Special Abilities of Microbes and Their Application in Agro-Biology

Special Abilities of Microbes and Their Application in Agro-Biology

Editors

Anuj Ranjan
Vishnu D. Rajput
Abhishek Chauhan

MDPI Basel • Beijing • Wuhan • Barcelona • Belgrade • Novi Sad • Cluj • Manchester

Editors

Anuj Ranjan
Academy of Biology
and Biotechnology
Southern Federal University
Rostov-on-Don
Russia

Vishnu D. Rajput
Academy of Biology
and Biotechnology
Southern Federal University
Rostov-on-Don
Russia

Abhishek Chauhan
Amity Institute of
Environmental Toxicology,
Safety and Management
Amity University
Noida
India

Editorial Office
MDPI AG
Grosspeteranlage 5
4052 Basel, Switzerland

This is a reprint of articles from the Special Issue published online in the open access journal *Microorganisms* (ISSN 2076-2607) (available at: www.mdpi.com/journal/microorganisms/special_issues/418T0MFKBN).

For citation purposes, cite each article independently as indicated on the article page online and as indicated below:

Lastname, A.A.; Lastname, B.B. Article Title. *Journal Name* **Year**, *Volume Number*, Page Range.

ISBN 978-3-7258-1704-7 (Hbk)
ISBN 978-3-7258-1703-0 (PDF)
doi.org/10.3390/books978-3-7258-1703-0

Contents

About the Editors

Daniela Fico

Daniela Fico obtained her PhD in Analytical Chemistry in 2015 at the Department of Cultural Heritage of the University of Salento (Lecce, Italy). From 2016 to 2020, she worked at the Department of Cultural Heritage (Lecce, Italy) dealing with the morphological and chemical characterization of materials, the development of innovative non-toxic and environmentally friendly coatings, and the development of innovative analytical protocols for the analysis of organic compounds using spectroscopic and chromatographic techniques, provenance studies, and statistical data processing. From 2021 to 2022, she worked at the Department of Engineering for Innovation of the University of Salento (Lecce, Italy), carrying out research activities related to material processing up to complete characterization (morphological–structural, chemical, physical–mechanical, rheological analysis), by means of a multidisciplinary approach (differential scanning calorimetry, mechanical analysis, permeability, colorimetry, modelling and 3D printing, etc.), and to the development of innovative materials (polymers, composites, biomaterials), with particular attention given to topics such as Sustainability and Circular Economy. Daniela Fico is currently working at the National Research Council-Institute of Heritage Science (CNR-ISPC, Lecce, Italy), and she is mainly involved in diagnostic and chemical analyses of Cultural Heritage, the study of manufacturing processes, alterations, conservation, and preservation, through the use of state-of-the-art instrumentation for advanced diagnostics and archaeometry (mapping and imaging spectroscopies, high-resolution mass spectrometry). She has participated in national and international projects, has supervised several theses, and is the author of several scientific articles in international Scopus-indexed journals. She currently serves as a Guest Editor for MDPI publishing house and as a reviewer for several international journals.

Preface

Today, ecological problems have led to the limited production of plastics and non-biodegradable materials and their replacement by materials with a low environmental impact. Biodegradable macromolecules and their composites are desirable candidates for a wide range of applications to overcome the difficulties of waste disposal. Moreover, one of the most crucial trends in current research on the development of new materials is associated with the use of raw waste materials or industrial by-products. This approach takes into account not only ecological issues, but also economic ones, since recycled and waste materials are significantly cheaper than virgin raw materials and their use produces low-cost end products. The substitution in industries, of raw materials that must be disposed of, by-products and waste with renewable and recyclable raw materials, constitutes an important transition to sustainable development and the circular economy: topics that many countries have already introduced into their environmental agenda through the creation of specific legislation.

This Reprint aims to highlight advanced research on the development of new eco-friendly materials and new technologies for sustainability and the circular economy.

Topics discussed by several experts in the field include, but are not limited to, the following:

- Biopolymers and biocomposites from natural raw materials;
- Innovative materials from recycled waste and industrial by-products;
- Synthesis, preparation and processing, and applications;
- Characterization, properties and potential of new biodegradable and eco-friendly materials;
- Studies on durability and biodegradability under different conditions and environments;
- Life cycle assessment of new materials;
- The circular economy, sustainability, and innovative and green materials.

Anuj Ranjan, Vishnu D. Rajput, and Abhishek Chauhan
Guest Editors

microorganisms

MDPI

Editorial

Editorial for "Special Abilities of Microbes and Their Application in Agro-Biology"

Anuj Ranjan [1,*], Vishnu D. Rajput [1] and Abhishek Chauhan [2]

[1] Academy of Biology and Biotechnology, Southern Federal University, Stachki 194/1, 344090 Rostov-on-Don, Russia; rvishnu@sfedu.ru

[2] Amity Institute of Environmental Toxicology, Safety and Management, Amity University, Noida 201301, India; akchauhan@amity.edu

* Correspondence: randzhan@sfedu.ru

Currently, climate change-related environmental issues associated with agriculture is alarming and poses a potential risk to global food security, causing significant global concern as a result [1]. Achieving the United Nations' Sustainable Development Goal of "Zero Hunger" necessitates the collective pursuit of sustainable agricultural practices [2]. Among these practices, the use of microbial-based biofertilizer has gained significant popularity due to its multifaceted role in agriculture. Microbes thriving in various environments are adapted and equipped with unique physiological and metabolic functionalities, which, upon their harnessing, could be useful for sustainable agricultural practices [3]. For instance, psychrophiles possessing plant growth-promoting traits could be useful for promoting soil fertility and crop production in geographically colder regions [4]. Similarly, thermophiles can be employed in a comparative tropical region where draught and soil–water stress are evident [5]. Therefore, this Special Issue, titled "Special Abilities of Microbes and Their Application in Agro-Biology", concerns the interaction between plants and beneficial microbes, and its scope covers the development of possibilities for enhancing agricultural productivity while mitigating environmental stressors. The studies reported in this editorial reveal the potential of beneficial microbes in agriculture.

Considering their evident adaptability to harsh environments and the usefulness of this special ability, one study on microbes in agriculture reports on anthropogenically polluted soil harboring microbes that have evolved the ability to survive, hence being useful for agricultural purposes, as exemplified by the species of *Bacillus*, *Brevibacterium*, and *Pseudomonad*, which exhibited exceptional plant growth-promoting traits [6]. Resilience to copper (Cu)-induced stress and the strengthening of the resilience of *Brassica napus* L. have been witnessed by using the metallotolerant *Bacillus altitudinis* strain TF16a, where Cu treatment was found to be accumulated in the roots and shoots and also elevated malondialdehyde (MDA) content by 20%. The application of a biofertilizer (prepared with biochar and *B. altitudinis* strain TF16a) with Cu decreased its accumulation by 20% for the shoots and 28% for the roots, maintaining MDA content similar to the control group, and both biofertilizer treatments, with and without Cu, increased chlorophyll a and b content, as well as non-enzymatic antioxidants, and led to increased biomass in the shoots and roots [7]. Improved phosphate mineralization and ex vitro acclimatization of *Musa acuminata* var. Valery using *Rahnella aquatilis* AZO16M2 was also evident when using microbes as a PGPR. In a solid medium with $Ca_3(PO_4)_2$, *R. aquatilis* AZO16M2 exhibited a solubilization index (SI) of 3.77 at 28 °C, and in a liquid medium, it produced 29.6 mg/L of soluble P at a pH of 4.4 [8]

In addition to this topic, this Special Issue attempts to provide details on microbial ecology, where the dynamics of soil microbiota come to light, such as in a case in which endophytic bacteria *Enterobacter* sp. ABk36 and HSTU-ABk39 were found to mineralize chlorpyrifos and support the health and growth of rice [9]. Similarly, phyllospheric bacteria isolated from *Coffea arabica* were found to be effective against coffee rust. The isolates

Citation: Ranjan, A.; Rajput, V.D.; Chauhan, A. Editorial for "Special Abilities of Microbes and Their Application in Agro-Biology". *Microorganisms* **2024**, *12*, 1179. https://doi.org/10.3390/microorganisms12061179

Received: 3 May 2024
Revised: 28 May 2024
Accepted: 5 June 2024
Published: 11 June 2024

CRRFLT7 and TRFLT8 showed urediniospore germination inhibition at rates of 81% and 82% [10], demonstrating the impact of microbial symbiosis in the protection of crop health and enhancing agricultural productivity. Through the transformative power of solid-state fermentation (SSF), another review article discusses approaches to unlock the underlying potential of waste materials to yield value-added agricultural formulations such as biostimulants and biopesticides [11], and it presents an evaluation of enhanced Brewer's spent grain's nutritional value and safety for use in animal feed [12]. By harnessing the principles of the circular economy, such innovative approaches promise to revolutionize agricultural production while minimizing environmental impact, heralding a new era of sustainable agriculture.

Additionally, this Special Issue includes an article that concerns precision agriculture, where nanotechnology emerges as a tool in the mitigation of plant pathogens and environmental stressors. Through the utilization of nano-minerals as alternatives to chemical fungicides, the researchers of the article demonstrated that the mycelial growth of *A. alternata* was inhibited by 85.1% with the application of 100 ppm nano-Se and that combining Se with SiO_2 at half doses resulted in a slightly lower efficacy rate of 77.8%, which could be significantly useful for crop protection strategies in sustainable crop management [13]. Lastly, this Special Issue also concerns ecosystem management, where a balance between pollinators, pathogens, and agricultural ecosystems comes into focus. In a separate study, the impact of probiotics on honeybees fed with a probiotic EM® for bees (TH2; TH3) showed a significant reduction in pathogen *Nosema* spp. spore counts from 25.18% to 96% on average across different sampling days, while the control groups (TH1, TH4) exhibited a continuous increase in infection levels along with improvement in gut microbiota [14]. To unravel the interactions between plants, diseases, and arthropods in open-field conditions, another study highlighted the impact of *Trichoderma harzianum* T22 on zucchini plants in open-field conditions, revealing increased attractiveness to aphids and *Hymenoptera parasitoids* but ineffectiveness against zucchini pathogens [15]. Similarly, a different study assessed the sensitivity of *Plasmopara halstedii* isolates to mefenoxam in sunflowers by analyzing host responses, including disease severity, growth reduction, and tissue reactions [16], to support biodiversity and the adoption of ecological resilience, which is the foundation of a more sustainable and resilient agroecosystem for the future.

In conclusion, this Special Issue stands as evidence of the potential of microbes with special abilities for utilization in innovative and collaborative efforts that address the multifaceted challenges facing modern agriculture. As we come across the complexities of rapidly changing climatic conditions, such collective endeavors bring us closer to a future where agriculture can thrive in harmony with nature.

Acknowledgments: V.D.R. would like to acknowledge the support from the Strategic Academic Leadership Program of the Southern Federal University ("Priority 2030").

Conflicts of Interest: The authors declare no conflicts of interest.

References

1. Rajpal, V.R.; Dhingra, Y.; Khungar, L.; Mehta, S.; Minkina, T.; Rajput, V.D.; Husen, A. Exploring Metal and metal-oxide nanoparticles for nanosensing and biotic stress management in plant systems. *Curr. Res. Biotechnol.* **2024**, *7*, 100219. [CrossRef]
2. Pânzaru, R.L.; Firoiu, D.; Ionescu, G.H.; Ciobanu, A.; Medelete, D.M.; Pîrvu, R. Organic Agriculture in the Context of 2030 Agenda Implementation in European Union Countries. *Sustainability* **2023**, *15*, 10582. [CrossRef]
3. Nadarajah, K.; Abdul Rahman, N.S.N. The microbial connection to sustainable agriculture. *Plants* **2023**, *12*, 2307. [CrossRef] [PubMed]
4. Ranjan, A.; Rajput, V.D.; Prazdnova, E.V.; Gurnani, M.; Sharma, S.; Bhardwaj, P.; Shende, S.S.; Mandzhieva, S.S.; Sushkova, S.; Minkina, T.; et al. Augmenting abiotic stress tolerance and root architecture: The function of phytohormone-producing PGPR and their interaction with nanoparticles. *S. Afr. J. Bot.* **2024**, *167*, 612–629. [CrossRef]
5. Singh, C.; Chauhan, A.; Arora, J.; Ranjan, A.; Singh Tuli, H.; Shahwan, M.; Rajput, V.D.; Minkina, T.; Venkat Eswaran, S.; Jindal, T. Exploration and Profiling of Potential Thermo-alkaliphilic Bacillus licheniformis and Burkholderia sp. from varied Soil of Delhi region, India and their Plant Growth-Promoting Traits. *J. Exp. Biol. Agric. Sci.* **2024**, *12*, 60–75. [CrossRef]

6. Barbaccia, P.; Gaglio, R.; Dazzi, C.; Miceli, C.; Bella, P.; Lo Papa, G.; Settanni, L. Plant Growth-Promoting Activities of Bacteria Isolated from an Anthropogenic Soil Located in Agrigento Province. *Microorganisms* **2022**, *10*, 2167. [CrossRef] [PubMed]
7. Kumar, A.; Borisova, G.; Maleva, M.; Tripti; Shiryaev, G.; Tugbaeva, A.; Sobenin, A.; Kiseleva, I. Biofertilizer Based on Biochar and Metal-Tolerant Plant Growth Promoting Rhizobacteria Alleviates Copper Impact on Morphophysiological Traits in *Brassica napus* L. *Microorganisms* **2022**, *10*, 2164. [CrossRef] [PubMed]
8. Acuña, D.L.; Toro, M.; Mendoza, R.S.; Dávila, D.Z. Role of *Rahnella aquatilis* AZO16M2 in phosphate solubilization and ex vitro acclimatization of *Musa acuminata* var. Valery. *Microorganisms* **2023**, *11*, 1596. [CrossRef] [PubMed]
9. Prodhan, M.Y.; Rahman, M.B.; Rahman, A.; Akbor, M.A.; Ghosh, S.; Nahar, M.N.-E.-N.; Simo; Shamsuzzoha, M.; Cho, K.M.; Haque, M.A. Characterization of Growth-Promoting Activities of Consortia of Chlorpyrifos Mineralizing Endophytic Bacteria Naturally Harboring in Rice Plants—A Potential Bio-Stimulant to Develop a Safe and Sustainable Agriculture. *Microorganisms* **2023**, *11*, 1821. [CrossRef] [PubMed]
10. Ogata-Gutiérrez, K.; Chumpitaz-Segovia, C.; Lirio-Paredes, J.; Zúñiga-Dávila, D. Antifungal Activity of Phyllospheric Bacteria Isolated from Coffea arabica against Hemileia vastatrix. *Microorganisms* **2024**, *12*, 582. [CrossRef] [PubMed]
11. Mattedi, A.; Sabbi, E.; Farda, B.; Djebaili, R.; Mitra, D.; Ercole, C.; Cacchio, P.; Del Gallo, M.; Pellegrini, M. Solid-State Fermentation: Applications and Future Perspectives for Biostimulant and Biopesticides Production. *Microorganisms* **2023**, *11*, 1408. [CrossRef] [PubMed]
12. Zeko-Pivač, A.; Bošnjaković, A.; Planinić, M.; Parlov Vuković, J.; Novak, P.; Jednačak, T.; Tišma, M. Improvement of the Nutraceutical Profile of Brewer's Spent Grain after Treatment with Trametes versicolor. *Microorganisms* **2022**, *10*, 2295. [CrossRef] [PubMed]
13. Taha, N.A.; Hamden, S.; Bayoumi, Y.A.; Elsakhawy, T.; El-Ramady, H.; Solberg, S.Ø. Nanofungicides with Selenium and Silicon Can Boost the Growth and Yield of Common Bean (*Phaseolus vulgaris* L.) and Control Alternaria Leaf Spot Disease. *Microorganisms* **2023**, *11*, 728. [CrossRef] [PubMed]
14. Tlak Gajger, I.; Nejedli, S.; Cvetnić, L. Influence of Probiotic Feed Supplement on *Nosema* spp. Infection Level and the Gut Microbiota of Adult Honeybees (*Apis mellifera* L.). *Microorganisms* **2023**, *11*, 610. [CrossRef] [PubMed]
15. Forlano, P.; Mang, S.M.; Caccavo, V.; Fanti, P.; Camele, I.; Battaglia, D.; Trotta, V. Effects of Below-Ground Microbial Biostimulant Trichoderma harzianum on Diseases, Insect Community, and Plant Performance in *Cucurbita pepo* L. under Open Field Conditions. *Microorganisms* **2022**, *10*, 2242. [CrossRef] [PubMed]
16. Nisha, N.; Vinogradov, S.; Körösi, K.; Berisha, A.; Bán, R. Assessing the Sensitivity of Plasmopara halstedii Isolates to Mefenoxam through Host Responses. *Microorganisms* **2023**, *11*, 821. [CrossRef] [PubMed]

microorganisms

MDPI

Article

Plant Growth-Promoting Activities of Bacteria Isolated from an Anthropogenic Soil Located in Agrigento Province

Pietro Barbaccia [1], Raimondo Gaglio [1], Carmelo Dazzi [1], Claudia Miceli [2], Patrizia Bella [1], Giuseppe Lo Papa [1] and Luca Settanni [1,*]

1 Dipartimento Scienze Agrarie, Alimentari e Forestali, Università Degli Studi di Palermo, 90128 Palermo, Italy
2 Council for Agricultural Research and Economics, Plant Protection and Certification Centre, 90121 Palermo, Italy
* Correspondence: luca.settanni@unipa.it

Abstract: Bacteria producers of plant growth-promoting (PGP) substances are responsible for the enhancement of plant development through several mechanisms. The purpose of the present work was to evaluate the PGP traits of 63 bacterial strains that were isolated from an anthropogenic soil, and obtained by modification of vertisols in the Sicily region (Italy) seven years after creation. The microorganisms were tested for the following PGP characteristics: indole acetic acid (IAA), NH_3, HCN and siderophore production, 1-aminocyclopropane-1-carboxylate deaminase activity (ACC) and phosphate solubilization. The results of principal component analysis (PCA) showed that *Bacillus tequilensis* SI 319, *Brevibacterium frigoritolerans* SI 433, *Pseudomonas lini* SI 287 and *Pseudomonas frederiksbergensis* SI 307 expressed high levels of IAA and production of ACC deaminase enzyme, while for the rest of traits analyzed the best performances were registered with *Pseudomonas* genus, in particular for the strains *Pseudomonas atacamensis* SI 443, *Pseudomonas reinekei* SI 441 and *Pseudomonas granadensis* SI 422 and SI 450. The in vitro screening provided enough evidence for future in vivo growth promotion tests of these eight strains.

Keywords: anthrosoils; cultivable bacteria; plant growth promoters; soil bacteria

Citation: Barbaccia, P.; Gaglio, R.; Dazzi, C.; Miceli, C.; Bella, P.; Lo Papa, G.; Settanni, L. Plant Growth-Promoting Activities of Bacteria Isolated from an Anthropogenic Soil Located in Agrigento Province. *Microorganisms* **2022**, *10*, 2167. https://doi.org/10.3390/microorganisms10112167

Academic Editors: Anuj Ranjan, Vishnu D. Rajput and Abhishek Chauhan

Received: 14 October 2022
Accepted: 29 October 2022
Published: 31 October 2022

1. Introduction

Unlike soils created by natural processes, anthropogenic soils (anthrosoils) have been affected, altered, or created by human activity. These soil types are generally found in different continents, and they are typically divided into four categories: urban, agricultural, mine related and archaeological soils [1]. The dumping of various materials for agricultural uses brings the soil to time zero, from the pedogenetic perspective; indeed, such events are seen as catastrophic [2]. In southern Italy, a lot of pedotechniques are used to improve the economic value of soils; these types of soil management are used in some areas of the Sicily region (Italy). Often, the original soils are covered with marly limestone, and subsequently plowed in order to improve the suitability of the areas for table grape cultivation [3]. Changes in the chemical composition of soils influence biological activities [4]. Thus, bacterial communities are subjected to new equilibriums that can affect plant growth.

Plant growth-promoting bacteria (PGPB) represent a huge and heterogenous group of bacteria that can be found as free-living in bulk soil or in rhizosphere, interacting in a mutualistic relationship with a huge variety of plant species [5,6]. They are involved to varying extents in the improvement of plant growth through several mechanisms [7]. Furthermore, PGPB are able to colonize all types of natural environments; in studies carried out by Antoun and Kloepper [8], around 5% of the root microflora is composed by PGPB. In particular, rhizobacteria possess different modes of action; these mechanisms are split directly and indirectly and provoke the improvement of plant physiology and the defense against phytopathogens [9].

PGPB can act as biofertilizers; in fact, they can increase plant growth thanks to the solubilization of some elements (mainly P, K and Zn), nitrogen fixation and production of siderophores (small molecules able to improve iron uptake capacity) [10–13]. PGPB also influence plant growth through the production of a series of organic substances, namely phytostimulators or plant growth regulators [14]. These compounds include the most important plant hormones: indole acetic acid (IAA), cytokinins, gibberellins and ethylene enzyme suppressors [15–18]. Furthermore, some of these bacteria might be considered valid alternatives to common pesticides; indeed, they have the capacity to produce antibiotics, HCN and hydrolytic enzymes that directly contrast the phytopathologies, but they are also able to compete with a high number of plant pathogens through indirect methods consisting in the search for radical exudates competing with the pathogens in order to obtain the nutrients [19,20]. PGPB are commonly applied in bioremediation strategies in order to remove or immobilize soil pollutants such as herbicides, pesticides, solvents, organic compounds and heavy metals [21]. Finally, these microorganisms can be employed to help plants overcome stresses of a biotic and abiotic nature [22–24].

The characterization of soil bacteria for their useful contributions in stimulating plant growth is of paramount importance in evaluating the positive traits of the natural microbial communities of soils subject to human modification. To this end, the PGP aptitude of several bacterial strains, detected at dominating levels from an anthropogenic soil of the Sicily region, were tested with the main aim of determining the state of health of the soils from a microbiological point of view. All strains were screened for IAA, NH_3, hydrogen cyanide, 1-aminocyclopropane-1-carboxylate deaminase (ACC), siderophore production and phosphate solubilization in order to estimate the positive functional role of the native bacterial community of these soils in supporting plant growth.

2. Materials and Methods

2.1. Plant Growth-Promoting Ability Assays

Bacteria used for the following assays were isolated and genetically characterized from an anthropogenic soil modified from Typic Haploxererts, located in the district of Giordano area within the Palma di Montechiaro (Agrigento, Italy) countryside, which is characterized by a Mediterranean climate [25].

The quantification of IAA was performed applying the method of Wholer [26]. The IAA was used for the construction of a standard curve in a range between 0 and 20 mg/L of water, using the Jenway Ltd. model 6400 (Dunmow, UK) spectrophotometer at 535 nm. All bacterial strains, previously stored at −80 °C, were cultured in nutrient broth overnight (Oxoid, Milan, Italy); afterwards, by centrifuging the culture media at 7000 rpm for five minutes, the cells were recollected and then cultured for 24 h at 37 °C in 3 mL of phosphate buffer (pH 7.5) containing 1% (*w*/*v*) of glucose and tryptophan. After incubation, cell suspensions were transferred into a solution of 2 mL of 5% (*v*/*v*) trichloroacetic acid and 1 mL of 0.5 M $CaCl_2$. The solution was filtered through Whatman No. 2 filters (Whatman International Ltd., Maidstone, UK) and 3 mL of filtrate was added along with 2 mL of Salper solution (2 mL 0.5 M $FeCl_3$ and 98 mL 35% (*v*/*v*) perchloric acid). The absorbance was measured at 535 nm after an incubation of half an hour in the dark at 25 °C.

In order to assess the ability to generate NH_3, all bacteria were grown in peptone water for 72 h at 30 °C. Nessler's reagent was added to each tube (0.5 mL), and the test was considered positive for NH_3 production if broth color turned yellow-brown [27].

Hydrogen cyanide production was tested in Petri dishes using a modified nutrient agar (4.4 g/L of glycine). A filter paper Whatman no. 1 (Whatman International Ldt) was dipped in a solution prepared with 2% (*w*/*v*) sodium carbonate and 0.5% (*v*/*v*) picric acid, and was laid onto the surface of the agar medium; 10 μL of microbial solution was taken by a refresh tube with a concentration of 10^9 CFU/mL, and was spread in a petri dish with the modified media and cultivated at 30 °C for 4 d. After that, colonies that acquired an orange or red color were considered positive for HCN production [5].

The synthesis of siderophores was carried out on a Chrome azurol S agar medium. Bacterial spots were transferred directly from growth plates, and after 48 h of incubation at 30 °C, the appearance of a bright orange halo surrounding the colonies indicated that siderophores had been produced in each single strain [28].

The method of Honma and Shimura [29] was modified to determine ACC-deaminase activity of bacterial strains. Pellets of bacterial cells were obtained as reported above. The cells were then resuspended in 5 mL of 0.1 mol/L Tris-HCL at pH 7.6, and after centrifugation at 16,000 rpm for 5 min, the pellets were further resuspended in 2 mL of 0.1 mol/L Tris-HCl at pH 8.5. The cell suspensions were added with 30 µL of toluene and vortexed for 30 s, and 200 µL of each cell suspension was transferred into a microtube, adding 20 µL of 0.5 mol/L ACC, and incubated at 30 °C for 15 min. After that, 1 mL of 0.56 mol/L HCl was added into a microtube that had been previously incubated, and the mix was homogenized and centrifuged for 5 min at 16,000 rpm at room temperature; 1 mL of supernatant was taken and mixed with 800 µL of 0.56 mol/L HCL, and 2 mL of 2,4-dinitrophenylhydrazine reagent was added to the mixture, vortexed and incubated at 30 °C for 30 min. Finally, 2 mL of 2 mol/L NaOH was added to the solution, and the absorbance was read spectrophotometrically at 540 nm. The measurement of α-ketobutyrate after hydrolysis of ACC is the basis of this method. The values obtained by this protocol was used to estimate the amount of µmol of α-ketobutyrate produced by the tested strains; these values were compared to a standard curve, obtained by adding 2 mL of 2,4-dinitrophenylhydrazine to each standard in a range between 0.1 and 1 µmol of α- ketobutyrate; the solution was vortexed and cultured at 30 °C for half an hour. The absorbance of the solution was measured at 540 nm after the addition of 2 mL 2 mol/L NaOH.

Phosphate solubilization was tested on a Pikovskaya medium (PVK). This medium had the following composition: 10 g/L glucose; 5 g/L $Ca3(PO_4)2$; 0.5 g/L $(NH_4)2SO_4$; 0.2 g/L NaCl; 0.1 g/L $MgSO_4 \cdot 7\ H_2O$; 0.2 g/L KCl; 0.5 g/L yeast extract; 0.5 g/L $MnSO_4 \cdot H_2O$; and 0.002 g/L $FeSO_4 \cdot 7\ H_2O$ [30]. After 15 d, the width of the halo around the colonies was measured, and colony diameter was subtracted from the total diameter. The phosphate dissolution rate was calculated using the formula (size of colony + size of clear zone)/diameter of colony.

Unless otherwise indicated, all chemicals and reagents were purchased from Sigma-Aldrich (Milan, Italy).

2.2. Statistical Analysis

IAA and ACC data from the screening of microorganisms were analyzed using the One-Way Variance Analysis (ANOVA). Version 7.5.2 of the XLStat software for Excel was used for the analysis (Addinsoft, New York, NY, USA). The various strains put through the tests were compared using Tukey's test. P values below 0.05 were considered statistically significant and are denoted by different letters.

To evaluate the correlation between the microorganisms and the parameters measured with the tests, the principal component analysis (PCA) was used. The number of major factors with eigen values greater than 1.00 were chosen using the Kaiser criterion [31]. The statistical significance within the dataset was examined with Barlett's sphericity test [32].

3. Results

3.1. Plant Growth-Promoting Ability Assays

Results obtained from the PGPB screening are reported in Table 1. Statistical treatment of results of IAA production generated 20 different groups. The three largest genera of bacteria analyzed (*Brevibacterium*, *Bacillus* and *Pseudomonas*) showed a capacity to produce highly variable IAA. In particular, the *Brevibacterium* group included strains with no ability to generate IAA (strain SI 325) and strains with a high IAA production, until 7.37 mg L^{-1} was registered for *Brevibacterium frigoritolerans* SI 433. A similar variability was observed for the *Bacillus* genus with *Bacillus halotolerans* strain SI 339 unable to express this character until 6.94 mg L^{-1} was displayed by *Bacillus megaterium* SI 404 and *Bacillus cabrialesii* SI 428.

Regarding *Pseudomonas*, which included 14 different bacterial strains, the range recorded was narrower than those of the two previous genera, from 0 mg L^{-1} of *Pseudomonas granadiensis* SI 450 to 6.50 mg L^{-1} of *Pseudomonas reinekei* SI 441. In addition to these three genera, there are interesting bacteria belonging to other genera such as *Streptomyces, Micrococcus, Sinorhizobium* and *Stenotrophomonas*, which possess the ability to produce consistent amounts of IAA.

Only 21 strains resulted positive for the NH_3 production assay by turning the medium color to yellow-brown (Figure 1A). Regarding the major taxonomic bacterial groups, only three strains of *Br. frigoritolerans* (SI 264, SI 312, and SI 400) and three strains of *Bacillus* (*B. megaterium* SI 408, *B. halotolerans* SI 339 and *B. cabrialesi* SI 428) resulted positive for this test, while nine *Pseudomonas* strains, belonging to five different species, generated NH_3. This character also registered positive for *Lysobacter soli*.

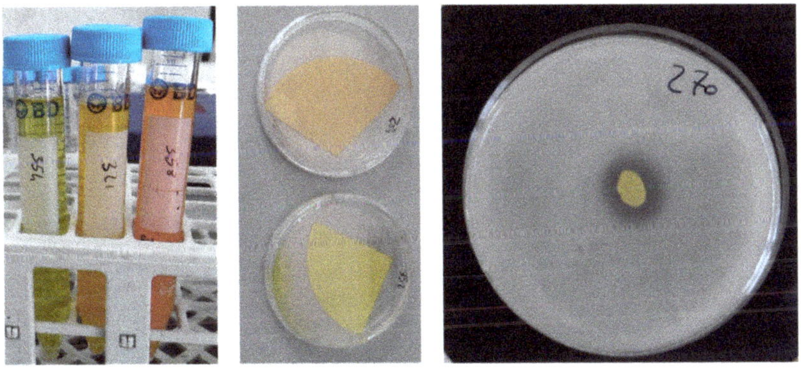

(a) (b) (c)

Figure 1. Visual results of plant-growth promoting (PGP) tests: (**a**) NH_3 production; (**b**) HCN production; (**c**) halo generated by phosphate solubilization.

Only eight strains among the totality of the screened bacteria resulted positive in the HCN test by turning the filter paper color from yellow to orange-red (Figure 1B). All these bacteria were *Pseudomonas*. In particular, the species able to generate HCN were: *Pseudonomas brassicacearum, Pseudomonas frederiksbergensis, Ps. reinekei, Pseudomonas atacamensis, Ps. granadensis* and *Pseudomonas lini*.

A higher percentage of strains resulted positive for siderophore production. Thirty-one strains, including all *Pseudomonas* of the collection, produced siderophores. Among the strains belonging to other taxonomic groups, this capacity was shown by four *Br. frigoritolerans* strains and five *Bacillus* belonging to the species *B. megaterium, Bacillus tequilensis* and *Bacillus halotolerans*.

The results of the ACC test are reported in Table 1. Statistical analysis showed a great variability of data, indicating 28 different groups. The strain that showed the highest production of α-ketobutyrate was *Br. frigoritolerans* SI 433 with 80.58 nmol. *Bacillus* genus showed a high percentage of positive strains, with 9 out of 14 strains producing α-ketobutyrate after the hydrolysis of ACC. Furthermore, the *Pseudomonas* genus displayed a high percentage of positive strains in this test, with 11 out of 14 strains tested. Within the *Pseudomonas* group, all strains that tested negative belonged to the species *Pseudomonas lini*, even though the strain showing the highest α ketobutyrate production is *Ps. lini* SI 287, with 62.28 nmol. Furthermore, all strains of *Peribacillus* displayed production of this acid.

Results highlighted that all bacteria able to solubilize phosphate belonged to the *Pseudomonas* genus, and the biggest halo diameter (8 mm) for phosphate solubilization was recorded for *Ps. lini* SI 270 (Figure 1C). The other three strains showing this character were *Streptomyces silaceus* SI 332, *Sinorizhobium melitoti* SI 240 and *Variovorax paradoxus* SI 435.

Table 1. Main plant-growth promoting (PGP) traits for which the bacteria were screened.

Strains	Species	IAA Production (mg/L)	NH_3 Production	HCN Production	ACC Deaminase Activity (nmol α-Ketobutyrate/g h)	Siderophore Production	Phospate Solubilization
SI 257	*Br. frigoritolerans*	2.50 ± 0.3 ^{ghijk}	–	–	0 ^{w}	–	–
SI 264	*Br. frigoritolerans*	2.50 ± 0.4 ^{ghijk}	+	–	0 ^{w}	–	–
SI 325	*Br. frigoritolerans*	0 ^{k}	–	–	37.17 ± 2 ^{hijk}	+	–
SI 312	*Br. frigoritolerans*	1.76 ± 0.2 ^{ijk}	+	–	22.72 ± 5 ^{mnopqr}	+	–
SI 385	*Br. frigoritolerans*	3.46 ± 0.2 ^{defghij}	–	–	9.05 ± 3 ^{stuvw}	–	–
SI 333	*Br. frigoritolerans*	3.75 ± 0.61 ^{cdefghij}	–	–	40.60 ± 6 ^{fghi}	–	–
SI 349	*Br. frigoritolerans*	6.72 ± 0.65 ^{ab}	–	–	28.91 ± 6 ^{jklm}	–	–
SI 400	*Br. frigoritolerans*	1.76 ± 0.1 ^{ijk}	+	–	0 ^{w}	+	–
SI 433	*Br. frigoritolerans*	7.37 ± 0.5 ^{a}	–	–	80.58 ± 4 ^{a}	–	–
SI 387	*Br. frigoritolerans*	6.72 ± 0.6 ^{ab}	–	–	16.18 ± 2 ^{opqrst}	+	–
SI 293	*R. erythropolis*	2.14 ± 0.2 ^{hijk}	–	–	72.56 ± 4 ^{ab}	–	–
SI 250	*R. equi*	3.15 ± 0.15 ^{efghij}	–	–	17.52 ± 7 ^{nopqrs}	+	–
SI 271	*N. globerula*	4.81 ± 0.2 ^{abcdefgh}	–	–	0 ^{w}	–	–
SI 279	*Str. mauvecolor*	3.15 ± 0.32 ^{efghij}	–	–	4.20 ± 1 ^{uvw}	–	–
SI 332	*Str. Silaceus*	5.33 ± 0.34 ^{abcdefg}	–	–	0 ^{w}	+	3.01
SI 362	*M. hydrocarboxydans*	5.57 ± 0.52 ^{abcdef}	–	–	34.85 ± 5 ^{ijkl}	–	–
SI 371	*M. oxydans*	4.83 ± 0.38 ^{abcdefgh}	–	–	23.98 ± 5 ^{lmnopq}	–	–
SI 295	*A. nitrophenolicus*	5.08 ± 0 ^{abcdeg}	–	–	55.96 ± 6 ^{cde}	–	–
SI 429	*P. aurescens*	3.46 ± 0.14 ^{defghij}	+	–	0 ^{w}	–	–
SI 236	*I. cucumis*	5.81 ± 0.31 ^{abcde}	–	–	9.05 ± 3 ^{stuvw}	–	–
SI 254	*Pb. simplex*	6.50 ± 0.37 ^{abc}	–	–	20.15 ± 3 ^{mnopqrs}	–	–
SI 397	*Pb. simplex*	4.03 ± 0.07 ^{bcdefghij}	–	–	7.51 ± 2 ^{tuvw}	+	–
SI 259	*Pb. simplex*	1.33 ± 0.3 ^{jk}	–	–	49.49 ± 5 ^{efg}	+	–
SI 306	*B. tequilensis*	3.15 ± 0.22 ^{efghij}	–	–	0 ^{w}	–	–
SI 296	*B. tequilensis*	3.46 ± 0.35 ^{defghij}	–	–	25.23 ± 4 ^{lmnop}	+	–
SI 319	*B. tequilensis*	6.27 ± 0.3 ^{abcd}	–	–	51.66 ± 5 ^{def}	+	–
SI 354	*B. tequilensis*	6.72 ± 0.1 ^{ab}	–	–	0 ^{w}	–	–
SI 305	*B. megaterium*	5.57 ± 0.3 ^{abcdef}	–	–	27.69 ± 4 ^{jklmn}	+	–
SI 404	*B. megaterium*	6.94 ± 0.88 ^{ab}	–	–	7.51 ± 0.7 ^{tuvw}	+	–
SI 408	*B. megaterium*	5.08 ± 0.11 ^{abcdefg}	+	–	38.32 ± 6 ^{ghij}	–	–
SI 470	*B. megaterium*	4.83 ± 0.04 ^{abcdefgh}	–	–	14.81 ± 2 ^{pqrstu}	–	–
SI 266	*B. megaterium*	6.04 ± 0.1 ^{abcde}	–	–	0 ^{w}	–	–
SI 339	*B. halotolerans*	0 ^{k}	+	–	34.85 ± 3 ^{ijkl}	–	–
SI 419	*B. halotolerans*	4.83 ± 0 ^{abcdefgh}	–	–	27.69 ± 4 ^{jklmn}	+	–
SI 297	*B. mohavensis*	1.33 ± 0.13 ^{jk}	–	–	7.51 ± 2 ^{tuvw}	–	–
SI 311	*B. cabrialensis*	4.83 ± 0.4 ^{abcdefgh}	–	–	0 ^{w}	–	–
SI 428	*B. cabrialesii*	6.94 ± 0.04 ^{ab}	+	–	0 ^{w}	–	–
SI 356	*T. saccharophilus*	6.27 ± 0.21 ^{abcd}	–	–	64.36 ± 6 ^{bc}	–	–
SI 243	*E. adherens*	2.50 ± 0.3 ^{ghijk}	+	–	0 ^{w}	–	–
SI 240	*Sn. meliloti*	5.57 ± 0.3 ^{abcdef}	–	–	0 ^{w}	+	2.96
SI 235	*Sn. meliloti*	6.50 ± 0.3 ^{abc}	–	–	9.05 ± 1 ^{stuvw}	–	–
SI 420	*S. quinivorans*	0 ^{k}	+	–	17.52 ± 3 ^{nopqrs}	+	–
SI 237	*C. respiraculi*	0 ^{k}	–	–	18.85 ± 3 ^{mnopqrs}	–	–
SI 435	*V. paradoxus*	5.33 ± 0.4 ^{abcdefg}	–	–	0 ^{w}	–	3.1
SI 439	*V. paradoxus*	5.33 ± 0.42 ^{abcdefg}	–	–	0 ^{w}	–	–
SI 321	*St. indicatrix*	6.50 ± 0.3 ^{abc}	+	–	26.47 ± 4 ^{klmno}	+	–
SI 358	*L. soli*	4.57 ± 0.37 ^{abcdefghij}	+	–	37.17 ± 3 ^{hijk}	–	–
SI 357	*L. soli*	2.14 ± 0.02 ^{hijk}	+	–	0 ^{w}	–	–
SI 377	*St. rhizophila*	4.57 ± 0.23 ^{abcdefghi}	–	–	18.85 ± 2 ^{mnopqrs}	+	–
SI 367	*Ps. Plecoglossicida*	6.50 ± 0.28 ^{abc}	–	–	20.15 ± 3 ^{mnopqrs}	+	–
SI 247	*Ps. brassicacearum*	3.75 ± 0.6 ^{cdefghij}	–	+	16.18 ± 2 ^{opqrst}	+	–
SI 307	*Ps. frederiksbergensis*	3.15 ± 0.2 ^{efghij}	+	+	48.40 ± 5 ^{efgh}	+	3.80
SI 441	*Ps. reinekei*	6.50 ± 1 ^{abc}	+	+	45.09 ± 6.4 ^{efghi}	+	–
SI 443	*Ps. atacamensis*	2.14 ± 0.2 ^{ghijk}	+	+	10.54 ± 1 ^{stuvw}	+	3.85
SI 422	*Ps. granadensis*	5.08 ± 1 ^{abcdefg}	+	+	16.18 ± 3 ^{opqrst}	+	3.2
SI 450	*Ps. granadensis*	0 ^{k}	+	+	12.00 ± 1 ^{rstuv}	+	–
SI 434	*Ps. moorei*	4.83 ± 0.65 ^{abcdefgh}	–	–	14.81 ± 3 ^{pqrstu}	+	–
SI 270	*Ps. lini*	2.83 ± 0.1 ^{fghijk}	+	–	0 ^{w}	+	4.99
SI 285	*Ps. lini*	2.83 ± 0.22 ^{fghijk}	+	+	0 ^{w}	+	–
SI 276	*Ps. lini*	2.14 ± 0.12 ^{hijk}	–	–	13.42 ± 2.4 ^{qrstu}	+	–
SI 284	*Ps. lini*	3.46 ± 0.2 ^{defghij}	+	+	0 ^{w}	+	–
SI 288	*Ps. lini*	6.27 ± 0.34 ^{abcd}	+	–	16.2 ± 3 ^{opqrst}	+	4
SI 287	*Ps. lini*	3.15 ± 0.1 ^{efghij}	–	–	62.28 ± 4.8 ^{bcd}	+	–

Abbreviations are as follows: IAA, indole acetic acid; ACC, 1-aminocyclopropane-1-carboxylate deaminase activity; *A., Arthrobacter*; *B., Bacillus*; *Br., Brevibacterium*; *C., Cupriavidus*; *E., Ensifer*; *I., Isoptericola*; *L., Lysobacter*; *M., Microbacterium*; *N., Nocardia*; *P., Paenarthrobacter*; *Pb., Peribacillus*; *Ps., Pseudomonas*; *R., Rhodococcus*; *S., Serratia*; *Sn., Sinorhizobium*; *St., Stenotrophomonas*; *Str., Streptomyces*; *T., Terribacillus*; *V., Variovorax*. Data within a column followed by the same letter are not significantly different for $p \leq 0.05$ according to Tukey's test.

3.2. Statistical Analysis

The correlation of the tested bacteria with PGP abilities, evaluated by production of IAA, siderophore, hydrogen cyanide, NH_3, expression of ACC deaminase activity and solubilization of phosphate, was analysed by PCA (Figure 2). The results highlighted how, for the first 2 components (PC1 and PC2), the eigen value reached 2.03 and 1.10, respectively.

The 33.83% of total variability was expressed by the first component, while PC2 accounted for the 18.38%; thus, PC1 and PC2 together accounted for 52.22% of total variability. The graphical biplot shows that the first component (F1) had a strong influence on IAA and production of the ACC deaminase enzyme, while PC2 showed an influence on the other characteristics evaluated.

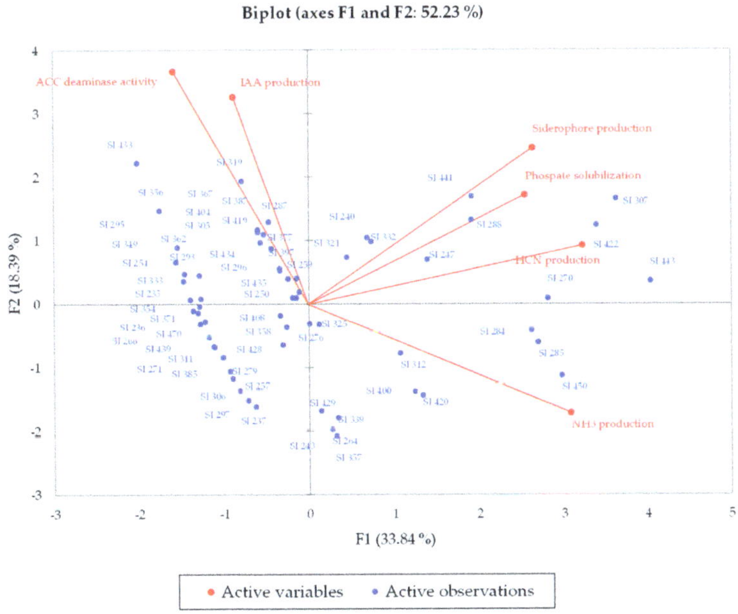

Figure 2. Principal component analysis (PCA) among the screened bacterial strains and the PGP traits.

The graphical distribution of microorganisms showed that the strains *Ps. atacamensis* (SI 443), *Ps. granadensis* (SI 422), *Ps. reinekei* (SI 441) and *Ps. granadensis* (SI 450) had the best PGP performances for siderophore production, phosphate solubilization and HCN and NH_3 production, whereas *B. tequilensis* (SI 319), *Ps. lini* (SI 287), *Br. frigoritolerans* (SI 433) and *Ps. frederiksbergensis* (SI 307) highlighted the best performances for IAA production and ACC deaminase activity.

4. Discussion

Microbial diversity is one of the main factors characterizing natural ecosystems; soil is considered one of the best storehouses of useful microorganisms in the world. Although the role of most of these microorganisms is still unknown, scientific progress is providing a better comprehension of the specific ecological functions of soil microorganisms [33]. The microbial community encountered in soil includes bacteria, molds and protozoa; some of them are free-living, while others live in symbiotic form with various species of plants. These microorganisms can be in different types of relationship with the plants, since their role can be indifferent, harmful or favorable [33].

In order to evaluate the PGP abilities of the indigenous bacteria present in an anthropogenic soil, in this work 63 soil bacteria, belonging to three different *phyla* (Actinobacteria, Firmicutes and Proteobacteria) and isolated by a modified Sicilian soil [24], were tested in vitro for their PGP abilities. The scope of this research was to establish if the natural bacterial community resident in this site was able to support plant growth. Indeed, anthropogenic soils are not cultivated soon after modification, in order to give the microbial community a certain period of time to find a new equilibrium after the addition of exogenous material. Practical observations in the area of Palma di Montechiaro (Sicily) under study suggested around five years as the optimal time before starting grape plant cultivation.

In general, Actinobacteria are one of the richest phyla of PGPB; bacteria belonging to *Frankia* genus are involved in symbiosis with plants. Other Actinobacteria, especially *Arthrobacter, Micrococcus* and *Streptomyces,* are considered plant growth boosters, although they do not take part in symbiotic relationships [34,35]. Firmicutes represent the most important phylum involved in PGP. In particular, *Bacillus* was thoroughly proven to exert positive effects in soil, which is directly related to plant growth [36–40]. These bacteria use the broadest range of PGP mechanisms, such as production of siderophores and IAA, ACC-deaminase activity and phosphate dissolution [41]. Proteobacteria are also counted as PGPB. Among these, Alphaproteobacteria include 13 different genera, especially *Ensifer* and *Cupriavidus,* that are recognized as symbiotic organisms with legumes [33]; Gammaproteobacteria include the genus *Pseudomonas* whose species might be plant pathogenic, but also PGP, especially by producing auxins, gibberellins, cytokinin, and ethylene, as well as by asymbiotic nitrogen fixation and mineral solubilization [42,43]. Some strains of *Pseudomonas aurantiaca* are also involved in HCN and siderophore production, and solubilization of phosphate [44]. Among phytohormones, one of the most important groups is undoubtedly composed of auxins. They influence many cellular functions [45]; despite the fact that a number of naturally occurring auxins have been identified, IAA has received the greatest attention by the scientific community, and the terms auxin and IAA are commonly used synonymously. In plants, IAA is typically found in conjugated forms that are primarily involved in IAA catabolism transport, storage and protection [45,46]. Tryptophan, a common precursor in root exudates, is widely converted in nature into IAA by plants and PGPB through the metabolic processes of transamination and decarboxylation [47]. It has been proposed that IAA produced by PGPB may shield cells from the harmful effects of environmental stresses [48]. Furthermore, it was demonstrated by several authors that the bacterial IAA promoted lateral and adventitious root growth, improving mineral and nutrient uptake [45]. Auxin is widely produced by soil bacteria, with an estimated 80% of soil bacteria showing this characteristic. In fact, numerous strains of soil bacteria, as well as *Alcaligenes, Azotobacter, Azospirillum, Enterobacter, Klebsiella, Pantoea, Pseudomonas, Rhizobium* and *Streptomyces*, have been found to express this property [45,47]. Not all isolates tested in this study produced IAA, even though the majority of them produced IAA amounts in the range of those described in previous works; for example, levels between 29 and 71 mg L^{-1} were reported by Tara and Saharan [49] for *Br. frigoritolernas* strains, and in this study, *Br. frigoritolernas* showed values between 0 and 7.37 mg L^{-1}. According to Wahyudi et al. [50], 5 strains of *Streptomyces* that were obtained from soybean rhizosphere produced IAA in the range of 5.25–12.04 mg L^{-1}, which are values closest to those found for our *Streptomyces* strains (*Streptomyces mauvecolor* and *Streptomyces silaceus*), with 3.15 and 5.33 mg L^{-1}, respectively. Our results showed that several *Bacillus* species were able to produce IAA, although their levels were quite variable. It is among *Bacillus* genus that our investigation found the highest IAA production, and this could be explained by the high efficiency of this genus to utilize nutrients supplied by the plant through exudates [51]. Regarding *Pseudomonas,* our results were comparable to those found by some authors [51,52], while other authors reported higher values than ours [33,51,53,54]; this heterogeneity in IAA production is attributable to multiple biochemical pathways, genetic control, and environmental influences [55].

The production of ammonia is another notable aspect related to PGPB. In particular, this compound indirectly influences plant development. In this study, not all isolates were able to produce ammonia. Plants use released ammonia as a source of nutrients. Furthermore, in nitrogen-rich soils, an accumulation of ammonia can cause the soil to become alkaline; these soil conditions prevent the growth of some fungi [56,57]. According to Joseph et al. [58], a high percentage of bacteria belonging to the *Pseudomonas* genus resulted positive to ammonia tests; but with regards to the *Bacillus* genus, the results are not comparable to those found in the previous study, because only 3 out of 14 strains were positive in the NH_3 test.

HCN production is of particular importance in soil, because its overproduction might suppress plant fungal infections [59]. Furthermore, the generation of hydrogen cyanide is positively correlated with nitrogen accumulation, root elongation, biomass production, and shoot elongation [60]. *Bacillus, Pseudomonas, Serratia, Arthrobacter* and *Stenotrophomonas* are considered PGPB and are involved in HCN production [61]. Although a wide variety of bacterial genera are recognized as HCN producers, in our study only eight strains, all belonging to the genus *Pseudomonas*, were found to be producers of this volatile substance.

Several proteins involved in a variety of both microbial and plant processes need iron as a cofactor. Thus, iron is essential for plant growth and development. The fourth most common element in the crust of the earth is iron [62]. Unfortunately, a relatively little amount of this element is in the ferric ion (Fe^{3+}) form that is assimilated by living organisms [63]. This obstacle is overcome by several bacteria, especially siderophores, tiny organic compounds produced by microbes in iron-limited environments that increase the capability to absorb iron [64,65]. Moreover, the presence of siderophores allows plants to absorb iron despite the presence of other metals such as cadmium and nickel [66]. Producers of siderophores, in addition to chelating iron, can also adsorb other heavy metals such as lead arsenic, aluminum, magnesium, zinc, copper, cobalt and strontium [62]; for this reason, these microorganisms can be used as bioremediators. Our results demonstrate that a high number of bacteria tested were positive in the siderophores test. In particular, 31 strains belonging to nine different genera were seen to be producers of these organic compounds. Among these, four strains were *Br. frigoritolerans*; indeed, the same species resulted positive to this test in the work of Rasool et al. [67], demonstrating the good aptitude of this species as PGP. *Bacillus megaterium* was one of the best siderophore producers in our study, and similar findings were reported by Wani and Khan [68]. Also, a *Serratia* strain tested positive for this character, confirming what had already been reported by Koo and Cho [69]. Finally, *Pseudomonas* are widely utilized as bioremediators thanks to their ability to produce siderophores [70–72]. Our results showed that all *Pseudomonas* strains tested tested positive for siderophores, showing their important role in soil.

The production of ethylene is an important strategy developed by plants to induce a rapid protective response in reaction to external stress [73]. Basically, plant response consists of two phases: the first phase is characterized by a small peak of ethylene production, while in cases of chronic or intense stress, plants react with a huge production of ethylene that can lead to a variety of processes, including aging, chlorosis and defoliation, which impede plant growth [74]. As described by several authors, the production of ethylene in higher plants is regulated by three enzymes: S-adenosyl-L-methionine (SAM), 1-aminocyclopropane-1-carboxylic acid (ACC) and ACC oxidase [45]. Some microorganisms possess a particular enzyme (ACC deaminase) that is able to split the precursors of ethylene ACC into ammonia and α-ketobutyrate [75,76], thus reducing the amount of ethylene formed. In our work, 19 out of 63 bacteria did not show the presence of the ACC deaminase enzyme. Among those positive for this character, *Br. frigoritolerans* (the strain SI 433) showed the highest value of α-ketobutyrate with 80.58 nmol /g protein h, while the other bacteria within the *Brevibacterium* genus (ranging between 9.05 and 40.60 nmol/g protein h of α-ketobutyrate) behaved similarly to the brevibacteria isolated and screened by Tiryaki et al. [77]. In our study, strains of the *Pseudomonas* genus showed a high percentage of positivity to ACC deaminase activity. In particular, *Ps. Lini* (SI 287) showed the highest value of α-ketobutyrate among this genus (62.28 nmol/g protein h), and similar values for the same bacteria species were reported by Palacio Rodríguez [78]. Regarding *Bacillus*, values of α ketobutyrate synthetized by ACC deaminase of our strains were similar to those found by Misra et al. [79].

The last PGP test performed in this work was phosphate solubilization. Phosphorus, as well as nitrogen, is one of the most important elements involved in plant nutrition [80]. In particular, phosphorus has a role in every major metabolic function, including energy transmission, signal transduction, respiration, macromolecular biosynthesis and photosynthesis. Despite being one of the elements most present in soils, both in organic and inorganic form, 95–99% of the phosphate is contained in the insoluble, immobilized, and precipitated forms, making absorption by plants quite difficult. For this reason, solubilization and mineralization of phosphate is one of the most important characteristics of PGPB. These bacteria have a low rhizosphere pH thanks to their secretion of different organic acids, such as carboxylic acid and succinic acids, which causes the bound forms of phosphate like $Ca_3(PO_4)_2$ to be released in calcareous soils [81,82]. Inorganic phosphate solubilization has also been linked to the release of H^+ [83] and the creation of chelating agents [84,85]. Furthermore, phosphorous biofertilizers can increase the nitrogen fixation and implement the availability of substances like iron and zinc [62]. Only eight bacterial strains in our study tested positive in the phosphate solubilization test, and the genus mainly represented was *Pseudomonas*. As reported by several works, several bacteria expressing this character and known as PGP belong to this genus [82,86–89]. The biggest halo (8 mm) was observed for the strain *Ps. lini* SI 27. Zhang et al. [90] measured wider halos for the same species. *Streptomyces silaceus* SI 332 showed a solubilization halo of barely 4 mm, the smallest one registered in the screening. *Streptomyces* genus is actually active in solubilizing soil phosphate [91–94]. None of the *Bacillus* strains solubilized phosphates, although it is reported as a genus particularly active from this perspective [95–98].

Finally, all PGP traits of the bacteria tested were analyzed by multivariate statistical analysis to better individuate the strains characterized by the best PGP characteristics. For PCA, it emerged that two technological traits, i.e., IAA production and ACC deaminase activity, were positively related to each other, in contrast to those reported by Castellano-Hinojosa et al. [98], while according to the same authors, siderophore production was positively related to phosphate solubilization. The strains *Ps. Atacamensis* (SI 443), *B. Tequilensis* (SI 319), *Ps. Lini* (SI 287), *Br. frigoritolerans* (SI 433), *Ps. frederiksbergensis* (SI 307), *Ps. granadensis* (SI 422), *Ps. Reinekei* (SI 441) and *Ps. granadensis* (SI 450) had the largest contributions to the total variance, according to PCA analysis, so by this analysis we can say that these eight bacterial strains possessed the best PGP performances, and they could be used for single or consortium inoculations in vivo in order to test their abilities as PGPB, as reported by several works [99–102].

5. Conclusions

In conclusion, the PGP screening showed that all bacteria analyzed displayed positivity to at least one of the tests applied; these findings highlight that the microbial biodiversity present in the anthropogenic soil seven years after creation reached a certain capacity to provide support for plant growth functions. In addition, eight bacterial strains distributed among *Pseudomonas, Bacillus* and *Brevibacteria* genera were recognized as excellent producers of PGP substances. Additional research will be needed to evaluate the in vivo PGP performance of these microorganisms in fields cultivated for table grapes.

Author Contributions: Investigation, P.B. (Pietro Barbaccia) and C.M.; data curation, P.B. (Patrizia Bella), R.G. and C.M.; visualization, P.B. (Pietro Barbaccia) and G.L.P.; writing—original draft preparation, P.B. (Pietro Barbaccia); conceptualization, C.D. and L.S.; resources, R.G.; supervision, C.D.; writing—review and editing, L.S. and G.L.P.; funding acquisition, R.G.; Project administration, L.S. All authors have read and agreed to the published version of the manuscript.

Funding: This research has received co-funding from the European Commission's ERASMUS+ Programme under grant agreement No 2017-1-SE01-KA203-034570.

Data Availability Statement: Not applicable.

Acknowledgments: Not applicable.

Conflicts of Interest: The authors declare no conflict of interest.

References

1. Howard, J. *Anthropogenic soils*; Springer: Berlin, Germany, 2017.
2. Papa, G.L.; Palermo, V.; Dazzi, C. Is land-use change a cause of loss of pedodiversity? The case of the Mazzarrone study area, Sicily. *Geomorphology* **2011**, *135*, 332–342. [CrossRef]
3. Papa, G.L.; Antisari, L.V.; Vianello, G.; Dazzi, C. Soil interpretation in the context of anthropedogenic transformations and pedotechniques application. *Catena* **2018**, *166*, 240–248. [CrossRef]
4. Wang, A.S.; Angle, J.S.; Chaney, R.L.; Delorme, T.A.; McIntosh, M. Changes in soil biological activities under reduced soil pH during *Thlaspi caerulescens* phytoextraction. *Soil Biol. Biochem.* **2006**, *38*, 1451–1461. [CrossRef]
5. Ahmad, F.; Ahmad, I.; Khan, M.S. Screening of free-living rhizospheric bacteria for their multiple plant growth promoting activities. *Microbiol. Res.* **2008**, *163*, 173–181. [CrossRef]
6. Wu, S.C.; Cao, Z.H.; Li, Z.G.; Cheung, K.C.; Wong, M.H. Effects of biofertilizer containing N-fixer, P and K solubilizers and AM fungi on maize growth: A greenhouse trial. *Geoderma* **2005**, *125*, 155–166. [CrossRef]
7. Haghighi, B.J.; Alizadeh, O.; Firoozabadi, A.H. The role of plant growth promoting rhizobacteria (PGPR) in sustainable agriculture. *Adv. Environ. Biol.* **2011**, *5*, 3079–3083.
8. Antoun, H.; Kloepper, J.W. Plant growth promoting rhizobacteria. In *Encyclopedia of Genetics*; Brenner, S., Miller, J.H., Eds.; Academic Press: New York, NY, USA, 2001; pp. 1477–1480.
9. Zakry, F.A.A.; Shamsuddin, Z.H.; Rahim, K.A.; Zakaria, Z.Z.; Rahim, A.A. Inoculation of *Bacillus sphaericus* UPMB-10 to young oil palm and measurement of its uptake of fixed nitrogen using the 15N isotope dilution technique. *Microbes Environ.* **2012**, *27*, 257–262. [CrossRef] [PubMed]
10. Vansuyt, G.; Robin, A.; Briat, J.F.; Curie, C.; Lemanceau, P. Iron acquisition from Fe-pyoverdine by Arabidopsis thaliana. *Mol. Plant-Microbe Interact.* **2007**, *20*, 441–447. [CrossRef]
11. Yazdani, M.; Bahmanyar, M.A.; Pirdashti, H.; Esmaili, M.A. Effect of phosphate solubilization microorganisms (PSM) and plant growth promoting rhizobacteria (PGPR) on yield and yield components of corn (*Zea mays* L.). *WASET* **2009**, *49*, 90–92.
12. Sandhya, V.Z.A.S.; SK Z., A.; Grover, M.; Reddy, G.; Venkateswarlu, B.S.S.S. Alleviation of drought stress effects in sunflower seedlings by the exopolysaccharides producing *Pseudomonas putida* strain GAP-P45. *Biol. Fertile. Soils* **2009**, *46*, 17–26. [CrossRef]
13. Weyens, N.; Truyens, S.; Dupae, J.; Newman, L.; Taghavi, S.; van der Lelie, D.; Carleer, R.; Vangronsveld, J. Potential of the TCE-degrading endophyte *Pseudomonas putida* W619-TCE to improve plant growth and reduce TCE phytotoxicity and evapotranspiration in poplar cuttings. *Environ. Pollut.* **2010**, *158*, 2915–2919. [CrossRef] [PubMed]
14. Damam, M.; Kaloori, K.; Gaddam, B.; Kausar, R. Plant growth promoting substances (phytohormones) produced by rhizobacterial strains isolated from the rhizosphere of medicinal plants. *Int. J. Pharm. Scie. Rev. Res.* **2016**, *37*, 130–136.
15. Kumar, A.; Singh, M.; Singh, P.P.; Singh, S.K.; Singh, P.K.; Pandey, K.D. Isolation of plant growth promoting rhizobacteria and their impact on growth and curcumin content in *Curcuma longa* L. *Biocatal. Agric.* **2016**, *8*, 1–7. [CrossRef]
16. García de Salamone, I.E.; Hynes, R.K.; Nelson, L.M. Cytokinin production by plant growth promoting rhizobacteria and selected mutants. *Can. J. Microbiol.* **2001**, *47*, 404–411. [CrossRef]
17. Noel, T.C.; Sheng, C.; Yost, C.K.; Pharis, R.P.; Hynes, M.F. *Rhizobium leguminosarum* as a plant growth-promoting rhizobacterium: Direct growth promotion of canola and lettuce. *Can. J. Microbiol.* **1996**, *42*, 279–283. [CrossRef]
18. Gutiérrez-Mañero, F.J.; Ramos-Solano, B.; Probanza, A.N.; Mehouachi, J.R.; Tadeo, F.; Talon, M. The plant-growth-promoting rhizobacteria *Bacillus pumilus* and Bacillus licheniformis produce high amounts of physiologically active gibberellins. *Physiol. Plant* **2001**, *111*, 206–211. [CrossRef]
19. Lugtenberg, B.; Kamilova, F. Plant-growth-promoting rhizobacteria. *Annu. Rev. Microbiol.* **2009**, *63*, 541–556. [CrossRef]
20. Singh, V.K.; Singh, A.K.; Kumar, A. Disease management of tomato through PGPB: Current trends and future perspective. *3 Biotech* **2017**, *7*, 1–10.
21. Uqab, B.; Mudasir, S.; Nazir, R. Review on bioremediation of pesticides. *J. Bioremed. Biodegr.* **2016**, *7*, 343.
22. Vejan, P.; Abdullah, R.; Khadiran, T.; Ismail, S.; Nasrulhaq Boyce, A. Role of plant growth promoting rhizobacteria in agricultural sustainability—A review. *Molecules* **2016**, *21*, 573. [CrossRef]
23. Foyer, C.H.; Rasool, B.; Davey, J.W.; Hancock, R.D. Cross-tolerance to biotic and abiotic stresses in plants: A focus on resistance to aphid infestation. *J. Exp. Bot.* **2016**, *67*, 2025–2037. [CrossRef] [PubMed]
24. Consentino, B.B.; Sabatino, L.; Vultaggio, L.; Rotino, G.L.; La Placa, G.G.; D'Anna, F.; Leto, C.; Iacuzzi, N.; De Pasquale, C. Grafting Eggplant Onto Underutilized Solanum Species and Biostimulatory Action of *Azospirillum brasilense* Modulate Growth, Yield, NUE and Nutritional and Functional Traits. *Horticulturae* **2022**, *8*, 722. [CrossRef]
25. Barbaccia, P.; Dazzi, C.; Franciosi, E.; Di Gerlando, R.; Settanni, L.; Lo Papa, G. Microbiological Analysis and Metagenomic Profiling of the Bacterial Community of an Anthropogenic Soil Modified from Typic Haploxererts. *Land* **2022**, *11*, 748. [CrossRef]
26. Wohler, I. Auxin-indole derivatives in soils determined by a colorimetric method and by high performance liquid chromatography. *Microbiol. Res.* **1997**, *152*, 399–405. [CrossRef]
27. Cappuccino, J.C.; Sherman, N. Negative staining. In *Microbiology: A Laboratory Manual*, 3rd ed.; Cappuccino, J.C., Sherman, N., Eds.; Pearson: London, UK, 1992; pp. 125–179.
28. Schwyn, B.; Neilands, J.B. Universal chemical assay for the detection and determination of siderophores. *Anal. Biochem.* **1987**, *160*, 47–56. [CrossRef]
29. Honma, M.; Shimomura, T. Metabolism of 1-aminocyclopropane-1-carboxylic acid. *Agric. Biol. Chem.* **1978**, *42*, 1825–1831.

30. Surange, S.; Wollum Ii, A.G.; Kumar, N.; Nautiyal, C.S. Characterization of Rhizobium from root nodules of leguminous trees growing in alkaline soils. *Can. J. Microbiol.* **1997**, *43*, 891–894. [CrossRef]
31. Jolliffe, I.T. *Principal Component Analysis for Special Types of Data*; Springer: New York, NY, USA, 2002.
32. Mazzei, P.; Francesca, N.; Moschetti, G.; Piccolo, A. NMR spectroscopy evaluation of direct relationship between soils and molecular composition of red wines from Aglianico grapes. *Anal. Chim. Acta* **2010**, *673*, 167–172. [CrossRef]
33. Ahmad, I.; Pichtel, J.; Hayat, S. (Eds.) *Plant-Bacteria Interactions: Strategies and Techniques to Promote Plant Growth*; John Wiley & Sons: Hoboken, NJ, USA, 2008.
34. Siddiqui, Z.A.; Mahmood, I. Role of bacteria in the management of plant parasitic nematodes: A review. *Bioresour. Technol.* **1999**, *69*, 167–179. [CrossRef]
35. Gray, E.J.; Smith, D.L. Intracellular and extracellular PGPR: Commonalities and distinctions in the plant–bacterium signaling processes. *Soil Biol. Biochem.* **2005**, *37*, 395–412. [CrossRef]
36. Molina, L.; Constantinescu, F.; Michel, L.; Reimmann, C.; Duffy, B.; Défago, G. Degradation of pathogen quorum-sensing molecules by soil bacteria: A preventive and curative biological control mechanism. *FEMS Microbiol.* **2003**, *45*, 71–81. [CrossRef]
37. Compant, S.; Duffy, B.; Nowak, J.; Clément, C.; Barka, E.A. Use of plant growth-promoting bacteria for biocontrol of plant diseases: Principles, mechanisms of action, and future prospects. *Appl. Environ. Microbiol.* **2005**, *71*, 4951–4959. [CrossRef] [PubMed]
38. Spadaro, D.; Gullino, M.L. Improving the efficacy of biocontrol agents against soilborne pathogens. *Crop Prot.* **2005**, *24*, 601–613. [CrossRef]
39. Rodríguez, H.; Fraga, R. Phosphate solubilizing bacteria and their role in plant growth promotion. *Biotechnol. Adv.* **1999**, *17*, 319–339. [CrossRef]
40. Bargabus, R.L.; Zidack, N.K.; Sherwood, J.E.; Jacobsen, B.J. Screening for the identification of potential biological control agents that induce systemic acquired resistance in sugar beet. *Biol. Control* **2004**, *30*, 342–350. [CrossRef]
41. Berkeley, R.; Heyndrickx, M.; Logan, N.; De Vos, P. (Eds.) *Applications and Systematics of Bacillus and Relatives*; John Wiley & Sons: Hoboken, NJ, USA, 2008.
42. Belimov, A.A.; Kunakova, A.M.; Kozhemiakov, A.P.; Stepanok, V.V.; Yudkin, L.Y. Effect of associative bacteria on barley grown in heavy metal contaminated soil. In Proceedings of the International Symposium on Agro-Environmental Issues and Future Strategies: Towards the 21st Century, Faisalakad, Pakistan, May 1998.
43. Antoun, H.; Prévost, D. Ecology of plant growth promoting rhizobacteria. In *PGPR: Biocontrol and Biofertilization*; Siddiqui, Z.A., Ed.; Springer: Berlin, Germany, 2005; pp. 1–38.
44. Rosas, S.; Rovera, M.; Andrés, J.A.; Pastor, N.A.; Guiñazú, L.B.; Carlier, E.; Correa, N.S. Characterization of Pseudomonas aurantiaca as biocontrol and PGPR agent. Endophytic properties. In Proceedings of the Prospects and Applications for Plant Associated Microbes, 1st International Conference on Plant–Microbe Interactions: Endophytes and Biocontrol Agents, Finland, Lapland, 18–22 April 2005.
45. Gamalero, E.; Glick, B.R. Mechanisms used by plant growth-promoting bacteria. In *Bacteria in Agrobiology: Plant Nutrient Management*; Springer: Berlin/Heidelberg, Germany, 2011; pp. 17–46.
46. Seidel, C.; Walz, A.; Park, S.; Cohen, J.D.; Ludwig-Muller, J. Indole-3-acetic acid protein conjugates: Novel players in auxin homeostasis. *Plant Biol.* **2006**, *8*, 340–345. [CrossRef]
47. Apine, O.A.; Jadhav, J.P. Optimization of medium for indole-3-acetic acid production using *Pantoea agglomerans* strain PVM. *J. Appl. Microbiol.* **2011**, *110*, 1235–1244. [CrossRef]
48. Egamberdieva, D. Alleviation of salt stress by plant growth regulators and IAA producing bacteria in wheat. *Acta Physiol. Plant* **2009**, *31*, 861–864. [CrossRef]
49. Tara, N.; Saharan, B.S. Plant growth promoting traits shown by bacteria *Brevibacterium frigrotolerans* SMA23 isolated from *Aloe vera* rhizosphere. *Agric. Sci. Dig.-A Res. J.* **2017**, *37*, 226–231.
50. Wahyudi, A.T.; Priyanto, J.A.; Afrista, R.; Kurniati, D.; Astuti, R.I.; Akhdiya, A. Plant growth promoting activity of actinomycetes isolated from soybean rhizosphere. *Online J. Biol. Sci.* **2019**, *19*, 1–8. [CrossRef]
51. Susilowati, D.N.; Sudiana, I.M.; Mubarik, N.R.; Suwanto, A. Species and functional diversity of rhizobacteria of rice plant in the coastal soils of Indonesia. *Indones. J. Agric. Sci.* **2015**, *16*, 39–50. [CrossRef]
52. Zahid, M.; Abbasi, M.K.; Hameed, S.; Rahim, N. Isolation and identification of indigenous plant growth promoting rhizobacteria from Himalayan region of Kashmir and their effect on improving growth and nutrient contents of maize (*Zea mays* L.). *Front. Microbiol.* **2015**, *6*, 207. [CrossRef] [PubMed]
53. Verma, J.P.; Yadav, J.; Tiwari, K.N. Enhancement of nodulation and yield of chickpea by co-inoculation of indigenous *mesorhizobium* spp. and Plant Growth–Promoting Rhizobacteria in Eastern Uttar Pradesh. *Commun. Soil Sci. Plant Anal.* **2012**, *43*, 605–621. [CrossRef]
54. Verma, J.P.; Yadav, J.; Tiwari, K.N.; Kumar, A. Effect of indigenous *Mesorhizobium* spp. and plant growth promoting rhizobacteria on yields and nutrients uptake of chickpea (*Cicer arietinum* L.) under sustainable agriculture. *Ecol. Eng.* **2013**, *51*, 282–286. [CrossRef]
55. Duca, D.; Lorv, J.; Patten, C.L.; Rose, D.; Glick, B.R. Indole-3-acetic acid in plant–microbe interactions. *Anton. Leeuw.* **2014**, *106*, 85–125. [CrossRef]
56. Jha, B.; Gontia, I.; Hartmann, A. The roots of the halophyte *Salicornia brachiata* are a source of new halotolerant diazotrophic bacteria with plant growth-promoting potential. *Plant Soil* **2012**, *356*, 265–277. [CrossRef]

57. Howell, C.R.; Beier, R.C.; Stipanovic, R.D. Production of ammonia by Enterobacter cloacae and its possible role in the biological control of Pythium preemergence damping-off by the bacterium. *Phytopathology* **1988**, *78*, 1075–1078. [CrossRef]
58. Joseph, B.; Patra, R.R.; Lawerence, R. Characterization of plant growth promoting rhizobacteria with chickpea (*Cicer arietinum* L.). *Int. J. Plant Prod.* **2007**, *1*, 141–151.
59. Flaishman, M.A.; Eyal, Z.A.; Zilberstein, A.; Voisard, C.; Hass, D. Suppression of *Septoria tritici* blotch and leaf rust of wheat by recombinant cyanide producing strains of *Pseudomonas putida*. *Mol. Plant Microbe Int.* **1996**, *9*, 642–645. [CrossRef]
60. Marques, A.P.; Pires, C.; Moreira, H.; Rangel, A.O.; Castro, P.M. Assessment of the plant growth promotion abilities of six bacterial isolates using *Zea mays* as indicator plant. *Soil Biol. Biochem.* **2010**, *42*, 1229–1235. [CrossRef]
61. Kanchiswamy, C.N.; Malnoy, M.; Maffei, M.E. Chemical diversity of microbial volatiles and their potential for plant growth and productivity. *Front. Plant Sci.* **2015**, *6*, 151. [CrossRef] [PubMed]
62. Prasad, M.; Srinivasan, R.; Chaudhary, M.; Choudhary, M.; Jat, L.K. Plant growth promoting rhizobacteria (PGPR) for sustainable agriculture: Perspectives and challenges. In *PGPR Amelioration in Sustainable Agriculture*; Singh, A.K., Kumar, A., Singh, P.K., Eds.; Elsevier: Amsterdam, The Netherland, 2019; pp. 129–157.
63. Ammari, T.; Mengel, K. Total soluble Fe in soil solutions of chemically different soils. *Geoderma* **2006**, *136*, 876–885. [CrossRef]
64. Whipps, J.M. Microbial interactions and biocontrol in the rhizosphere. *J. Exp. Bot.* **2001**, *52*, 487–511. [CrossRef]
65. Li, M.; Ahammed, G.J.; Li, C.; Bao, X.; Yu, J.; Huang, C.; Yin, H.; Zhou, J. Brassinosteroid ameliorates zinc oxide nanoparticles-induced oxidative stress by improving antioxidant potential and redox homeostasis in tomato seedling. *Front. Plant Sci.* **2016**, *7*, 615. [CrossRef]
66. Beneduzi, A.; Ambrosini, A.; Passaglia, L.M. Plant growth-promoting rhizobacteria (PGPR): Their potential as antagonists and biocontrol agents. *Genet. Mol. Biol.* **2012**, *35*, 1044–1051. [CrossRef]
67. Rasool, A.; Mir, M.I.; Zulfajri, M.; Hanafiah, M.M.; Unnisa, S.A.; Mahboob, M. Plant growth promoting and antifungal asset of indigenous rhizobacteria secluded from saffron (*Crocus sativus* L.) rhizosphere. *Microb. Pathog.* **2021**, *150*, 104734. [CrossRef] [PubMed]
68. Wani, P.A.; Khan, M.S. *Bacillus* species enhance growth parameters of chickpea (*Cicer arietinum* L.) in chromium stressed soils. *Food Chem. Toxicol.* **2010**, *48*, 3262–3267. [CrossRef] [PubMed]
69. Koo, S.Y.; Cho, K.S. Isolation and characterization of a plant growth-promoting rhizobacterium, *Serratia* sp. SY5. *J. Microbiol. Biotechnol.* **2009**, *19*, 1431–1438. [PubMed]
70. Sriprang, R.; Hayashi, M.; Ono, H.; Takagi, M.; Hirata, K.; Murooka, Y. Enhanced accumulation of Cd21 by a *Mesorhizobium* sp. transformed with a gene from *Arabidopsis thaliana* coding for phytochelatin synthase. *Appl. Environ. Microbiol.* **2003**, *69*, 1791–1796. [CrossRef] [PubMed]
71. Rajkumar, M.; Freitas, H. Influence of metal resistant-plant growth-promoting bacteria on the growth of *Ricinus communis* in soil contaminated with heavy metals. *Chemosphere* **2008**, *71*, 834–842. [CrossRef]
72. Ma, Y.; Rajkumar, M.; Freitas, H. Improvement of plant growth and nickel uptake by nickel resistant-plant-growth promoting bacteria. *J. Hazard. Mater.* **2009**, *166*, 1154–1161. [CrossRef] [PubMed]
73. Van Loon, L.C.V.; Glick, B.R. Increased plant fitness by rhizobacteria. In *Molecular Ecotoxicology of Plants*; Sabderman, H., Jr., Ed.; Springer: Berlin/Heidelberg, Germany, 2004; Volume 170, pp. 177–205.
74. Glick, B.R.; Cheng, Z.; Czarny, J.; Duan, J. Promotion of plant growth by ACC deaminase-producing soil bacteria. In *New Perspectives and Approaches in Plant Growth-Promoting Rhizobacteria Research*; Bakker, P.A., Raaijmakers, J.M., Bloemberg, G., Höfte, M., Lemanceau, P., Cooke, B.M., Eds.; Springer: Berlin/Heidelberg, Germany, 2007; pp. 329–339.
75. Bayliss, C.; Bent, E.; Culham, D.E.; MacLellan, S.; Clarke, A.J.; Brown, G.L.; Wood, J.M. Bacterial genetic loci implicated in the *Pseudomonas putida* GR12-2R3-canola mutualism: Identification of an exudate-inducible sugar transporter. *Can. J. Microbiol.* **1997**, *43*, 809–818. [CrossRef] [PubMed]
76. Penrose, D.M.; Glick, B.R. Levels of ACC and related compounds in exudate and extracts of canola seeds treated with ACC deaminase-containing plant growth-promoting bacteria. *Can. J. Microbiol.* **2001**, *47*, 368–372. [CrossRef]
77. Tiryaki, D.; Aydın, İ.; Atıcı, Ö. Psychrotolerant bacteria isolated from the leaf apoplast of cold-adapted wild plants improve the cold resistance of bean (*Phaseolus vulgaris* L.) under low temperature. *Cryobiology* **2019**, *86*, 111–119. [CrossRef] [PubMed]
78. Palacio-Rodríguez, R.; Coria-Arellano, J.L.; López-Bucio, J.; Sánchez-Salas, J.; Muro-Pérez, G.; Castañeda-Gaytán, G.; Sáenz-Mata, J. Halophilic rhizobacteria from *Distichlis spicata* promote growth and improve salt tolerance in heterologous plant hosts. *Symbiosis* **2017**, *73*, 179–189. [CrossRef]
79. Misra, S.; Dixit, V.K.; Khan, M.H.; Mishra, S.K.; Dviwedi, G.; Yadav, S.; Lehri, A.; Chauhan, P.S. Exploitation of agro-climatic environment for selection of 1-aminocyclopropane-1-carboxylic acid (ACC) deaminase producing salt tolerant indigenous plant growth promoting rhizobacteria. *Microbiol. Res.* **2017**, *205*, 25–34. [CrossRef] [PubMed]
80. Anand, K.; Kumari, B.; Mallick, M. Phosphate solubilizing microbes: An effective and alternative approach as biofertilizers. *Int. J. Pharm. Pharm.* **2016**, *8*, 37–40.
81. Sharma, S.B.; Sayyed, R.Z.; Trivedi, M.H.; Gobi, T.A. Phosphate solubilizing microbes: Sustainable approach for managing phosphorus deficiency in agricultural soils. *Springer Plus* **2013**, *2*, 587. [CrossRef]
82. Chen, Y.P.; Rekha, P.D.; Arun, A.B.; Shen, F.T.; Lai, W.A.; Young, C.C. Phosphate solubilizing bacteria from subtropical soil and their tricalcium phosphate solubilizing abilities. *Appl. Soil Ecol.* **2006**, *34*, 33–41. [CrossRef]

83. Illmer, P.; Schinner, F. Solubilization of inorganic calcium phosphates-solubilization mechanisms. *Soil Biol. Biochem.* **1995**, *27*, 265–270. [CrossRef]
84. Sperber, J.I. The incidence of apatite-solubilizing organisms in the rhizosphere and soil. *Aust. J. Aric. Res.* **1958**, *9*, 778–781. [CrossRef]
85. Duff, R.B.; Webley, D.M. 2-Ketogluconic acid as a natural chelator produced by soil bacteria. *Chem. Ind.* **1959**, *1959*, 1376–1377.
86. Di Simine, C.D.; Sayer, J.A.; Gadd, G.M. Solubilization of zinc phosphate by a strain of *Pseudomonas fluorescens* isolated from a forest soil. *Biol. Fert. Soils* **1998**, *28*, 87–94. [CrossRef]
87. Gulati, A.; Rahi, P.; Vyas, P. Characterization of phosphate-solubilizing fluorescent *Pseudomonas* from the rhizosphere of seabuckthorn growing in the cold deserts of Himalayas. *Curr. Microbiol.* **2008**, *56*, 73–79. [CrossRef] [PubMed]
88. Park, K.H.; Lee, C.Y.; Son, H.J. Mechanism of insoluble phosphate solubilization by *Pseudomonas fluorescens* RAF15 isolated from ginseng rhizosphere and its plant growth-promoting activities. *Lett. Appl. Microbiol.* **2009**, *49*, 222–228. [CrossRef] [PubMed]
89. Malboobi, M.A.; Behbahani, M.; Madani, H.; Owlia, P.; Deljou, A.; Yakhchali, B.; Moradi, M.; Hassanabadi, H. Performance evaluation of potent phosphate solubilizing bacteria in potato rhizosphere. *World J. Microbiol. Biotechnol.* **2009**, *25*, 1479–1484. [CrossRef]
90. Zhang, M.; Yang, L.; Hao, R.; Bai, X.; Wang, Y.; Yu, X. Drought-tolerant plant growth-promoting rhizobacteria isolated from jujube (*Ziziphus jujuba*) and their potential to enhance drought tolerance. *Plant Soil* **2020**, *452*, 423–440. [CrossRef]
91. Molla, M.A.Z.; Chowdhury, A.A.; Islam, A.; Hoque, S. Microbial mineralization of organic phosphate in soil. *Plant Soil* **1984**, *78*, 393–399. [CrossRef]
92. Mba, C.C. Rock phosphate solubilizing and cellulolytic actinomycetes isolates of earthworm casts. *Environ. Manag.* **1994**, *18*, 257–261. [CrossRef]
93. Hamdali, H.; Hafidi, M.; Virolle, M.J.; Ouhdouch, Y. Rock phosphate-solubilizing Actinomycetes: Screening for plant growth-promoting activities. *World J. Microbiol. Biotechnol.* **2008**, *24*, 2565–2575. [CrossRef]
94. Chang, C.H.; Yang, S.S. Thermo-tolerant phosphate-solubilizing microbes for multi-functional biofertilizer preparation. *Bioresour. Technol.* **2009**, *100*, 1648–1658. [CrossRef] [PubMed]
95. de Freitas, J.R.; Banerjee, M.R.; Germida, J.J. Phosphate-solubilizing rhizobacteria enhance the growth and yield but not phosphorus uptake of canola (*Brassica napus* L.). *Biol. Fertil. Soils* **1997**, *24*, 358–364. [CrossRef]
96. Toro, M.; Azcon, R.; Barea, J. Improvement of arbuscular mycorrhiza development by inoculation of soil with phosphate-solubilizing rhizobacteria to improve rock phosphate bioavailability ((sup32) P) and nutrient cycling. *Appl. Environ. Microbiol.* **1997**, *63*, 4408–4412. [CrossRef] [PubMed]
97. Rojas, A.; Holguin, G.; Glick, B.R.; Bashan, Y. Synergism between *Phyllobacterium* sp. (N2-fixer) and *Bacillus licheniformis* (P-solubilizer), both from a semiarid mangrove rhizosphere. *FEMS Microbiol. Ecol.* **2001**, *35*, 181–187. [CrossRef] [PubMed]
98. Sahin, F.; Cakmakci, R.; Kantar, F. Sugar beet and barley yields in relation to inoculation with N2-fixing and phosphate solubilizing bacteria. *Plant Soil* **2004**, *265*, 123–129. [CrossRef]
99. Castellano-Hinojosa, A.; Pérez-Tapia, V.; Bedmar, E.J.; Santillana, N. Purple corn-associated rhizobacteria with potential for plant growth promotion. *J. Appl. Microbiol.* **2018**, *124*, 1254–1264. [CrossRef]
100. Fan, Z.Y.; Miao, C.P.; Qiao, X.G.; Zheng, Y.K. Diversity, distribution, and antagonistic activities of rhizobacteria of *Panax notoginseng*. *J. Ginseng Res.* **2016**, *40*, 97–104. [CrossRef]
101. Bender, R.R.; Haegele, J.W.; Ruffo, M.L.; Below, F.E. Nutrient uptake, partitioning, and remobilization in modern, transgenic insect-protected maize hybrids. *Agron. J.* **2013**, *105*, 161–170. [CrossRef]
102. Kifle, M.H.; Laing, M.D. Isolation and screening of bacteria for their diazotrophic potential and their influence on growth promotion of maize seedlings in greenhouses. *Front. Plant Sci.* **2016**, *6*, 1225. [CrossRef]

microorganisms

MDPI

Article

Characterization of Growth-Promoting Activities of Consortia of Chlorpyrifos Mineralizing Endophytic Bacteria Naturally Harboring in Rice Plants—A Potential Bio-Stimulant to Develop a Safe and Sustainable Agriculture

Md. Yeasin Prodhan [1,†], Md. Bokhtiar Rahman [1,†], Aminur Rahman [2,*], Md. Ahedul Akbor [3], Sibdas Ghosh [4], Mst. Nur-E-Nazmun Nahar [1], Simo [1], Md. Shamsuzzoha [5], Kye Man Cho [6] and Md. Azizul Haque [1,*]

1 Department of Biochemistry and Molecular Biology, Hajee Mohammad Danesh Science and Technology University, Dinajpur 5200, Bangladesh; bokhtiarbdj@gmail.com (M.B.R.); nnnbhstu@gmail.com (M.N.-E.-N.N.); tahminasimo148@gmail.com (S.)
2 Department of Biomedical Sciences, College of Clinical Pharmacy, King Faisal University, Al-Ahsa 31982, Saudi Arabia
3 Institute of National Analytical Research and Services (INARS), Bangladesh Council of Scientific and Industrial Research (BCSIR), Dhaka 1205, Bangladesh; akborbcsir@yahoo.com
4 Department of Biological Sciences, College of Arts and Sciences, Carlow University, 3333 Fifth Avenue, Pittsburgh, PA 15213, USA; sghosh@carlow.edu
5 Department of Chemistry, Hajee Mohammad Danesh Science and Technology University, Dinajpur 5200, Bangladesh; ms_zoha2006@yahoo.com
6 Department of Green Bio Science and Agri-Food Bio Convergence Institute, Gyeongsang National University, Jinju 52725, Republic of Korea; kmcho@gnu.ac.kr
* Correspondence: marahman@kfu.edu.sa (A.R.); helalbmb2016@hstu.ac.bd (M.A.H.); Tel.: +966-(0)-547757460 (A.R.); +880-1761106098 (M.A.H.)
† These authors contributed equally to this work and both should be considered as first authors.

Citation: Prodhan, M.Y.; Rahman, M.B.; Rahman, A.; Akbor, M.A.; Ghosh, S.; Nahar, M.N.-E.-N.; Simo; Shamsuzzoha, M.; Cho, K.M.; Haque, M.A. Characterization of Growth-Promoting Activities of Consortia of Chlorpyrifos Mineralizing Endophytic Bacteria Naturally Harboring in Rice Plants—A Potential Bio-Stimulant to Develop a Safe and Sustainable Agriculture. *Microorganisms* **2023**, *11*, 1821. https://doi.org/10.3390/microorganisms11071821

Academic Editors: Vishnu D. Rajput, Anuj Ranjan and Abhishek Chauhan

Received: 12 June 2023
Revised: 3 July 2023
Accepted: 12 July 2023
Published: 16 July 2023

Abstract: Eighteen pesticide-degrading endophytic bacteria were isolated from the roots, stems, and leaves of healthy rice plants and identified through 16S rRNA gene sequencing. Furthermore, biochemical properties, including enzyme production, dye degradation, anti-bacterial activities, plant-growth-promoting traits, including N-fixation, P-solubilization, auxin production, and ACC-deaminase activities of these naturally occurring endophytic bacteria along with their four consortia, were characterized. *Enterobacter cloacae* HSTU-ABk39 and *Enterobacter* sp. HSTU-ABk36 displayed inhibition zones of 41.5 ± 1.5 mm, and 29 ± 09 mm against multidrug-resistant human pathogenic bacteria *Staphylococcus aureus* and *Staphylococcus epidermidis*, respectively. FT-IR analysis revealed that all eighteen isolates were able to degrade chlorpyrifos *pesticide*. Our study confirms that pesticide-degrading endophytic bacteria from rice plants play a key role in enhancing plant growth. Notably, rice plants grown in pots containing reduced urea (30%) mixed with either endophytic bacterial consortium-1, consortium-2, consortium-3, or consortia-4 demonstrated an increase of 17.3%, 38.6%, 18.2%, and 39.1% yields, respectively, compared to the control plants grown in pots containing 100% fertilizer. GC–MS/MS analysis confirmed that consortia treatment caused the degradation of chlorpyrifos into different non-toxic metabolites, including 2-Hydroxy-3,5,6 trichloropyridine, Diethyl methane phosphonate, Phorate sulfoxide, and Carbonochloridic. Thus, these isolates could be deployed as bio-stimulants to improve crop production by creating a sustainable biological system.

Keywords: pesticide-degrading endophyte; growth promotion; MDR bacterial inhibition; synthetic consortia; GC–MS/MS analysis; rice plant; yields enhancement

1. Introduction

Agriculture remains one of the most important economic sectors in Bangladesh and plays a crucial role in the rural economy of the country. Thus, the performance of this sector is critical to irradicating poverty. Excessive use of pesticides discharging into surrounding water ultimately impregnates populations through direct and indirect contact, leading to fatal non-communicable human diseases [1]. Furthermore, the excessive importation of fertilizers and pesticides contributes to the interruption of reserve dollar in the national economy and an increase in the prevalence of non-communicable diseases. All in all, this interferes with the sustainable development goals of Bangladesh. Continuing to achieve a steady increase in food production along with human health requires favorable weather conditions and efficient use of fertilizers along with measured application of biological control systems for pests, insects, and fungi. One such example includes the use of growth-promoting endophytic bacteria with capabilities of degrading pesticides.

Endophytic bacteria ubiquitously colonize the internal tissues of plants and are found in nearly every plant worldwide, which can promote the growth of plants through increased germination rates, biomass, leaf area, chlorophyll content, nitrogen content, protein content, hydraulic activity, roots and shoot length, yield and tolerance to abiotic stresses, including but not limited to drought, flood, and salinity [2,3]. Furthermore, endophytic bacteria have been demonstrated to promote plant growth directly through biological nitrogen fixation, phytohormone production, phosphate solubilization, inhibition of ethylene biosynthesis under both biotic or abiotic stresses (induced systemic tolerance), or indirectly through inducing resistance to the pathogen [4]. Some of these endophytes are reported to have generated adaptation against fungicides, herbicides, insecticides, and pesticides [5] by their abilities to degrade and/or metabolize these organic-based synthetic control agents [6,7]. The biochemical mechanisms involving degradation of organophosphate pesticides include adsorption, hydrolysis of P–O alkyl and aryl bonds, photodegradation, and enzymatic mineralization. Microbe-specific enzymes, including esterase, organophosphorus hydrolase, amidohydrolase, carboxylesterase, phosphotriesterase [6,8], diisopropyl fluorophosphatase, parathion hydrolase, and paraoxonase [9,10], have been demonstrated to be involved in the degradation of insecticides.

Rice is the staple food grain for more than 3.5 billion people around the world, particularly in Asia, Latin America, and parts of Africa, and serves as an important source of fiber, energy, minerals, vitamins, bioactive compounds, among other biomolecules [11,12]. Due to high year-round demands, farmers extensively adopt high-yielding varieties for increased production. However, the commercial farming of this vital crop is under immense threat from pests, insects, and diseases along with both biotic and abiotic stresses, leading to annual loss of yield up to 50% globally [13]. According to the Bangladesh Bureau of Statistics (2021), 54,500 metric tons of pesticide was applied in 2020, of which 20,896 metric tons was insecticides, to combat pests and diseases. In 2022, the Government of Bangladesh subsidized 300,000 million takas (over US $3M) to import additional fertilizers, a severe threat to the national economy. Therefore, controlling pests and pathogens in rice cultivation with potential consortia of pesticide-degrading endophytic bacteria might be a bio-solution ensuring enhanced yield along with the security of food and human health.

The present study aims to isolate and identify naturally occurring pesticide-mineralizing endophytic bacteria in rice plants with potential growth-promoting activities. A consortium of the isolates is identified to be used as a biofertilizer to increase the yield of rice with the following five objectives: (i) isolate and identify pesticide-mineralizing endophytic bacteria from the roots, shoots, and leaves of BRRI-28 and Kalijera rice plants from different fields, (ii) select the isolates showing accelerated plant-growth-promoting traits using biochemical analysis, (iii) determine the pesticide-mineralizing activities of the isolates in vitro using FT-IR and GC–MS, (iv) identify the inhibitory activities of the isolates against multidrug-resistant human pathogenic bacteria, and (v) conduct pot experiments under field conditions to determine the impact of consortia comprising the isolates on rice yields.

2. Materials and Methods

2.1. Sample Collection and Processing

Varieties of rice plants, *Kalijeera* and BRRI-28, were collected from two different paddy fields near Basher Hat, Dinajpur, Bangladesh. These locations were exposed frequently to organophosphate pesticides chlorpyrifos and diazinon for several years prior to our collections of the rice plants in 2020. These two locations were strategically selected to collect naturally occurring endophytic bacterial strains resistant to chlorpyrifos or capable of mineralizing chlorpyrifos. Two months after transplantation, healthy rice plants exposed to chlorpyrifos pesticide at least twice were selected, carefully cut into small pieces, and washed with tap water to remove soil and dust. Tissue samples were then surface sterilized with 75% ethanol for 3 min, followed by shaking in 1.2% (*w*/*v*) sodium hypochlorite (NaOCl) solution for 20 min. Samples were then washed thrice with sterile distilled water by shaking for 20 min each time. Only samples that demonstrated no infections, as described in ref. [14], were used for further experiments. Juices were extracted from each sterilized root, shoot, and leaf tissue separately using mortar and pestle and collected into sterilized test tubes for further analyses.

2.2. Screening for Pesticide-Degrading Bacteria

Endophytic bacteria were screened as previously described by Gyaneshwar et al. (2001) [13]. Briefly, the aliquot of juices from specific plant tissues was centrifuged at 1300 rpm at room temperature under aseptic conditions for 10 min. The supernatants were serially diluted up to 10^{-5}, and each dilution was transferred to a pesticide-containing liquid medium in a conical flask and incubated at 37 °C with 130 rpm for 4 days. Then, the samples were cultured into the pesticide-containing medium as described [15]. The pesticide-degrading bacterial isolates were selected with the streak plate method containing 1 gm/100 mL of chlorpyrifos [7,15,16]. The isolates were selected based on distinct colony morphology and growth criteria, and this procedure was repeated several times until the pure colonies were achieved. Finally, the isolates of pure colonies were grown into tryptic soy broth liquid medium at 37 °C for 24 h and stored at 4 °C in the short term and at −20 °C in 50% (*w*/*v*) glycerol until further use.

2.3. Molecular Characterization and Phylogeny of Endophytic Bacteria

The genomic DNA of bacterial isolates was extracted as described by Haque et al. (2015) [14]. The amplification of 16S rRNA gene was performed using forward primer 27F 5′-AGA GTT TGA TCM TGG CTC AG-3′ and reverse primer 1492R 5′-GGT TAC CTT GTT ACG ACT T-3′ [15]. The primers, template DNA, and Taq DNA polymerase were added into the master mix right before loading the sample. After PCR reaction, the amplification was visualized by gel electrophoresis. After purification of the amplified 16S rRNA gene, the concentration was measured, diluted into 5 ng/µL, mixed with a ready reaction premix, and run for PCR sequences using genetic analyzer 3130 (Applied Biosystem, CL, Beverly, MA, USA). The resulting sequences were analyzed by BLASTn, submitted to NCBI, and the accession numbers were registered. The 16S rRNA sequences were compared to different 16S rRNA genes of other bacteria in the reference RNA sequences from NCBI nucleotide BLAST. The query sequences were used to create a phylogenetic tree using the neighbor-joining method, where the bootstrap test included 2000 replicates [16].

2.4. FT-IR for Pesticide-Degrading Activity Confirmation

The FT-IR spectroscopy was performed as described by Pourbabaee et al. (2018) [16] to obtain information about the qualitative changes of chlorpyrifos in bacteria-treated solution. Eighteen isolates were incubated in chlorpyrifos-containing liquid media for 14 days. From each culture, 5 mL was extracted using 10 mL n-hexane on a rotary shaker for 30 min, dehydrated via anhydrous sodium sulfate (Na_2SO_4), evaporated to dryness, subsequently diluted to a final volume of 5 mL with acetone, and analyzed by using FT-IR spectrophotometer in the range of 400–4000 cm^{-1} at 20 °C [17].

2.5. Biochemical Analysis

Biochemical tests were performed to characterize each isolate. Fresh bacterial isolates were grown on autoclaved nutrient agar media. The catalase and oxidase activities of the isolates were conducted as described [7]. Briefly, each isolate was incubated in Simmons Citrate agar medium, and a change of color from green to blue due to pH change indicated a positive reaction after incubation for 48 h at 37 °C [18]. The tests for Indole, Methyl red and Voges–Proskauer, urease, motility, triple sugar iron agar, glucose, maltose, lactose, and sucrose fermentation were performed as described [7,19]. The activities of cell wall hydrolytic enzymes, including cellulase, xylanase, pectinase, amylase, and protease, were performed in minimal nutrient agar media containing carboxymethylcellulose, oat-spelt xylan, pectin, starch, and casein powder (1%) as the sole source of carbon [15,20]. The lignin derivatives' degrading activity of the isolates was confirmed using their growth on aromatic-dye-enriched minimal nutrient media [20,21]. A test tube containing the pure culture of endophytic bacteria was inoculated with 10 mL of phenol red broth supplemented with 1 g/100 mL of various sources of carbohydrates and incubated for 24 h at 37 °C. A yellow color indicated positive reaction, and the bubbles trapped inside the Durham tube indicated gas production [18]. The pure colonies were subjected on specific substrate agar plates, as previously described [15].

2.6. Indole-3-Acetic Acid (IAA) and ACC Deaminase Production

The quantification method for IAA of the endophytes was adopted as previously described [22]. To measure the production of IAA, bacterial isolates were inoculated into 0.5 mg L-tryptophan/mL containing medium and at 37 °C with continuous shaking at 125 rpm for 48 h, as described [23]. Then, the 2 mL culture was centrifuged at 15,000 rpm for 1 min, and a 1 mL aliquot of the supernatant was mixed with 2 mL of Salkowski's reagent, incubated for 20 min in darkness at room temperature. The absorbance was measured on a spectrophotometer at 530 nm, and the concentration was determined using a standard curve of pure IAA, as previously described [22].

The ACC deaminase activity of the chlorpyrifos-degrading bacteria was determined according to the modified methods [24–26], which measure the amount of α-ketobutyrate produced upon the hydrolysis of ACC. The endophytic bacterial strains were separately grown in tryptic soy broth medium (TSB) for 18 h at 28 °C to determine the ACC deaminase activity as described in our study [21]. The cell suspension without ACC was used as a negative control, and the one with $(NH_4)_2SO_4$ (0.2% *w*/*v*) was used as a positive control. The number of μmol of α-ketobutyrate produced by this reaction was determined by comparing the absorbance at 540 nm of a sample to a standard curve of α-ketobutyrate ranging between 10 and 200 μmol [24,26].

2.7. Phosphate and Nitrogen Solubilization

The N-fixation ability of the endophytic bacteria was determined by streaking isolates on the Jensen's medium. A yellow halo zone around bacterial growth after 5–7 days incubation at 37 ± 2 °C indicated positive N-fixation activity [7]. The phosphate solubilization by isolates growing in 'National Botanical Research Institute's Phosphate Growth Medium' was observed with a halo zone [7,27].

2.8. Anti-Bacterial Activity against Multidrug-Resistant Bacteria

Four different multidrug-resistant human pathogenic bacteria, including *S. aureus, E. coli, Klebseilla* sp., *S. epidermidis,* were collected from Dhaka Central International Medical College and Hospital, Dhaka, Bangladesh. Sensitivity tests of all endophytes against human pathogenic bacteria were conducted using the procedures described [15]. The diameters of the inhibition zones were measured in millimeters after 16, 32, and 48 h of inoculation [15]. Since the multidrug-resistant (MDR) pathogenic bacteria did not respond to any known antibiotics or known strains within our capacity, we used *Bacillus* sp. Strain HSTU-10 (MG582603) as a negative control.

2.9. Rice Plant Growth-Promoting Effects of Endophytic Bacteria

The crew members of each consortium were deliberately selected to incorporate a diverse range of bacterial genera, aiming to maximize the combined effects of promoting plant growth in rice plants. The following four synthetic consortia were created using various combinations of the endophytic bacterial isolates to test the effectiveness of growth-promoting activities of the pesticide-degrading endophytic bacteria: consortium-1: *Klebsiella* sp. HSTU-Bk11, *Acinetobacter* sp. HSTU-Abk29, *Citrobacter* sp. HSTU-Abk30, and *Enterobacter cloacae* HSTU-Abk39; consortium-2: *Enterobacter cloacae* HSTU-Abk37, *Enterobacter* ludwigii HSTU-Abk40, *Acinetobacter baumannii* HSTU-ABK42, *Klebsiella* sp. HSTU-Abk31, *Acinetobacter* sp. HSTU-Bk12; consortium-3: *Pseudomonas* sp. HSTU-Bk13, *Citrobacter* sp. HSTU-Bk14, *Acinetobacter* sp. HSTU-Bk15, *Acinetobacter* sp. HSTU-Abk32, and *Burkholderia* sp. HSTU-ABK33; and consortium-4: *Acinetobacter* sp. HSTU-Abk34, *Enterobacter* HSTU-Abk36, *Enterobacter* sp. HSTU-Abk38, and *Serratia marcescens* HSTU-Abk41. All endophytes were tested for their compatibility for consortium preparation [28]. In short, each endophytic isolate was grown in nutrient broth for 24 h (10^6 CFU/mL for rice plant) and was used as inoculum. One loopful of each endophytic isolate was streaked on the opposite side of the medium in a Petri plate and then incubated at 30 °C for 48–72 h [28]. The endophytic bacterial effects on the growth promotion of rice plant genotype (*Shorumukhi*) were assayed at germination along with the vegetative-to-paddy yield stages. To this end, the germination test was performed with individual endophytes and their consortia in Petri plates. The effects of endophytes on the vegetative and reproductive stages were assessed with synthetic consortia.

2.10. Seed Germination Performance

Twenty-five rice seeds sterilized with 70% ethanol for 5 min and washed with distilled water five times in a septic condition were plated on a Petri dish containing sterilized 1% agar media. Autoclaved distilled water was used for all experiments. Bacterial treatment was provided individually at 100 µL/Petri dish after 24 h (approximately 10^6) CFU/mL, except for the control plate. Following the sowing of seeds, germination was recorded at 24 h intervals and continued up to 6 days, when the seed was considered germinated, as the plumule and radicle were >2 mm long [29]. The bacterial inoculum was prepared following a previously published method [30]. The effects of both the individual members of the bacterial consortia and the effects of each of the four consortia were recorded. The germination percentages of the seeds were calculated. The root and shoot lengths of individual seedlings were measured after 7 days of sowing [29], and the vigor index was calculated using percent germination multiplied by seedling length (shoot length + root length).

2.11. Effect of Endophytic Consortia on Growth of Rice Plant and Yield of Grains

The pot experiments under natural conditions were conducted in the paddy field of Hajee Mohamad Danesh Science and Technology University, Dinajpur, Bangladesh, following appropriate procedures, as published by Das et al. (2022) [7]. Dried, sterilized, and pulverized soil was used to prepare eleven pots, and the experiment layouts were as follows: Control = Only Soil; Fer = Soil + Fertilizer (100% urea); Com+ = Compost + Fer; Con-1 = Soil + Consortium-1; Con-2 = Soil + Consortium-2; Con-3 = Soil + Consortium-3; Con-4 = Soil + Consortium-4; Com + Con-1 = Com + Soil + Consortium-1; Com + Con-2 = Com + Soil + Consortium-2; Com + Con-3 = Com + Soil + Consortium-3; and Com+ Con-4 = Com + Soil + Consortium-4. The bacterial treatments were performed in triplicate, and their agronomic data were recorded.

2.12. Chlorophyll Content of Fresh Leaf

Prior to measuring the chlorophyll content, the fresh leaves of rice plants were weighed. Using the spectrophotometric approach, the chlorophyll was extracted in 80% acetone, centrifuged, and the absorption of the extracts at wavelengths of 663 nm (D663) and

645 nm (D645) was measured. Total chlorophyll (Chl-t), chlorophyll a (Chl-a), and chlorophyll b (Chl-b) concentrations were estimated as described (Zhang et al., 2009) [31].

2.13. Root Length, Shoot Length, and Plant Height

The root length was measured from the collar region to the tip of the longest root in centimeters, while the shoot length was measured from the collar region to the tip of the shoot, and mean shoot length was expressed in centimeters. The plant heights were measured from the ground level to the tip of the topmost leaf at early stages (15, 30, 60, 90 days), up to the tip of the main panicle at maturity, and the average height was expressed in centimeters [32].

2.14. Plant Dry Matter Production

The shoots, roots, and leaves were first washed and then air-dried in the shade for 24 to 36 h prior to weighing, and the average dry weight of the plant was expressed in grams [32].

2.15. Yield Parameters (Grain Yield per Plant)

The weight of the grains in the panicles per plant from five tillers selected from randomly labeled plants was recorded, and the mean was expressed in grams [32].

2.16. Harvesting and Observations

The paddy crops were harvested after 120 days of transplantation. The germination, seedling growth parameters, plant growth parameters, plant biomass production, and yield parameters were recorded [32].

2.17. GC–MS/MS Analysis of Chlorpyrifos Degradation by Each of the Four Consortia

To ensure the consortia chlorpyrifos degradation compatibility, the chlorpyrifos (1% as carbon sources) enriched media were treated with each of the four synthetic consortia for 14 days. To perform the GC–MS analysis, 5 mL of consortia treated broth was shifted to separating funnels. Next, 25 mL of deionized water and 5 mL of n-hexane were added to the separating funnel as described [7]. After 5–10 min of shaking, the n-hexane layer with the solvents, which appeared on the upper hexane layer, was kept for further analysis with the Shimadzu GCMS-QP2010 Ultra (Japan) mass detector connected with a capillary column of Rxi-5ms, 30 m long, 0.25 mm i.d., 0.25 µm film thickness. One microliter of the sample was injected in a splitless mode, and the analyses were performed in a full scan mode, ranging from m/z 50 to 400. The compounds were detected after analyzing the mass spectrum of each component using the NIST11 library [33].

2.18. Statistical Analysis

The data were analyzed (frequency, homogeneity of variances, and LSD ($\rho < 0.05$)) and visualized (graph and bar chart) using SPSS, Microsoft Excel, and R language statistical software. In the bar charts, the means and error bars depict standard errors, while each letter indicates significant ($\rho < 0.05$) differences in plant growth parameters between treatments.

3. Results

3.1. Isolation and Selection of Chlorpyrifos-Degrading Endophytic Bacteria

Forty isolates obtained from the roots, shoots, and leaves of *Kalijeera* (indigenous variety) and BRRI-28 rice plants were screened based on their capabilities of utilizing chlorpyrifos as their sole carbon source—a characteristic of pesticide-degrading bacteria. Out of these forty, we selected eighteen isolates by analyzing morphological (size, shape, color) data and biochemical test results. As depicted in Figure 1, all strains demonstrated noticeable growth over 12 days compared to that observed in control.

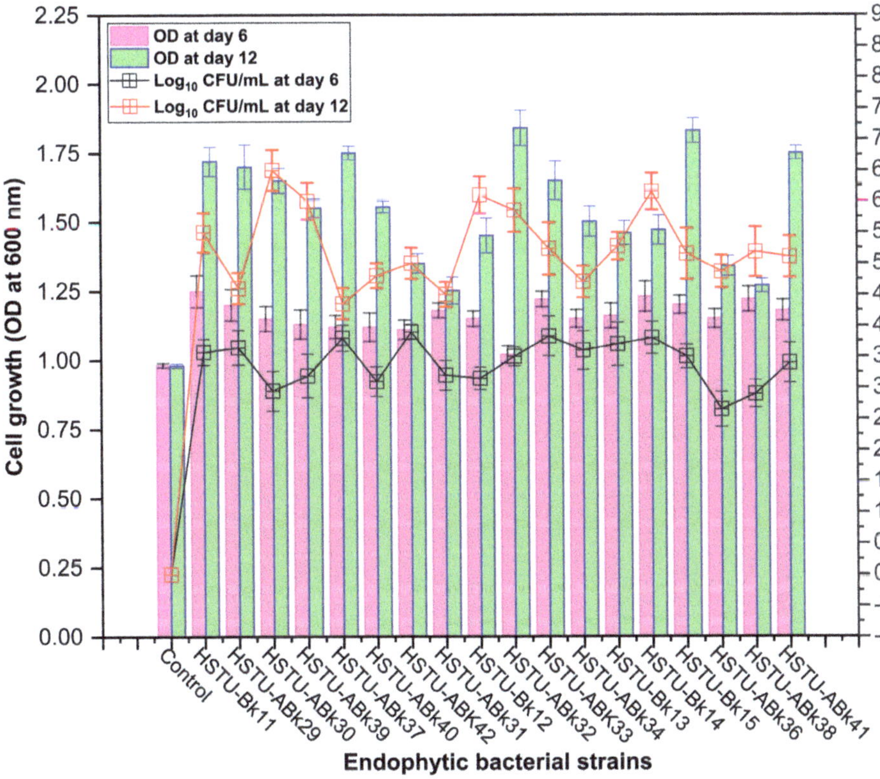

Figure 1. Growth performance of endophytic strains in minimal nutrient media (MSM) with chlorpyrifos (1 g/100 mL) as the carbon source.

3.2. Biochemical Characterization of the Pesticide-Degrading Endophytic Bacteria

The biochemical properties of each strain are summarized in Table 1. All eighteen strains demonstrated positive tests for oxidase, catalase, citrate utilization test, triple sugar iron (TSI), as well as lactose, sucrose, dextrose, and maltose fermentation. Furthermore, all strains except HSTU-Abk29, HSTU-Abk32, HSTU-Abk33, and HSTU-Bk14 showed positive motility. Except for strains HSTU-Abk30, HSTU-Abk31, HSTU-Bk12, HSTU-Bk15, HSTU-Abk33, and HSTU-Abk36, all strains tested positive for urease. Strains HSTU-Abk30, HSTU-Abk33, and HSTU-Bk15 were positive in the Voges–Proskauer test but negative in the methyl red (MR) test. Cell wall hydrolytic enzymes, amylase, proteases, and xylanase were secreted by all strains except strain HSTU-Bk12, which appeared to be negative for xylanase. The cellulase enzyme was secreted by all strains except HSTU-Abk34, HSTU-Bk13, HSTU-Bk14, and HSTU-Bk15. In addition, the ligninolytic enzyme secretion by all strains was assayed against several dye compounds enriched with minimal nutrient media. The assays revealed that all strains had the abilities of degrading dyes, including trypan blue, congo red, toluidine blue, avitera blue, and bromothymol blue, except strains HSTU-Abk29 and HSTU-Abk32, which were unable to degrade bromothymol blue and toluidine blue, respectively (Table 1).

Table 1. Summary of biochemical characterization of the endophytic bacteria isolated from rice plant genotypes Kalijera and Shonamukhi.

Isolates	Oxi	Cit	Cat	MIU	Mot	Ure	VP	MR	TSI	Lac	Suc	Dex	Mal	CMC	Xy	Amy	Pro	CR	TB	BTB	AB	TB
Klebsiella sp. HSTU-Bk11	+	+	+	+	+	+	+	+	+	+	+	+	+	+	+	+	+	+	+	+	+	+
Acinetobacter sp. HSTU-Abk29	+	+	+	-	-	+	+	-	+	+	+	+	+	+	+	+	+	+	+	-	+	+
Citrobacter sp. HSTU-Abk30	+	+	+	+	+	-	-	+	+ (Fe)	+	+	+	+	+	+	+	+	+	+	+	+	+
Enterobacter cloacae HSTU-Abk39	+	+	+	+	+	+	+	-	+	+	+	+	+	+	+	+	+	+	+	+	+	+
Enterobacter cloacae HSTU-Abk37	+	+	+	+	+	+	+	-	+	+	+	+	+	+	+	+	+	+	+	+	+	+
Enterobacter ludwigii HSTU-Abk40	+	+	+	+	+	+	+	-	+	+	+	+	+	+	+	+	+	+	+	+	+	+
Acinetobacter baumannii HSTU-ABK42	+	+	+	+	+	+	+	-	+	+	+	+	+	+	+	+	+	+	+	+	+	+
Klebsiella sp. Strain HSTU-Abk31	+	+	+	+	+	-	+	+	+	+	+	+	+	+	+	+	+	+	+	+	+	+
Acinetobacter sp. Strain HSTU-Bk12	+	+	+	+	+	-	+	-	+	+	+	+	+	+	-	+	+	+	+	+	+	+
Acinetobacter sp. HSTU-Abk32	+	+	+	-	-	+	+	-	+	+	+	+	+	+	+	+	+	+	+	+	+	-
Burkholderia sp. HSTU-ABK33	+	+	+	-	-	-	-	-	+	+	+	+	+	+	+	+	+	+	+	+	+	+
Acinetobacter sp. HSTU-Abk34	+	+	+	+	+	+	+	-	+	+	+	+	+	-	+	+	+	+	+	+	+	+
Pseudomonas sp. HSTU-Bk13	+	+	+	+	+	+	+	+	+	+	+	+	+	-	+	+	+	+	+	+	+	+
Citrobacter sp. HSTU-Bk14	+	+	+	-	-	+	+	-	+ (Fe)	+	+	+	+	-	+	+	+	+	+	+	+	+
Acinetobacter sp. HSTU-Bk15	+	+	+	+	+	-	-		+	+	+	+	+	-	+	+	+	+	+	+	+	+
Enterobacter sp. HSTU-Abk36	+	+	+	+	+	-	+	-	+	+	+	+	+	+	+	+	+	+	+	+	+	+
Enterobacter sp. HSTU-Abk38	+	+	+	+	+	+	+	-	+	+	+	+	+	+	+	+	+	+	+	+	+	+
Serratia marcescens HSTU-Abk41	+	+	+	+	+	+	+	-	+	+	+	+	+	+	+	+	+	+	+	+	+	+

Oxi, oxidase; Cit, citrate; Cata, catalase; MIU, motility indole and urease; Mot, motility; Ure, urease; VP, Voges–Proskauer; MR, methyl red; TSI, triple sugar iron agar; Lac, lactose; Suc, sucrose; Dex, dextrose; Mal, maltose; CMC, carboxymethyl cellulose; Xy, xylanase; Amy, Amylase; Pro, protease; Fe, iron; CR, congo red; TB, Trypan blue; BTB, bromothymol blue; AB, Avitera blue; TB, trypan blue. The symbol "+" indicates positive and "-" indicates negative.

3.3. Molecular Characterization of the Pesticide-Degrading Endophytic Bacteria

The vigorous mineralizing capabilities of chlorpyrifos (CPF) by five endophytic bacteria, including *Klebsiella*, *Enterobacter*, *Citrobacter*, *Serratia*, and *Acinetobacter* isolated from the roots of rice plants, are illustrated in Figure 2A. Two strains of *Enterobacter* sp. HSTU-Abk38 and HSTU-Abk36 showed substantial genetic distances and occupied different taxa in the phylogenetic tree (Figure 2A). In particular, the strain HSTU-Abk38 was placed in a separate node located between the *Klebsiella pneumoniae* and *Klebsiella aerogenes*. Similarly, the *Citrobacter* sp. Strain HSTU-Abk30 was deviated to a single node from the *Citrobacter freundi* and *Enterobacter ludwigi* (Figure 2A). A very similar observation was recorded for the *Serratia marcescens* strain HSTU-Abk41 and *Acinetobacter* sp. Strain HSTU-Abk29, suggesting genetic diversities of these strains.

Four genera, *Acinetobacter*, *Pseudomonas*, *Citrobacter*, and *Enterobacter,* isolated from the shoots of rice plants (Figure 2B), were not placed in the same node or sister taxa with reference type strains, except the *Acinetobacter* sp. strain HSTU-Bk15, which showed 100% similarity with the *Acinetobacter soli* strain B1. Notably, the *Citrobacter* sp. strain HSTU-Bk14 was placed in sister taxa with the *Enterobacter cloacae* strain HSTu-ABk39, which formed a different node from the *Citrobacter* and *Enterobacter* (Figure 2B). Similarly, the *Pseudomonas* sp. strain HSTU-Bk13 occupied a single node, which was closer to the

Pseudomonas aeruginosa strain DSM50071 and *Pseudomonas aeruginosa* strain ATCC10145. Overall, these results suggest that the chlorpyrifos mineralizing endophytes isolated from the shoots exhibit diversities but are dominated by the *Enterobacter* species.

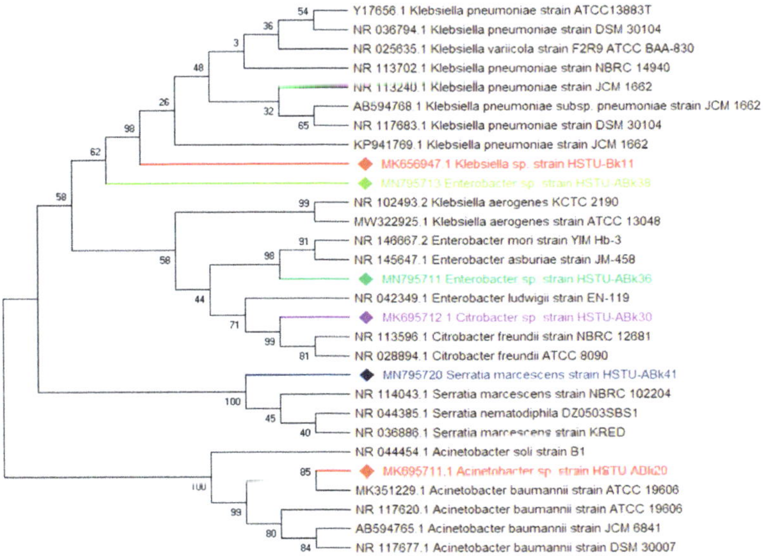

(**A**)

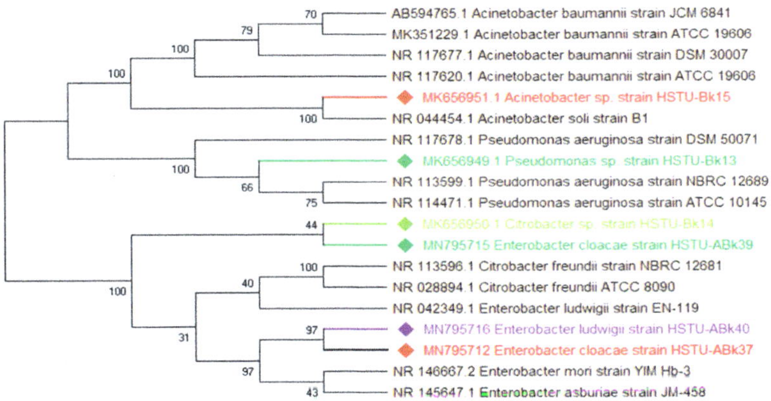

(**B**)

Figure 2. *Cont.*

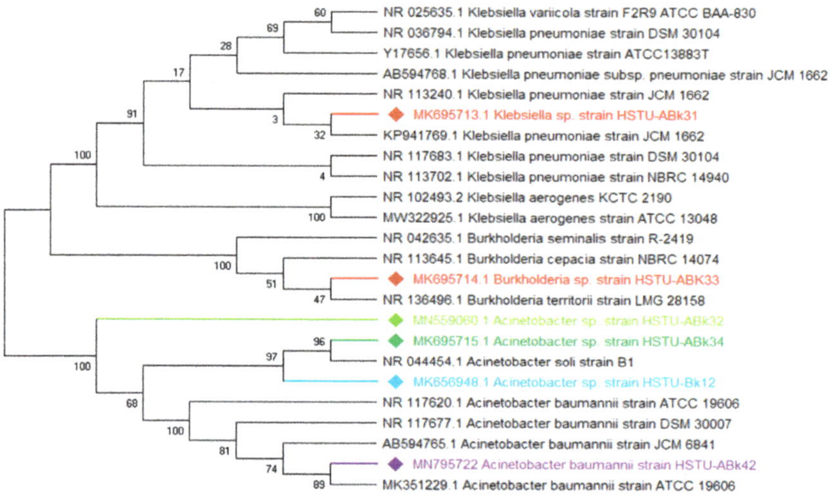

(C)

Figure 2. (**A**) Diversity of chlorpyrifos-resistant endophytic bacteria isolated from two different rice fields. The fields were exposed to massive pesticides for rice cultivation over the last 10 years. (**A**) Phylogeny of root endophytic bacteria. (**B**) Diversity of chlorpyrifos-resistant endophytic bacteria isolated from two different rice fields. The fields were exposed to massive pesticides for rice cultivation over the last 10 years. (**B**) Phylogeny of shoot endophytic bacteria. (**C**) Diversity of chlorpyrifos-resistant endophytic bacteria isolated from two different rice fields. The fields were exposed to massive pesticides for rice cultivation over the last 10 years. (**C**) Phylogeny of leaf endophytic bacteria. Black color strains are reference bacteria whether different colors are isolated strains in this study.

A total of three genera of endophytic bacteria, i.e., *Klebsiella*, *Burkholderia*, and *Acinetobacter*, were obtained from the leaves of rice plants (Figure 2C). Four strains were placed with the same taxon of *Klebsiella*, *Burkholderia*, and *Acinetobacter* species. The *Acinetobacter baumannii* strain HSTu-ABk42 was placed in sister taxa with the *Acinetobacter baumannii* strain ATCC 19606, while the *Acinetobacter* sp. strain HSTu-ABk32, *Acinetobacter* sp. strain HSTu-ABk34, and *Acinetobacter* sp. strain HSTU-Bk12 were separately placed in different nodes of the same clade. Furthermore, the *Burkholderia* sp. strain HSTu-ABk33 occupied a sister taxon with the *Burkholderia territorii* strain LMG28158, which greatly deviated from the other *Burkholderia* nodes (Figure 2B). In aggregate, these results demonstrate that the chlorpyrifos mineralizing leaf endophytes of rice plants were dominated by the *Acinetobacter* species.

3.4. Chlorpyrifos Biodegradation Confirmation Using FT-IR Analysis

Figure 3 illustrates the FT-IR spectrum of chlorpyrifos biodegradation observed after 14 days of incubation of the endophytic strains with minimal nutrient media (MSN) enriched with chlorpyrifos, when the C–H asymmetric vibration bond belonging to the typical methyl at 2870–2960 cm^{-1} and 772 cm^{-1} completely disappeared after bacterial treatment. The peak around 1370–1462 cm^{-1} indicates the C=C and C=N bonds along with the peak at 1220 cm^{-1}, representing the C-N bonding vibration observed in control but one that disappeared in the case of all endophyte treatments. Moreover, the peak at 1024 cm^{-1} assigned for C=O appeared only in untreated control samples (Figure 3). Notably, some new peaks around 640–650 cm^{-1}, 1100–1120 cm^{-1}, and 3200–3250 cm^{-1} were recorded for the samples treated with endophytic strains. These results indicate that all endophytic strains belonging to the genera of *Klebsiella*, *Enterobacter*, *Citrobacter*, *Serratia*, *Acinetobacter*,

Pseudomonas, and *Burkholderia* isolated from rice plants were capable of using chlorpyrifos as a carbon source.

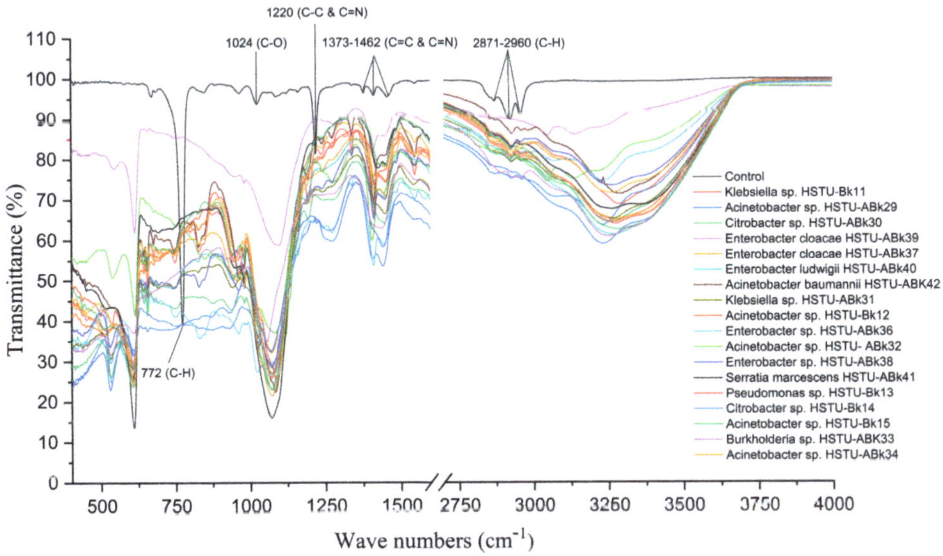

Figure 3. FT-IR spectra of chlorpyrifos degrading evidence of the endophytic strains in the MSM with chlorpyrifos (1 g/100 mL). The strains were grown for 14 days.

3.5. Plant-Growth-Promoting Traits of the Pesticide-Degrading Endophytic Bacteria

3.5.1. N-Fixation and PO_4- Solubilization Activity

The nitrogen fixation activities of the pesticide-degrading endophytic bacteria were assayed in nitrogen-free Jensen's growth media, as presented in Figure 4. Among the root endophytes, the strains *Klebsiella* sp. HSTU-Bk11, *Acinetobacter* sp. HSTu-ABk29, *Enterobacter* sp. HSTu-ABk36, *Serratia marcescens* HSTu-ABk41 were grown well in Jensen's media, which was further evidenced by the creation of a holo zone of 11–15.8 mm in diameter. However, the expansion of the holo zone was limited to 5.5–7.75 mm in *Enterobacter* sp. HSTu-ABk38 and *Citrobacter* sp. HSTu-ABk30.

In the case of shoot endophytes, three strains, *Enterobacter cloacae* HSTu-ABk39, *Enterobacter cloacae* HSTu-ABk37, *Enterobacter ludwigii* HSTu-ABk40, formed holo zones ranging from 6.48 to 7.11 mm. Notably, three other shoot endophytic strains, *Pseudomonas* sp. HSTU-Bk13, *Citrobacter* sp. HSTU-Bk14, *Acinetobacter* sp. HSTU-Bk15, demonstrated a wider holo zone spanning 10.80–14.10 mm (Figure 4) compared to those formed by the root endophytes. Interestingly, the leaf endophytes showed the best levels of nitrogen fixation capacity, as evident by the maximum holo zone formation recorded as 25.15 mm for *Acinetobacter* sp. HSTu-ABk34, followed by 15.11 mm, 14.99 mm, 13.84 mm, 12.57 mm holo zone formation in *Burkholderia* sp. HSTu-ABk33, *Acinetobacter* sp. HSTU-Bk12, *Acinetobacter baumannii* HSTu-ABk42, and *Klebsiella* sp. HSTu-ABk31, respectively. Collectively, these findings suggest that the leaf endophytes showed superior levels of nitrogen fixation capacity from the atmosphere without symbiotic association with the plants (Figure 4).

Tolubilizeization activities of the endophytes were also recorded (Supplementary Figure S1). All strains showed phosphate solubilization activities in PVK agar media with formation of holo zones in the range of 7–18 mm in diameter. Notably, the root endophyte strains *Enterobacter* sp. HSTu-ABk36 and *Acinetobacter* sp. HSTU-Bk15, the leaf endophyte strains *Acinetobacter* sp. HSTU-Bk12 and *Acinetobacter* sp. HSTu-ABk34, and the shoot endophyte strain *Pseudomonas* sp. HSTU-Bk13 demonstrated a high level of phosphate solubilization (Supplementary Figure S1).

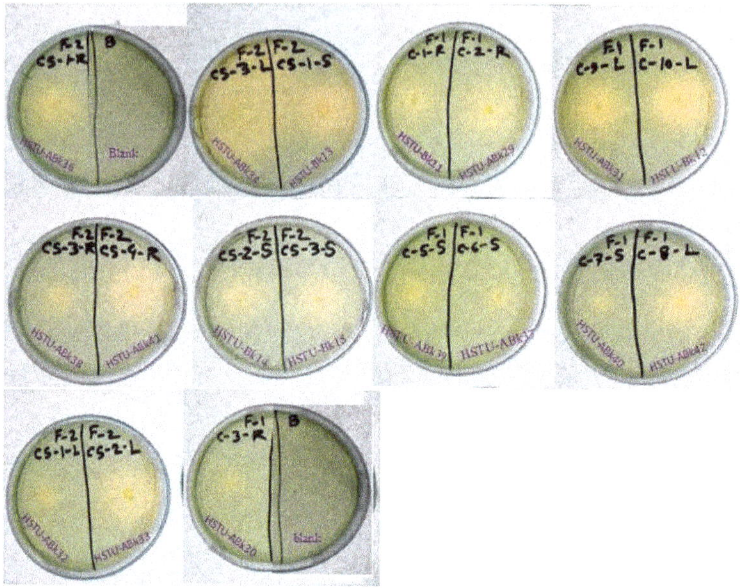

Figure 4. Plant-growth-promoting traits of the strains. Nitrogen fixation abilities of the endophytic strains on the nitrogen-free Jensen's media. The diameter of the holo zone indicates the spectrum of anti-microbial activities.

3.5.2. IAA and ACC-Deaminase Activity

Figure 5 illustrates the varying production capacities of indole-3-acetic acid (IAA) by the endophytes isolated from the roots, shoots, and leaves of rice plants. Eight strains, including *Acinetobacter* sp. HSTU-Bk12, *Acinetobacter* sp. HSTU-Bk15, *Klebsiella* sp. HSTu-ABk31, *Enterobacter* sp. HSTu-ABk36, *Enterobacter cloacae* HSTu-ABk37, *Enterobacter* sp. HSTu-ABk38, *Enterobacter cloacae* HSTu-ABk39, and *Enterobacter ludwigi* HSTu-ABk40, produced nearly 7.5 μg/mL of IAA, which was the highest amount compared to that produced by the remaining ten other strains. Except for the strain *Serratia marcescens* HSTu-ABk41, which produced approximately 6.0 μg/mL of IAA, the remaining nine strains produced IAA below 3.0 μg/mL.

Similarly, the ACC-deaminase production varied greatly among the eighteen endophytic strains (Figure 5). The maximum activity of 0.037~0.048 μM/mg/h was recorded for the strains *Serratia marcescens* HSTU-ABk41, *Citrobacter* sp. HSTU-Bk14, *Klebsiella* sp. HSTU-Bk11, *Acinetobacter* sp. HSTU-ABk29, *Enterobacter cloacae* HSTU-ABk39, while the lowest amount of ACC-deaminase activity of 0.005 μM/mg/h was found in the strain *Citrobacter* sp. HSTU-ABk30. The remaining strains showed a moderate level of ACC-deaminase production, ranging from 0.02 to 0.035 μM/mg/h.

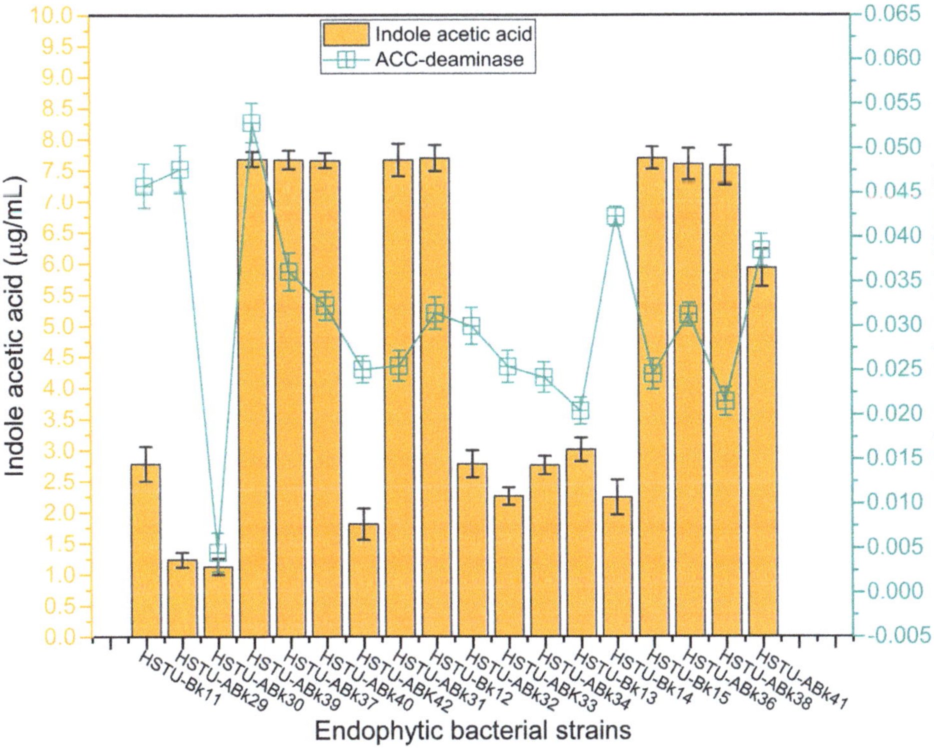

Figure 5. Plant-growth-promoting traits of the strains. Auxin (Indole acetic acid) and ACC-deaminase activities of the endophytic strains.

3.6. Anti-Bacterial Activity against Multidrug-Resistant Human Pathogenic Bacteria

The growth inhibition activity of the endophytic bacterial isolates against four multidrug-resistant human pathogenic bacteria was observed (Table 2; Supplementary Figure S2). The results revealed that 40% of the endophytic isolates possessed anti-bacterial activities against *S. aureus* after 16 h of treatment, which rose to 55% and 65% after 32 h and 48 h, respectively. The highest inhibition zones of 41.5 ± 1.5 and 26 ± 0.6 mm were created by *Enterobacter cloacae* HSTu-ABk39 and *Acinetobacter* sp. HSTu-ABk34, respectively, against *S. aureus* after 32 h of treatment. It is noteworthy that 35% of the endophytic isolates produced inhibition zones against *E. coli* within 16 h of treatment, which increased to 55% after 32 h; however, these were less effective compared to those treatments observed against *S. aureus*. A similarly poor activity was demonstrated by these isolates when they were tested against *Klebsiella* sp. The second highest inhibition zone of 29 ± 0.9 mm was created by *Enterobacter* HSTu-ABk36 after 48 h of treatment with *S. epidermidis*. While 55% of the isolates were unable to demonstrate any activity against *S. epidermidis*, four strains, including *Acinetobacter* sp. ABk32, *Acinetobacter* sp. ABk34, and *Pseudomonas* sp. HSTU-Bk13, showed inhibitory activities against all four pathogenic strains (*S. aureus*, *E. coli*, *Klebsiella* sp., and *S. epidermidis*).

Table 2. Minimum inhibitory concentration in mm of treated bacteria against human pathogenic bacteria.

Isolates	Multidrug-Resistant Human Pathogenic Bacteria											
	S. aureus (mm ± SE)			*E. coli* (mm ± SE)			*Klebshilla* (mm ± SE)			*S. epidermidis* (mm ± SE)		
	16 h	32 h	48 h	16 h	32 h	48 h	16 h	32 h	48 h	16 h	32 h	48 h
Klebsiella sp. HSTU-Bk11	12.75 ± 0.25	15 ± 0.2	20.5 ± 0.5	-	8.5 ± 0.5	7.5 ± 0.5	-	-	-	13.5 ± 3.5	17 ± 5.0	18.5 ± 1.5
Acinetobacter sp. HSTU-ABk29	-	6.5 ± 1.5	6.5 ± 1.5	-	-	-	-	-	-	9.5 ± 1.5	13 ± 2.0	12 ± 0.0
Citrobacter sp. HSTU-ABk30	- -	11 ± 0.0	10.5 ± 0.5	- -	- -	- -	- -	- -	- -	- -	- -	- -
Enterobacter cloacae HSTU-ABk39	40 ± 2	**41.5 ± 1.5 ****	40 ± 0.0	-	-	-	-	-	9 ± 1.0	-	-	9.5 ± 1.5
Enterobacter cloacae HSTU-ABk37	-	-	10 ± 0.0	6 ± 0.0	6 ± 0.0	6 ± 0.0	-	-	-	-	9.5 ± 1.5	10 ± 0.0
Enterobacter ludwigii HSTU-ABk40	-	-	-	6 ± 0.0	6 ± 0.0	6 ± 0.0	-	-	8 ± 0.0	-	-	-
Acinetobacter baumannii HSTU-ABK42	-	-	-	-	-	-	8.5 ± 1.5	-	-	-	-	-
Klebsiella sp. HSTU-ABk31	-	-	-	-	7 ± 0.0	7 ± 0.0	8.5 ± 0.5	-	-	-	-	-
Acinetobacter sp. HSTU-Bk12	-	-	-	-	5.5 ± 0.5	6 ± 0.0	6.5 ± 0.5	-	-	-	-	-
Acinetobacter sp. HSTU-ABk32	16.5 ± 0.5	16.5 ± 0.5	17 ± 0.0	8.5 ± 0.5	8.5 ± 0.5	10 ± 0.0	-	7 ± 0.0	15 ± 0.0	9 ± 0.0	10.5 ± 0.5	9.5 ± 0.5
Burkholderia sp. HSTU-ABK33	-	-	-	-	-	-	-	5.5 ± 0.5	5.5 ± 0.5	-	-	-
Acinetobacter sp. HSTU-ABk34	18 ± 2.0	**26 ± 0.6 ****	24.5 ± 4.5	12.5 ± 0.5	12.5 ± 0.5	12.5 ± 0.5	9.5 ± 1.5	9.5 ± 1.5	10.5 ± 0.5	10 ± 0.0	11 ± 0.0	10.5 ± 0.5
Pseudomonas sp. HSTU-Bk13	15 ± 1.0	16 ± 2.0	17.5 ± 0.5	8 ± 1.0	7.5 ± 0.5	7.5 ± 0.5		6.5 ± 0.5		18.5 ± 1.5	18.5 ± 1.5	19 ± 1.0
Citrobacter sp. HSTU-Bk14	10 ± 2.0	11.5 ± 1.5	11.5 ± 1.5	-	-	-	-	-	-	-	-	-
Acinetobacter sp. HSTU-Bk15	11.5 ± 1.5	16.5 ± 0.5	23.5 ± 1.5	-	-	-	17 ± 1.0	16 ± 2.0	16 ± 3.0	16 ± 1.0	16 ± 1.0	17.5 ± 2.5
Enterobacter HSTU-ABk36	-	-	-	6 ± 0.0	11.5 ± 2.5	14.5 ± 0.5	10 ± 1.0	10.5 ± 0.5	9 ± 1.0	25 ± 10.0	28.5 ± 10.5	**29 ± 9.0 ****
Enterobacter sp. HSTU-ABk38	-	11.5 ± 0.5	11.5 ± 0.5	-	-	-	-	6 ± 0.0	7 ± 0.0	-	-	-
Serratia marcescens HSTU-ABk41	-	-	-	-	-	-	-	6 ± 0.0	6.5 ± 0.5	-	-	-

The bold and ** means high activity was achieved.

3.7. *Rice Plant Growth-Promoting Effect*

3.7.1. Effects of Individual and Consortia of Endophytes on Germination and Seedling Growth

The germination of Shunamukhi genotype rice seeds was tested after the inoculation of bacterial endophytes individually and consortia of endophytes (Table 3). There was a significant difference among the treated and control samples after 8 days. The root and shoot lengths were significantly increased after 8 and 12 days of treatment of individual endophytes compared to those treated in the control (no endophytes). The LSD value of seed germination (9.55%), shoot lengths after 8 days (1.40), shoot lengths after 12 days (2.02), root lengths after 8 days (2.73), root lengths after 12 days (2.78), and the vigor index (354.61) were recorded for plants treated with individual endophytes (Table 3; Supplementary Figure S2). The germination percentages were not significantly different when treated with any consortia of endophytes. However, consortium-1 and consortium-2 treatments displayed the highest shoot growth compared with those treated with either no bacterial endophytes (control) or consortium-3 and consortium-4. In addition, the most increased vigor activity was observed for consortium-1 treated samples (Table 3; Supplementary Figure S2).

Table 3. Individual and consortium effects on seedling and growth promotion.

Treatment	Germination % (Mean ± SE)	Shoot Length after 8 Days (Mean ± SE)	Shoot Length after 12 Days (Mean ± SE)	Root Length after 8 Days (Mean ± SE)	Root Length After 12 Days (Mean ± SE)	Vigor Index (Mean ± SE)
Klebsiella sp. HSTU-Bk11	95.56 ± 2.22^{a}	5.00 ± 0.50^{abcd}	6.77 ± 1.01^{abc}	6.73 ± 1.92^{ab}	7.43 ± 0.46^{abc}	1116.67 ± 119.46^{abc}
Acinetobacter sp. HSTU-ABk29	91.11 ± 5.88^{ab}	5.40 ± 0.31^{abcd}	7.37 ± 0.33^{abc}	6.43 ± 1.32^{abcd}	7.50 ± 0.76^{abc}	1068.89 ± 79.08^{abc}
Citrobacter sp. HSTU-ABk30	95.56 ± 2.22^{a}	4.77 ± 0.12^{bcd}	5.57 ± 0.58^{bcd}	6.33 ± 1.20^{abcd}	$6.77 + 1.65^{abc}$	1056.00 ± 83.44^{abc}
Enterobacter cloacae HSTU-ABk39	88.89 ± 2.22^{ab}	4.73 ± 0.39^{bcd}	7.30 ± 0.80^{abc}	3.90 ± 0.56^{dc}	4.97 ± 0.93^{c}	771.56 ± 105.30^{cd}
Enterobacter cloacae HSTU-ABk37	93.33 ± 0.0^{a}	5.33 ± 0.22^{abcd}	5.50 ± 0.53^{cd}	7.67 ± 0.33^{ab}	6.27 ± 1.50^{bc}	1213.33 ± 51.40^{ab}
Enterobacter ludwigii HSTU-ABk40	93.33 ± 3.85^{a}	5.33 ± 0.38^{abcd}	7.33 ± 0.45^{abc}	6.83 ± 0.69^{ab}	7.83 ± 0.44^{ab}	1137.11 ± 105.13^{ab}
Acinetobacter baumannii HSTU-ABK42	88.89 ± 2.22^{ab}	5.17 ± 0.17^{abcd}	6.50 ± 0.29^{abcd}	7.17 ± 0.88^{ab}	7.50 ± 1.04^{abc}	1098.89 ± 101.99^{abc}
Klebsiella sp. HSTU-ABk31	95.56 ± 2.22^{a}	5.90 ± 0.31^{abc}	6.53 ± 0.30^{abcd}	7.27 ± 0.67^{ab}	7.07 ± 1.38^{abc}	1256.67 ± 16.51^{ab}
Acinetobacter sp. HSTU-Bk12	93.33 ± 0.00^{a}	5.10 ± 0.10^{abcd}	7.27 ± 0.59^{abc}	6.07 ± 0.70^{bcd}	6.67 ± 0.17^{abc}	1042.22 ± 71.76^{abc}
Acinetobacter sp. HSTU-ABk32	93.33 ± 0.00^{a}	5.27 ± 0.15^{abcd}	6.63 ± 0.93^{abc}	6.10 ± 0.31^{abcd}	8.03 ± 0.27^{ab}	1060.89 ± 41.86^{abc}
Burkholderia sp. HSTU-ABK33	95.56 ± 2.22^{a}	6.27 ± 0.46^{a}	7.00 ± 0.50^{abc}	6.67 ± 0.41^{abc}	9.07 ± 0.64^{a}	1237.56 ± 87.10^{ab}
Acinetobacter sp. HSTU ABk34	88.89 ± 2.22^{ab}	6.00 ± 0.40^{ab}	7.57 ± 0.35^{ab}	7.83 ± 0.60^{ab}	8.10 ± 0.47^{ab}	1228.89 ± 86.47^{ab}
Pseudomonas sp. HSTU-Bk13	93.33 ± 3.85^{a}	5.43 ± 0.64^{abcd}	6.37 ± 1.10^{abcd}	7.33 ± 0.67^{ab}	6.70 ± 1.82^{abc}	1196.22 ± 138.18^{ab}
Citrobacter sp. HSTU-Bk14	91.11 ± 5.8^{ab}	4.83 ± 0.33^{bcd}	8.17 ± 0.64^{a}	7.50 ± 0.29^{ab}	7.33 ± 0.44^{abc}	1128.89 ± 111.31
Acinetobacter sp. HSTU-Bk15	97.78 ± 2.22^{a}	4.73 ± 0.26^{bcd}	6.97 ± 0.75^{abc}	5.20 ± 0.51^{bcd}	7.20 ± 0.85^{abc}	969.56 ± 62.34
Enterobacter sp. HSTU-ABk36	95.56 ± 2.22^{a}	5.77 ± 0.64^{abcd}	6.10 ± 0.32^{bcd}	8.80 ± 0.76^{a}	7.27 ± 0.93^{abc}	1386.00 ± 100.34^{a}
Enterobacter sp. HSTU-ABk38	97.78 ± 2.22^{a}	4.40 ± 0.20^{de}	6.53 ± 0.62^{abcd}	6.73 ± 0.23^{ab}	9.17 ± 0.17^{a}	1087.11 ± 18.72^{abc}
Serratia marcescens HSTU-ABk41	93.33 ± 3.85^{a}	4.77 ± 0.37^{bcd}	7.33 ± 0.73^{abc}	6.00 ± 1.53^{bcd}	7.83 ± 1.20^{ab}	1009.78 ± 188.26^{bc}
Control	82.22 ± 5.88^{b}	3.27 ± 0.43^{e}	4.60 ± 0.46^{d}	4.00 ± 0.29^{cd}	6.13 ± 0.52^{bc}	600.44 ± 80.22^{d}
LSD of individual bacteria	9.55	1.40	2.02	2.73	2.78	354.61
Consortia/Group-1	96.67 ± 3.3^{a}	6.45 ± 0.05^{a}	8.5 ± 1^{a}	6.65 ± 0.35^{a}	8.75 ± 0.75^{a}	1267.33 ± 72^{a}
Consortia/Group-2	93.33 ± 6.7^{a}	6.4 ± 0.2^{a}	8.7 ± 0.1^{a}	5.7 ± 0.3^{a}	8.15 ± 1.85^{a}	1128.67 ± 71^{ab}
Consortia/Group-3	93.33 ± 0.0^{a}	5.5 ± 0.5^{b}	8.8 ± 0.5^{a}	6.5 ± 0.5^{a}	7.15 ± 0.55^{ab}	1050.00 ± 116^{ab}
Consortia/Group-4	93.33 ± 00^{a}	9.75 ± 0.25^{ab}	7.5 ± 1^{ab}	7.1 ± 0.1^{a}	7.25 ± 0.25^{ab}	1176.00 ± 37^{ab}
Control	93.33 ± 0.0^{a}	4.55 ± 0.05^{b}	5.85 ± 0.35^{b}	6 ± 1.0^{a}	5.5 ± 0.0^{b}	984.67 ± 98^{b}
LSD of group treatment	8.93	1.55	2.39	1.88	2.56	273.50

Treatments with the same letter are not significantly different.

3.7.2. Effects of Consortia on Vegetative and Reproductive Stages and Yield

The effects of all four endophytic consortia on vegetative growth, reproductive growth, and yield of rice plants are presented in Figure 6.

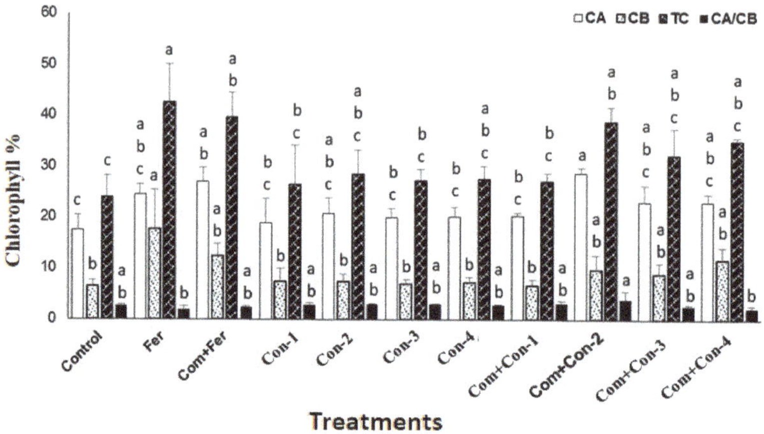

(**A**)

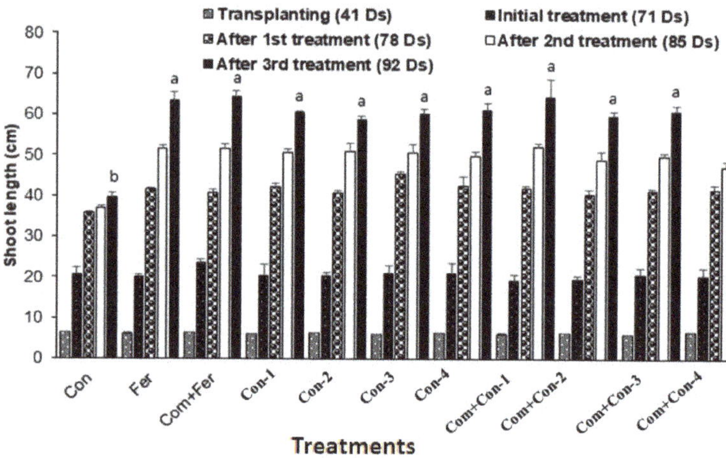

(**B**)

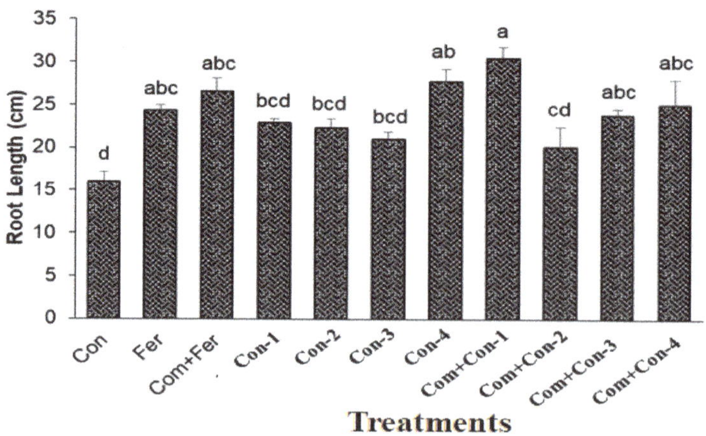

(**C**)

Figure 6. *Cont.*

(**D**)

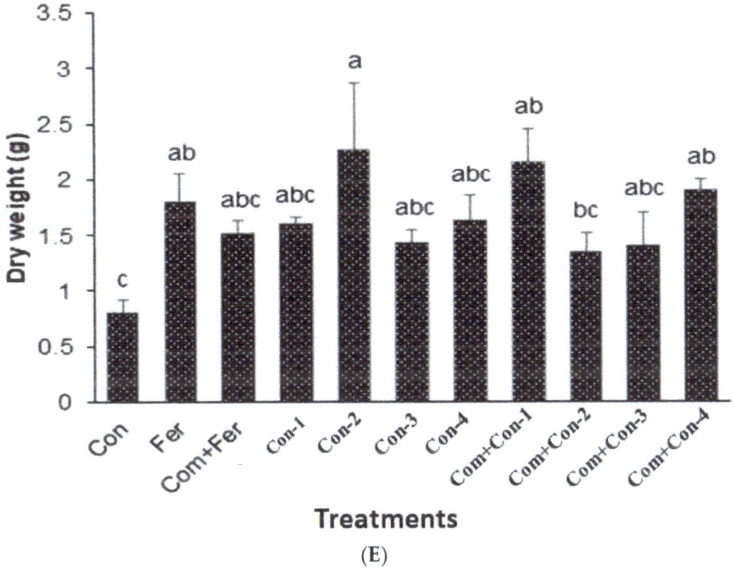

(**E**)

Figure 6. *Cont.*

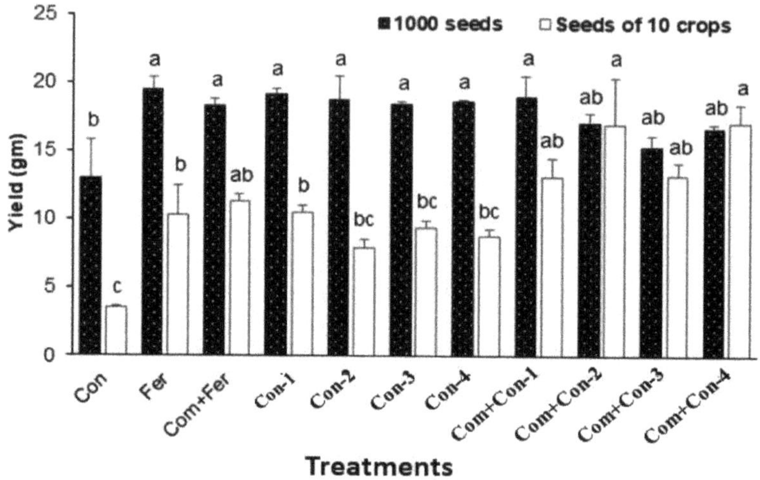

(F)

Figure 6. (**A**) Comparing chlorophyll content among the treated and untreated groups. (Fer, Fertilizer; Com, Compost; Bac, Bacteria; G, Bacterial consortium group). (**B**) Shoot length analysis among the endophytic bacteria in the treated and untreated groups in a time-dependent manner. (Con, Control; Fer, Fertilizer; Com, Compost; Bac, Bacteria; G, Bacterial consortium group). (**C**) Root length analysis of the endophytic bacteria. (**D**) Rice plant growth promotion with bacterial consortia at vegetative stage. (**E**) Mean of dry weights among different bacteria in the treated and untreated groups. (Con, Control; Fer, Fertilizer; Com, Compost; Bac, Bacteria; G, Bacterial consortium group). (**F**) Mean yield among different bacteria in the treated and untreated groups. (Con, Control; Fer, Fertilizer; Com, Compost; Bac, Bacteria; G, Bacterial consortium group).

3.7.3. Chlorophyll Content

Rice plants treated with compost mixed with consortium-2 produced the largest amount of chlorophyll a. On the other hand, the highest levels of chlorophyll b and total chlorophyll contents were found in the fertilizer treated rice plants (control). However, the highest ratio of chlorophyll a and chlorophyll b was observed in the com + consortium-2 treated rice plants (Figure 6A). Although the urea application was reduced by 70% in consortia (1–4) treated rice plants, there were no significant differences in chlorophyll ratio between the consortia treated plants and fertilizer treated (control) plants. This result indicates that the bacterial consortia amended nitrogen from the atmosphere and/or fortified the nutrients from the added compost.

3.7.4. Root and Shoot Lengths

All four consortia treated rice plants demonstrated significant shoot growth compared to those of the control and fertilizer treated plants. Consortium-3 produced maximum shoot growth activity (Figure 6B) compared to those plants treated with other consortia-1, -2, and -4. When compost was added, no significant shoot elongation of rice plants was observed among all four bacterial consortia treated plants.

There was a significant difference in rice plant root length among consortia treated plants (Figure 6C). Again, consortia-3 produced maximum root growth compared with other consortia (-1, -2, and -4) treated rice plants, as among those observed for shoots. It is important to note that both compost and consortia (1–4) treated rice plants were much more greenish and disease free compared to the control, fertilizer, and compost + fertilizer treated plants (Figure 6D(a–d)).

3.7.5. Yield

Consortium-2 treated rice plants produced maximum dry weight at harvest (Figure 6E). Notably, fertilizer, compost + Fertilizer, consortium-1, consortium-2, consortium-3, and consortium-4 treated rice plants showed similar yields (1000 seeds) compared with that of the control plants (Figure 6F). In contrast, the grain weights of 10 tillers/crops were significantly higher for the compost mixed with consortium-2 treated rice plants harvested from pot experiments. In addition, pots containing compost mixed with either consortium-1, consortium-2, consortium-3, or consortium-4 increased the rice yields by 17.3%, 38.6%, 18.2%, and 39.1%, respectively, compared to those pots containing only compost mixed with fertilizer (urea).

3.8. Roles of Consortia of Endophytic Bacteria in Chlorpyrifos Biodegradation

The consortia treated chlorpyrifos biodegradation was further evidenced by GC–MS/MS analysis (Figure 7). The control (untreated) chlorpyrifos solution had a major peak at spectrum 88, while no mentionable peak was detected in consortium-1 treated extract. Consequently, unmatching compounds were aligned with the NIST11s library. In addition, the consortium-2 treated extracts' GC–MS spectra showed the existence of chlorpyrifos, Phorate sulfone, Chlorpyrifos-methyl, 2-Hydroxy-3,5,6-trichloropyridine, Carbofenothion sulfoxide, Oxydisulfoton, Carbonochloridic acid, Thionodemeton sulfone, dl-(2-Thienyl)-α-alanine, Chlorpyrifos Oxon, and Diethyl methanephosphonate (Table S1). In fact, consortium-3 treatment created several new fragments, including Phorate sulfoxide, Phosphoric acid, Acetamide, and 4-Pyridinol (Table 4). Consortium-4 treatment also generated several new fragments, including Thionodemeton sulfone, Phosphorodithioic acid, and Thiophene (Table S2). These results suggest that each consortium has different biodegradation and mineralization mechanisms of chlorpyrifos. This may be attributed to the different endophytic bacterial compositions of the consortia with varied capacities and enzymatic activities.

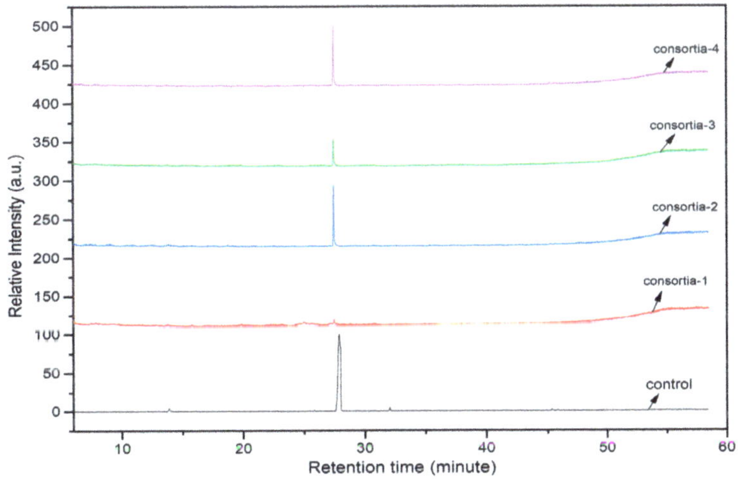

Figure 7. GC–MS spectra of chlorpyrifos treated with consortia-1, consortia-2, consortia-3, consortia-4.

Table 4. Biodegradation profile of consortium-3 treated chlorpyrifos (1 g/100 mL) enriched with minimal broth medium.

Similarity of Hit	Search Spectrum	Soft Ionization (SI)	Spectrum	Molecular Weight (Da)	Molecular Form	Molecular Structure
1,2,3,4,5,8	76,69,67,66,65,57	2921	88	2	Chlorpyrifos	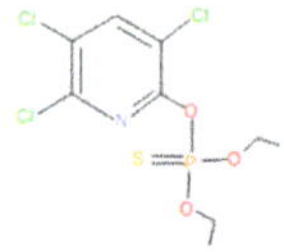
6	59	6515	38	4	2-Hydroxy-3,5,6 trichloropyridine	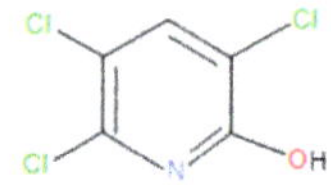
7	58	6515	38	4	2-Hydroxy-3,5,6-trichloropyridine	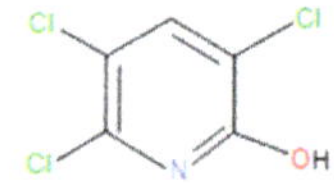
9	56	2588	3	6	Phorate sulfoxide	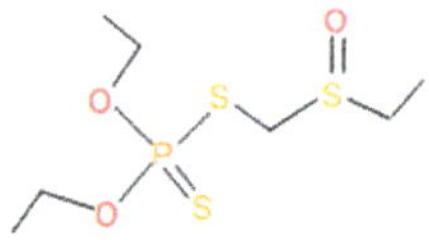
10	56	2588	3	6	Phorate sulfoxide	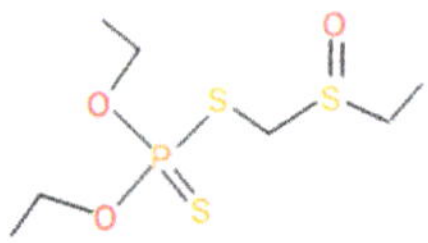
11	55	5598	13	0	Chloropyriphos-methyl	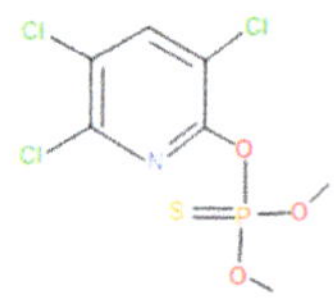
12	55	2588	3	6	Phorate sulfoxide	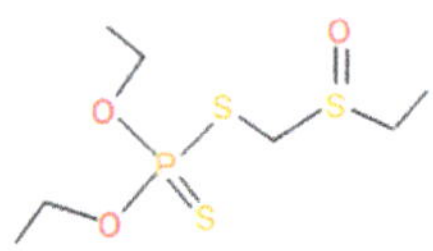
13	55	2588	4	7	Phorate sulfoxide	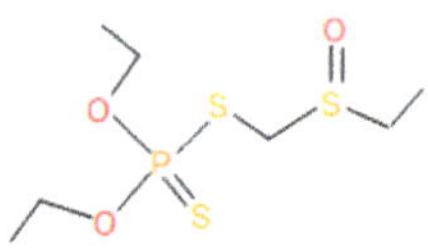
14	54	2921	88	2	Chlorpyrifos	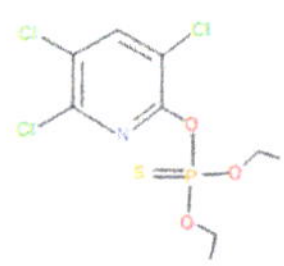
15	53	2588	4	7	Phorate sulfone	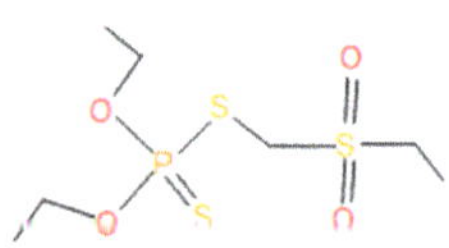

Table 4. *Cont.*

Similarity of Hit	Search Spectrum	Soft Ionization (SI)	Spectrum	Molecular Weight (Da)	Molecular Form	Molecular Structure
16	53	16,947	69	6	Carbonochloridic	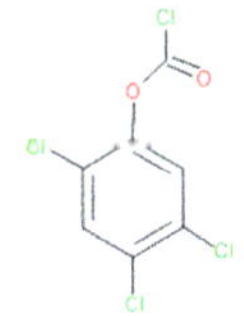
17	52	5598	15	2	Phosphoric acid	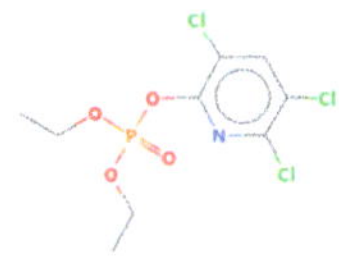
18	51	2021	58	1	dl-2-.beta.-Thienyl-.alpha.-alanine	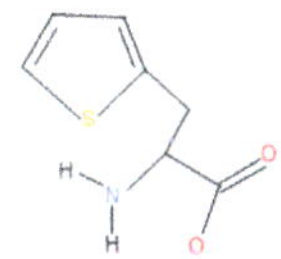
19	50	2497	7	6	Oxydisulfoton	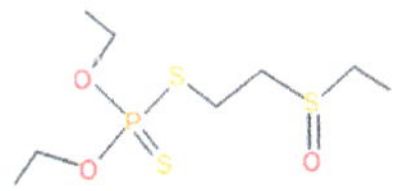
20	50	683	8	9	Diethyl methanephosphonate	
21	50	683	8	9	Diethyl methanephosphonate	
22	50	5598	13	0	Chloropyriphos-methyl	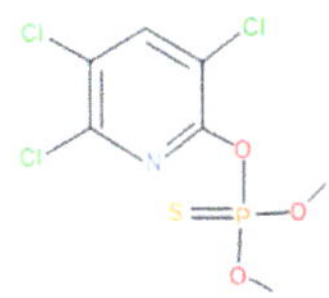
23	50	17,297	40	4	Carbofenothion sulfoxide	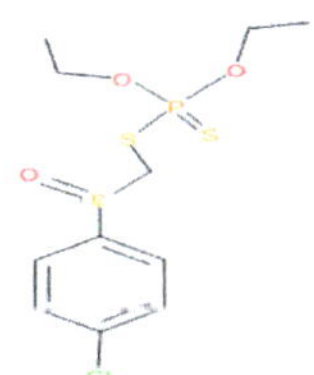
24	49	0	0	0	Acetamide,	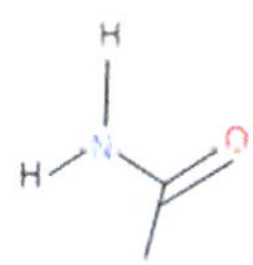

Table 4. *Cont.*

Similarity of Hit	Search Spectrum	Soft Ionization (SI)	Spectrum	Molecular Weight (Da)	Molecular Form	Molecular Structure
25	49	1970	40	7	4-Pyridinol,	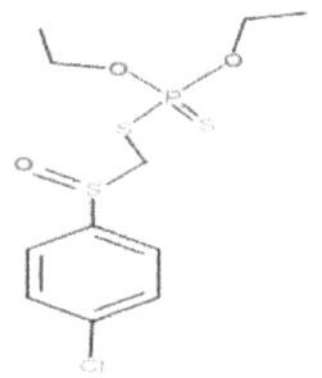

4. Discussion

Previously, organochlorine, organophosphorus, and carbamate group pesticides were widely used in agricultural fields in Bangladesh [34]. However, due to the banning of organochlorine group pesticides by the Bangladesh Environment Conservation Act 1995 [35], organophosphorus pesticides are widely used in agriculture. Chlorpyrifos (O,O-diethyl O-3,5,6-trichloro-2-pyridyl phosphorothioate) is one such widely used organophosphate pesticide employed to control a range of insects and pests in agriculture. The application of chlorpyrifos to increase crop yield poses a health risk to humans, animals, and other organisms alike. At high levels of chlorpyrifos exposure, this inherent neurotoxin can be lethal to humans [36]. To mitigate this issue, organophosphorus pesticide-degrading endophytic bacteria are much more desirable for biofertilizer components to develop a safe and sustainable agriculture practice. Endophytes exhibit cell wall hydrolytic and lignin-related dye-degrading enzyme activities, facilitating their penetration into host plants as symbionts. Additionally, their anti-microbial activities enable them to protect themselves and resist pathogenic strains during symbiosis, thus promoting healthier plants [20,21,37,38]. Furthermore, pesticide-degrading endophytes can contribute to the remediation of soil and plants by alleviating pesticide contamination [7,37,39]. These combined properties enable endophytes to fulfill their endophytic roles in host plants, fostering the development of resilient plants, which are resistant to pests and pathogens [20,40,41]. The current study demonstrated that the application of this synthetic consortium for rice cultivation can lead to healthy and high-yielding plants without the need for pesticides. This approach holds promise for promoting safe and sustainable agriculture practices.

In this paper, we report a total of eighteen endophytic bacterial strains isolated from *Kalijeera* (Field-1) and BRRI-28 (Field-2) genotypic rice plants capable of mineralizing chlorpyrifos. Employing a culture-dependent technique, these endophytic bacterial strains were grown and reproduced with chlorpyrifos serving as the only source of carbon, demonstrating a strong pesticide-degrading capability [7,8,15]. Furthermore, these strains were evaluated for their roles in the growth promotion of Shuna6+mukhi rice plant along with their inhibitory activities toward multidrug-resistant human pathogenic bacteria.

The 16S rRNA gene sequencing analysis of these strains indicated that the eighteen isolates belonged to six different species, including *Klebsiella* sp., *Acinetobacter* sp., *Pseudomonas* sp., *Citrobacter* sp., *Burkholderia* sp., and *Serratia* sp. While a wide range of chlorpyrifos mineralizing strains were found in the root samples, *Enterobacter* sp. (50%) and *Acinetobacter* sp. (50%) were found to be the most dominant among the shoot and leaf endophytic strains, respectively. Previously, several chlorpyrifos degrading bacteria, such as the *Enterobacter* strain B14 [42] and the *Klebsiella* sp. [43], were reported. Like the present study, endophytic strains of the *Enterobacter* sp. were previously isolated from the roots and grains of rice plants by Walitang et al. (2017) [44]. Other chlorpyrifos degrading endophytes, such as the *Pseudomonas aeruginosa* strain RRA, *Bacillus megaterium* strain RRB, *Sphingobacterium siyangensis* strain RSA, *Stenotrophomonas pavanii* strain RSB, and *Curtobacterium plantarum* strain RSC, were also isolated from chlorpyrifos treated rice plants grown in China [45]. All of these eighteen strains along with four different consortia included the strains reported in this paper (Figure 2A–C), demonstrating rice plant growth-promoting traits (Table 3; Figure 6), including enhanced germination rate, and increasing the root–shoot length and yield in the presence of only 30% urea application (Table 3; Figure 6D–F). In fact, a substantial amount of rice yields was obtained with consortia treatments, mimicking the yields under full-dose urea fertilizer (100%) applications. In the present study, a synthetic consortium consisting of diverse bacterial genera was utilized to enhance the growth and development of rice plants, enabling them to withstand environmental stressors, such as drought, heat, and pathogenic infections [38–41]. This endophytic consortium demonstrates the capacity to support plants in tolerating adverse conditions and resisting pathogen attacks, thus promoting their overall resilience and health. All four consortia significantly enhanced rice grain yields in pot experiments (Figure 6F) and degraded chlorpyrifos (Figure 7). In particular, the enhanced growth (4.3–6.5 log_{10} CFU/mL) in chlorpyrifos enriched with minimal nutrient media can be attributed to the abilities of these strains utilizing chlorpyrifos as their sole carbon source, as described above in Section 3.4. While consortia-2, -3, and -4 showed several degraded fragments of chlorpyrifos, consortium-3 presented fragments containing phorate sulfoxide, phosphoric acid, and acetamide, and consortium-4 showed thionodemeton sulfone, phosphorodithioic acid, and thiophene, which suggested their varied action toward chlorpyrifos in culture media (Figure 7; Table 4; Supplementary Tables S1 and S2). This probably occurred due to their different crew members in consortia exerting various sets of enzyme activities and metabolic routes. The existence of common compounds, such as TCP (2-Hydroxy-3,5,6 trichloropyridine), DEMP (Diethyl methanephosphonate), ensured that the consortia members degraded the phosphodiester bonds of chlorpyrifos. Haque et al. (2018, 2020) reported that organophosphorus hydrolase enzyme OpdA, OpdC, OpdD, OpdE of *Lactobacillus* species isolated from chlorpyrifos impregnated fermented food, e.g., kimchi samples degraded chlorpyrifos ester bonds to non-toxic compounds TCP and DETP [8,16]. However, the generation of several other chlorpyrifos derivatives detected for the first time in this study was due to the efficiency of the GC–MS techniques along with the NIST11 library search, which opens a new window for chlorpyrifos detoxification by endophytic consortia.

Almost all of the endophytic strains reported in this paper showed oxidase, catalase, xylanase, amylase, protease, and cellulase activities, which are crucial for endophytic competence. Catalase activity defends reactive oxygen species and is essential for the successful survival of colonizing endophytes during oxidative bursts by plants [46]. Previously, similar activities were reported for catalase in the *Klebsiella* sp. strain PS19 [47], for protease in the *Burkholderia* sp., *Pseudomonas* sp., *Enterobacter* sp., *Pseudomonas* sp. [48,49], and for amylase in the *Pseudomonas* sp. [48]. In our study, we found that all but one endophyte, the *Acinetobacter* sp. strain HSTU-Bk12, secreted xylanase. However, several studies revealed the *Acinetobacter* sp. as a cellulase producer, and the *Pseudomonas* sp. and *Enterobacter* sp. as cellulase and xylanase producers [7,15].

The bioremediation of dye by endophytic bacteria would be non-hazardous, environmentally friendly, and cost effective. In this study, all strains except the *Acinetobacter* sp. HSTU-ABk29 and *Acinetobacter* sp. HSTU-ABk32 degraded different dyes, including TPB, CR, TDB, ATB, and BTB. The *Acinetobacter* sp. strains HSTU-ABk29 and HSTU-ABk32 were unable to degrade BTB and BTB dyes, respectively. As reported previously (Ali et al., 2009) [50], we also found that the *Pseudomonas* sp. could decolorize dye solution or simu lated effluents. In addition, Tony et al. (2009) [51] reported that the consortia of microbes were capable of complete mineralization of azo dyes. However, endophytes harboring lignin-degrading activities might be presented with an opportunity to penetrate plants, forming symbionts [52]. An endophyte, which produces cellulase, protease, chitinase, and gelatinase, can hydrolyze fungal cell walls, and can inhibit the adhesion of fungal spores to plants, will have an advantage in mitigating fungal infection and be able to serve as a bio-fungicide.

The production of ammonia by endophytes is also a desirable trait for plant growth promotion, including the early establishment of seedlings, enhanced soil fertility, and increased phosphate solubilization [53]. Oteino et al. (2015) [54] observed that the inoculation with endophytes increased phosphate solubilization and significantly improved plant growth, which might be attributed to increased growth and development of rice plants treated with all four consortia. Auxin-producing endophytes *Burkoholderia vietnamiensis* promoted rice growth and yield [49]. Nitrogen-fixing bacterium *Pantoea agglomerans* [45] increased the growth of rice roots, shoots, flag leaves, and weights. Meanwhile, *Lysinibacillus sphaericus* produced ACC-deaminase and positively modulated the ethylene level in rice plants, consequently improving the number of panicles and grains per plant, straw, grain dry weight, and N and P uptake [55,56]. Moreover, additional Zinc (Zn) fortification was observed in rice treated with endophytes, such as the *Acinetobacter* sp., *Klebsiella* sp., *Enterobacter* sp., and *Bukholderia* sp. [57].

For the first time, we report that the *Enterobacter cloacae* strain HSTU-ABk39 and *Acinetobacter* sp. strain HSTU-ABk34 showed significant inhibitory activity against pathogenic *S. aureus*. Moreover, the *Enterobacter* sp. strain HSTU-ABk36 and *Citrobacter* sp. strain HSTU-ABk15 possessing growth inhibitor traits against *S. epidermidis* can serve as sources of novel antibiotics [58]. Bacterial strains, such as the *Bacillus* sp., *Lysinibacillus*, *Streptomyces*, *Streptomyces parvulus* Av-R5, also had anti-bacterial activity against *S. aureus* [59–63]. Endophytes are also the storehouse of several kinds of bioactive metabolites, including phenolics, alkaloids, quinones, steroids, saponins, tannins, and terpenoids, which makes them serve as potential anti-cancer, anti-malarial, anti-tuberculosis, anti-viral, anti-diabetic, anti-inflammatory, anti-arthritis, and immunosuppressive agents, as well as helping host plants become more resistant to abiotic and biotic stresses [64].

Bacterial consortia treatment of the Shunamukhi rice plant in a pot showed significant plant-growth-promoting effects in terms of root lengths, shoot lengths, yields, and chlorophyll contents compared with the control groups (Figure 6D,F). Higher cell numbers (10^{11} CFU/mL) of a growth-promoting endophytic bacterial consortium isolated from swamp soil have shown increased rice yield [65]. Previous research showed that nitrogen-fixing endophytic bacteria treatment directly influenced grain yield parameters in rice plants [66,67]. In a similar study, Yanni and Dazzo (2010) [68] reported that rice grain yield was augmented by 41% when they used endophytic *Rhizobium leguminosarum* as the inoculant, which is in support of our findings of enhanced yields when rice plants were treated with consortium-2 and consortium-4 (Figure 6F). Hence, the increased grain yield (38–39%), along with no changes in the total chlorophyll content even under the 70% reduced doses of nitrogen fertilizer, can be attributed to the cumulative effects of the plant-growth-promoting traits possessed by each of the crew members of all four consortia in this study. Several other studies reported similar findings, namely that plant-growth-promoting traits of endophytic bacteria augmented the nutrient uptake and enhanced rice yields [68–70] and yields of zucchini [71] and bermudagrass [72]. The endophytic strain of *Lysinibacillus sphaericus* [56] also demonstrated various aspects of plant-growth-promoting traits, such as those of the four consortia reported in the present study, which could accelerate plant growth along with fortifying the plant structure to provide resistance to phytopathogens. This finding might be indicative of the potential promises of these naturally occurring strains for sustainable rice production. Similarly, hormones such as auxin produced by endophytic consortia enhanced the root length and volume, augmented early seedling establishment, and increased nutrient intake from the soil (Figure 6C). Rice plants treated with compost mixed with any of the four consortia resulted in more tillers and longer spikelets, which collectively contributed to the increased (38–39%) grain weights (Figure 6F). The potential rice grain yield was affected by the increased rate of leaf photosynthesis, which could have an impact on dry matter production [73]. The fertilizer and consortium treatments showed a little difference in chlorophyll concentration, but the consortium treatments produced more rice. This might be attributed to other relevant factors, such as phytohormone synthesis and the furnishing of the crop structure toward immunity against biotic and abiotic stresses.

To date, there are no published reports on the effects of a consortium comprising naturally occurring endophytes on rice plants regarding the bioremediation of pesticides and urea reduction in field conditions. We assumed that the efficacy of consortia, comprising endophytic bacteria, as well as based biofertilizer harboring bioremediation properties, provided rice genotype *Shonamukhi* with resistance against both biotic and abiotic stresses in the field rather than in laboratory or greenhouse conditions (Figure 6D). In turn, the effects of these consortia comprising different species of endophytic bacteria on rice plants resulted in increased plant growth and tiller number (Figure 6D). This can be attributed to the abilities of members of the consortia to solubilize micronutrient and/or organic matter of the compost and make it available for root uptake. This was further confirmed by comparing the reduced growth and tillering performances of rice plants grown in compost mixed fertilizer only (Figure 6D). The significant growth-promoting role of pesticide-degrading rice endophytes along with their consortia and subsequent yield is a novel finding, which can be applied to manage a safe and sustainable agricultural practice. Furthermore, the reduced amount of nitrogen fertilizer (30%) along with endophytic consortia as a biofertilizer in rice cultivation may lower the demand for chemical fertilizers, such as urea, which, in turn, might have a beneficiary economic impact on farming in Grameen Bangladesh.

5. Conclusions

The tendency of a quick gain is increasing, resulting in uncontrolled and improper application of pesticides and fertilizers to enhance crop yield. However, the residues from excessive applications of pesticides and fertilizers emerge with their persistence in the environment and are responsible for the death of many endophytic species, leading to agricultural disaster through minimizing plant survival, growth, and subsequent yield. We isolated and identified eighteen endophytic bacterial isolates with high activities of growth-promoting auxin (IAA), ACC-deaminase, N-fixation, P-solubilization, and lignocellulolytic enzymes. In addition, we developed four consortia of these endophytic bacteria, which led to the successful growth and yield of rice at lower doses of urea (30%). The anti-bacterial along with plant-growth-promoting properties of these naturally occurring endophytic bacteria created the potential routes for their use as pesticide degraders and growth promoters in agriculture and as sources of anti-bacterial drugs in pharmaceuticals. The application of endophytic bacteria in agriculture as a microbial inoculant will reduce fertilizer utilization, as well as agricultural health hazards. These research outputs combined with bio-stimulant technologies will help us develop the inoculants of naturally occurring pesticide-degrading bacterial endophytes to substitute synthetic chemical fertilizers, creating a safe and sustainable agricultural practice, as articulated in the Sustainable Development Goals (also known as Global Goals) adopted by the United Nations in 2015.

Supplementary Materials: The following supporting information can be downloaded at: https://www.mdpi.com/article/10.3390/microorganisms11071821/s1: Figure S1: Phosphate solubilizing activities of the chlorpyrifos mineralizing endophytic bacteria; Figure S2: (a) Germinated rice plants after 8 days of endophytic bacterial treatment. (C) control, (R) replica; (b) Effect of endophytic consortium on seedling and growth parameter after 8 days. (C) control, (R) replica and (G) consortia; (c) Effect of endophytic consortium on Seedling and growth parameter after 12 days of treatment. (C) control, (R) replica and (G) consortia. Figure S3: (a–d) Anti-bacterial activity of endophytic bacteria against multidrug-resistant human pathogenic bacteria *S. aureus*, *E. coli*, *Klebsiella* sp., *S. epidermidis* after 32 h. Table S1: Biodegradation profile of chlorpyrifos (1 gm/100 mL) in minimal broth medium after treatment with endophytic bacterial synthetic consortium-2. Table S2: Biodegradation profile of chlorpyrifos (1 gm/100 mL) in minimal broth medium after treatment with endophytic bacterial synthetic consortium-4.

Author Contributions: M.Y.P., Experimentation, Data analysis, Writing, Resources; M.B.R., Experimentation, Data analysis, Writing; A.R., Conceptualization, Editing, Critical review, Data curation, Resources; M.A.A., GC–MS/MS Experimentation; S.G., Writing, Editing, Critical review, Data curation; M.N.-E.-N.N., Resources, Analysis; S., Experimentation, Data analysis; M.S., FT-IR Experimentation, Funding acquisition, Project writing; K.M.C., Conceptualization, Resources, Data analysis; M.A.H., Conceptualization, Funding acquisition, Data analysis, Data curation, Experimentation, Writing. All authors have read and agreed to the published version of the manuscript.

Funding: The authors extend their appreciation to the Deanship of Scientific Research, Vice Presidency Scientific Research, King Faisal University, Saudi Arabia (Researcher Track, GRANT 3636), and the Ministry of Science and Technology (Grant no. 39.00.0000.009.14.008.19/BS-304/315 in FY 2018–2019), Dhaka, Bangladesh.

Institutional Review Board Statement: Not applicable.

Informed Consent Statement: Informed consent was obtained from all subjects involved in the study.

Data Availability Statement: The data reported in this study are contained within the article. The 16S rRNA gene sequences of the bacterial strains were deposited in NCBI and will be available immediately after acceptance of the manuscript for publication. The underlying raw data are available on request from the corresponding author.

Conflicts of Interest: The authors declare no conflict of interest.

References

1. Chowdhury, A.Z.; Jahan, S.A.; Islam, M.N.; Moniruzzaman, M.; Alam, M.K.; Zaman, M.A.; Karim, N.; Gan, S.H. Occurrence of organophosphorus and carbamate pesticide residues in surface water samples from the Rangpur district of Bangladesh. *Bull. Environ. Contam. Toxicol.* **2012**, *89*, 202–207. [CrossRef] [PubMed]
2. Silambarasan, S.; Abraham, J. Ecofriendly method for bioremediation of chlorpyrifos from agricultural soil by novel fungus *Aspergillus terreus* JAS1. *Water Air Soil Pollut.* **2013**, *224*, 1369. [CrossRef]
3. Zhao, L.; Xu, Y.; Lai, X.H.; Shan, C.; Deng, Z.; Ji, Y. Screening and characterization of endophytic *Bacillus* and *Paenibacillus* strains from medicinal plant *Lonicera japonica* for use as potential plant growth promoters. *Brazil. J. Microbiol.* **2015**, *46*, 977–989. [CrossRef] [PubMed]
4. Bhattacharjee, R.B.; Singh, A.; Mukhopadhyay, S. Use of nitrogen-fixing bacteria as biofertiliser for non-legumes: Prospects and challenges. *Appl. Microbiol. Biotechnol.* **2008**, *80*, 199–209. [CrossRef] [PubMed]
5. Santoyo, G.; Moreno-Hagelsieb, G.; del Carmen Orozco-Mosqueda, M.; Glick, B.R. Plant growth-promoting bacterial endophytes. *Microbiol. Res.* **2016**, *183*, 92–99. [CrossRef] [PubMed]
6. Aislabie, J.; Lloyd-Jones, G. A review of bacterial-degradation of pesticides. *Soil Res.* **1995**, *33*, 925–942. [CrossRef]
7. Das, S.R.; Haque, M.A.; Akbor, M.A.; AbdullahAlMamun, M.; Debnath, G.C.; Hossain, M.S.; Hasan, Z.; Rahman, A.; Islam, M.A.; Hossain, M.A.; et al. Organophosphorus insecticides mineralizing endophytic and rhizospheric soil bacterial consortium infuence eggplant growth promotion. *Arch. Microbiol.* **2022**, *204*, 199. [CrossRef]
8. Haque, M.A.; Hong, S.Y.; Hwang, C.E.; Kim, S.C.; Cho, K.M. Cloning of an organophosphorus hydrolase (*opd*D) gene of *Lactobacillus sakei* WCP904 isolated from chlorpyrifos impregnated kimchi and hydrolysis activities of its gene product for organophosphorus pesticides. *Appl. Biol. Chem.* **2018**, *61*, 643–651. [CrossRef]
9. Zhongli, C.; Shunpeng, L.; Guoping, F. Isolation of methyl parathion-degrading strain M6 and cloning of the methyl parathion hydrolase gene. *Appl. Environ. Microbiol.* **2001**, *67*, 4922–4925. [CrossRef]
10. Kumar, S.; Kaushik, G.; Dar, M.A.; Nimesh, S.; Lopez-Chuken, U.J.; Villarreal-Chiu, J.F. Microbial degradation of organophosphate pesticides: A review. *Pedosphere* **2018**, *28*, 190–208. [CrossRef]
11. Narayanan, Z.; Glick, B.R. Secondary metabolites produced by plant growth promoting bacterial endophytes. *Microorganisms* **2022**, *10*, 2008. [CrossRef]
12. Rojas-Sánchez, B.; GuzmánGuzmán, P.; Morales-Cedeño, L.R.; Orozco-Mosqueda, M.C.; Saucedo Martínez, B.C.; Sánchez-Yáñez, J.M.; Fadiji, A.E.; Babalola, O.O.; Glick, B.R.; Santoyo, G. Bioencapsulation of microbial inoculants: Mechanisms, formulation types and application techniques. *Appl. Biosci.* **2022**, *1*, 198–220. [CrossRef]
13. FAO; IFAD; UNICEF; WFP; WHO. *The State of Food Security and Nutrition in the World 2020. Transforming Food Systems for Affordable Healthy Diets*; FAO: Rome, Italy, 2020.
14. Gyaneshwar, P.; James, E.K.; Mathan, N.; Reddy, P.M.; Reinhold-Hurek, B.; Ladha, J.K. Endophytic colonization of rice by a diazotrophic strain of *Serratia marcescens*. *J. Bacteriol.* **2001**, *183*, 2634–2645. [CrossRef] [PubMed]
15. Haque, M.A.; Cho, K.M.; Barman, D.N.; Kim, M.K.; Yun, H.D. A potential cellulose microfibril swelling enzyme isolated from *Bacillus* sp. AY8 enhances cellulose hydrolysis. *Process Biochem.* **2015**, *50*, 807–815. [CrossRef]

16. Haque, M.A.; Hwang, C.E.; Kim, S.C.; Cho, D.Y.; Lee, H.Y.; Cho, K.M.; Lee, J.H. Biodegradation of organophosphate insecticides by two organophosphorus hydrolase genes (*opd*A and *opd*E) from isolated *Leuconostoc mesenteriodes* WCP307 of kimchi origin. *Process Biochem.* **2020**, *94*, 340–348. [CrossRef]
17. Pourbabaee, A.; Soleymani, S.; Farahbakhsh, M.; Torabi, E. Biodegradation of diazinon by the *Stenotrophomonas maltophilia* PS: Pesticide dissipation kinetics and breakdown characterization using FTIR. *Int. J. Environ. Sci. Technol.* **2018**, *15*, 1073–1084. [CrossRef]
18. Aneja, K.R. *Experiments in Microbiology, Plant Pathology and Biotechnology*, 4th ed.; New Age International Publishers: New Delhi, India, 2007.
19. Seeley, H.W.; Vandemark, P.J.; Lee, J.J. *Microbes in Action: A laboratory Manual of Microbiology*, 4th ed.; W. H. Freeman: New York, NY, USA, 1990.
20. Abdullah-Al-Mamun, M.; Hossain, M.S.; Debnath, G.C.; Sultana, S.; Rahman, A.; Hasan, Z.; Das, S.R.; Ashik, M.A.; Prodhan, M.Y.; Aktar, S.; et al. Unveiling lignocellulytic traits of a goat omasum inhabitant *Klebsiella variicola* strain HSTU-AAM51 in light of biochemical and genome analyses. *Braz. J. Microbiol.* **2022**, *53*, 99–130. [CrossRef]
21. Haque, M.A.; Hossain, M.S.; Ahmad, I.; Akbor, M.A.; Rahman, A.; Manir, M.S.; Patel, H.M.; Cho, K.M. Unveiling chlorpyrifos mineralizing and tomato plant-growth activities of *Enterobacter* sp. strain HSTU-ASh6 using biochemical tests, field experiments, genomics, and in silico analyses. *Front. Microbiol.* **2022**, *13*, 1060554. [CrossRef]
22. Ullah, I.; Khan, A.R.; Park, G.S.; Lim, J.H.; Waqas, M.; Lee, I.J.; Shin, J.H. Analysis of phytohormones and phosphate solubilization in *Photorhabdus* spp. *Food Sci. Biotechnol.* **2013**, *22*, 25–31. [CrossRef]
23. Rahman, A.; Sitepu, I.R.; Tang, S.Y.; Hashidoko, Y. Salkowski's reagent test as a primary screening index for functionalities of rhizobacteria isolated from wild dipterocarp saplings growing naturally on medium-strongly acidic tropical peat soil. *Biosci. Biotechnol. Biochem.* **2010**, *74*, 2202–2208. [CrossRef]
24. Belimov, A.A.; Dodd, I.C.; Safronova, V.I.; Shaposhnikov, A.I.; Azarova, T.S.; Makarova, N.M.; Davies, W.J.; Tikhonovich, I.A. Rhizobacteria that produce auxins and contain 1-amino-cyclopropane-1- carboxylic acid deaminase decrease amino acid concentrations in the rhizosphere and improve growth and yield of well-watered and water-limited potato (*Solanum tuberosum*). *Ann. Appl. Biol.* **2015**, *167*, 11–25. [CrossRef]
25. Shahzad, S.; Khalid, A.; Arshad, M. Screening rhizobacteria containing ACC-deaminase for growth promotion of chickpea seedlings under axenic conditions. *Soil Environ.* **2010**, *29*, 38–46.
26. Shaharoona, B.; Riffat, B.; Muhammad, A.; Zahir, Z.; Ul-Hassan, Z. 1-Aminocylopropane-1- carboxylate (ACC)-deaminase rhizobacteria extenuates acc-induced classical triple response in etiolated pea seedlings. *Pak. J. Bot.* **2006**, *38*, 1491–1499.
27. Nautiyal, C.S. An efficient microbiological growth medium for screening phosphate solubilizing microorganisms. *FEMS Microbiol. Lett.* **1999**, *170*, 265–270. [CrossRef]
28. Fukui, R.; Schroth, M.; Hendson, M.; Hancock, J. Interaction between strains of pseudomonads in sugar beet spermospheres and their relationship to pericarp colonization by *Pythium ultimum* in soil. *Phytopathology* **1994**, *84*, 1322–1330. [CrossRef]
29. Krishnasamy, V.; Seshu, D. Germination after accelerated ageing and associated characters in rice varieties. *Seed Sci. Technol.* **1990**, *18*, 147–156.
30. Thompson, D.; Clarke, B.; Kobayashi, D. Evaluation of bacterial antagonists for reduction of summer patch symptoms in Kentucky bluegrass. *Plant Dis.* **1996**, *80*, 856–862. [CrossRef]
31. Zhang, J.; Han, C.; Liu, Z. Absorption spectrum estimating rice chlorophyll concentration: Preliminary investigations. *J. Plant Breed Crop Sci.* **2009**, *1*, 223–229.
32. Chadha, A.; Florentine, S.K.; Chauhan, B.S.; Long, B.; Jayasundera, M. Influence of soil moisture regimes on growth, photosynthetic capacity, leaf biochemistry and reproductive capabilities of the invasive agronomic weed; *Lactuca serriola*. *PLoS ONE* **2019**, *14*, e0218191. [CrossRef]
33. Zaman, N.R.; Chowdhury, U.F.; Reza, R.N.; Chowdhury, F.T.; Sarker, M.; Hossain, M.M.; Akbor, M.A.; Amin, A.; Islam, M.R.; Khan, H. Plant growth promoting endophyte *Burkholderia contaminans* NZ antagonizes phytopathogen *Macrophomina phaseolina* through melanin synthesis and pyrrolnitrin inhibition. *PLoS ONE* **2021**, *16*, e0257863. [CrossRef]
34. Dasgupta, S.; Meisner, C.; Huq, M. A pinch or a pint? Evidence of pesticide overuse in Bangladesh. *J. Agric. Econ.* **2007**, *58*, 91–114. [CrossRef]
35. Matin, M.; Malek, M.; Amin, M.; Rahman, S.; Khatoon, J.; Rahman, M.; Aminuddin, M.; Mian, A. Organochlorine insecticide residues in surface and underground water from different regions of Bangladesh. *Agric. Ecosyst. Environ.* **1998**, *69*, 11–15. [CrossRef]
36. Chowdhury, M.A.Z.; Banik, S.; Uddin, B.; Moniruzzaman, M.; Karim, N.; Gan, S.H. Organophosphorus and carbamate pesticide residues detected in water samples collected from paddy and vegetable fields of the Savar and Dhamrai Upazilas in Bangladesh. *Int. J. Environ. Res. Public Health* **2012**, *9*, 3318–3329. [CrossRef]
37. Haque, M.A.; Simo, T.; Prodhan, M.Y.; Ghosh, S.; Hossain, M.S.; Rahman, A.; Sharker, U.K.; Haque, M.A. Enhanced rice plant (BRRI-28) growth at lower doses of urea caused by diazinon mineralizing endophytic bacterial consortia and explorations of relevant regulatory genes in a *Klebsiella* sp. strain HSTU-F2D4R. *Arch. Microbiol.* **2023**, *205*, 231. [CrossRef]
38. Bargabus, R.L.; Zidack, N.K.; Sherwood, J.E.; Jacobsen, B.J. Characterisation of systemic resistance in sugar beet elicited by a non-pathogenic, phyllosphere-colonizing *Bacillus mycoides*, biological control agent. *Physiol. Mol. Plant Pathol.* **2002**, *61*, 289–298. [CrossRef]

39. Coombs, J.T.; Michelsen, P.P.; Franco, C.M. Evaluation of endophytic *Actinobacteria* as antagonists of *Gaeumannomyces graminis* var. tritici in wheat. *Biol. Control* **2004**, *29*, 359–366. [CrossRef]
40. Senthilkumar, M.; Govindasamy, V.; Annapurna, K. Role of antibiosis in suppression of charcoal rot disease by soybean endophyte *Paenibacillus* sp. HKA-15. *Curr. Microbiol.* **2007**, *55*, 25–29. [CrossRef]
41. Bakker, P.A.; Pieterse, C.M.; Van Loon, L.C. Induced systemic resistance by fluorescent *Pseudomonas* spp. *Phytopathology* **2007**, *97*, 239–243. [CrossRef] [PubMed]
42. Singh, B.K.; Walker, A.J.; Morgan, A.W.; Wright, D.J. Biodegradation of chlorpyrifos by *Enterobacter* strain B-14 and its use in bioremediation of contaminated soils. *Appl. Environ. Microbiol.* **2004**, *7*, 4855–4863. [CrossRef]
43. Ghanem, I.; Orfi, M.; Shamma, M. Biodegradation of chlorpyrifos by *Klebsiella* sp. isolated from an activated sludge sample of waste water treatment plant in Damascus. *Folia Microbiol.* **2007**, *52*, 423–427. [CrossRef]
44. Walitang, D.I.; Kim, K.; Madhaiyan, M.; Kim, Y.K.; Kang, Y.; Sa, T. Characterizing endophytic competence and plant growth promotion of bacterial endophytes inhabiting the seed endosphere of rice. *BMC Microbiol.* **2017**, *17*, 209. [CrossRef] [PubMed]
45. Feng, F.; Ge, J.; Li, Y.; He, S.; Zhong, J.; Liu, X.; Yu, X. Enhanced degradation of chlorpyrifos in rice (*Oryza sativa* L.) by five strains of endophytic bacteria and their plant growth promotional ability. *Chemosphere* **2017**, *184*, 505–513. [PubMed]
46. Taghavi, S.; Van Der Lelie, D.; Hofman, A.; Zhang, Y.B.; Walla, M.D.; Vangronsveld, J.; Newman, L.; Monchy, S. Genome sequence of the plant growth promoting endophytic bacterium *Enterobacter* sp. 638. *PLoS Genet* **2010**, *6*, e1000943. [CrossRef] [PubMed]
47. Ahemad, M.; Khan, M.S. Toxicological effects of selective herbicides on plant growth promoting activities of phosphate solubilizing *Klebsiella* sp. strain PS19. *Curr. Microbiol.* **2011**, *62*, 532–538. [CrossRef] [PubMed]
48. Carrim, A.J.I.; Barbosa, E.C.; Vieira, J.D.G. Enzymatic activity of endophytic bacterial isolates of Jacaranda decurrens Cham.(Carobinha-do-campo). *Braz. Arch. Biol. Technol.* **2006**, *49*, 353–359. [CrossRef]
49. Chimwamurombe, P.M.; Grönemeyer, J.L.; Reinhold-Hurek, B. Isolation and characterization of culturable seed-associated bacterial endophytes from gnotobiotically grown Marama bean seedlings. *FEMS Microbiol.* **2016**, *92*, fiw083. [CrossRef]
50. Ali, N.; Hameed, A.; Ahmed, S. Physicochemical characterization and bioremediation perspective of textile effluent, dyes and metals by indigenous bacteria. *J. Hazard. Mater.* **2009**, *164*, 322–328. [CrossRef]
51. Tony, B.D.; Goyal, D.; Khanna, S. Decolorization of Direct Red 28 by mixed bacterial culture in an up flow immobilized bioreactor. *J. Ind. Microbiol. Biotechnol.* **2009**, *36*, 955–960. [CrossRef]
52. Mattoo, A.J.; Nonzom, S. Endophytes in lignin valorization: A novel approach. *Front. Bioeng. Biotechnol.* **2022**, *10*, 895414. [CrossRef]
53. Li, X.; Geng, X.; Xie, R.; Fu, L.; Jiang, J.; Gao, L.; Sun, J. The endophytic bacteria isolated from elephant grass (*Pennisetum purpureum* Schumach) promote plant growth and enhance salt tolerance of Hybrid Pennisetum. *Biotechnol. Biofuels* **2016**, *9*, 190. [CrossRef]
54. Oteino, N.; Lally, R.D.; Kiwanuka, S.; Lloyd, A.; Ryan, D.; Germaine, K.J.; Dowling, D.N. Plant growth promotion induced by phosphate solubilizing endophytic *Pseudomonas* isolates. *Front. Microbiol.* **2015**, *6*, 745. [CrossRef] [PubMed]
55. Van, V.T.; Berge, O.; Ke, S.N.; Balandreau, J.; Heulin, T. Repeated beneficial effects of rice inoculation with a strain of *Burkholderia vietnamiensis* on early and late yield component in low fertility sulphate acid soils of Vietnam. *Plant Soil* **2000**, *218*, 273–284. [CrossRef]
56. Shabanamol, S.; Varghese, E.M.; Thampi, M.; Karthika, S.; Sreekumar, J.; Jisha, M.S. Enhancement of growth and yield of rice (*Oryza Sativa*) by plant probiotic endophyte, *Lysinibacillus sphaericus* under greenhouse conditions. *Commun. Soil Sci. Plant Anal.* **2020**, *51*, 1268–1282.
57. Tripathi, A.; Pandey, P.; Tripathi, S.N.; Kalra, A. Perspectives and potential applications of endophytic microorganisms in cultivation of medicinal and aromatic plants. *Front. Plant Sci.* **2022**, *13*, 985429. [CrossRef] [PubMed]
58. Mohamad, A.; Abdalla, O.; Li, L.; Ma, J.; Hatab, S.R.; Xu, L.; Guo, J.W.; Rasulov, B.A.; Liu, Y.H.; Hedlund, B.P.; et al. Evaluation of the antimicrobial activity of endophytic bacterial populations from Chinese traditional medicinal plant licorice and characterization of the bioactive secondary metabolites produced by *Bacillus atrophaeus* against *Verticillium dahliae*. *Front. Microbiol.* **2018**, *9*, 924. [CrossRef]
59. El-Deeb, B.; Fayez, K.; Gherbawy, Y. Isolation and characterization of endophytic bacteria from *Plectranthus tenuiflorus* medicinal plant in Saudi Arabia desert and their antimicrobial activities. *J. Plant Interact.* **2013**, *8*, 56–64. [CrossRef]
60. Chandrakar, S.; Gupta, A. Antibiotic potential of endophytic actinomycetes of medicinal herbs against human pathogenic bacteria. *Proc. Natl. Acad. Sci. India-B Biol. Sci.* **2017**, *87*, 905–915. [CrossRef]
61. Tian, B.; Zhang, C.; Ye, Y.; Wen, J.; Wu, Y.; Wang, H.; Li, H.; Cai, S.; Cai, W.; Cheng, Z.; et al. Beneficial traits of bacterial endophytes belonging to the core communities of the tomato root microbiome. *Agric. Ecosyst. Environ.* **2017**, *247*, 149–156. [CrossRef]
62. Diale, M.O.; Ubomba-Jaswa, E.; Serepa-Dlamini, M.H. The antibacterial activity of bacterial endophytes isolated from Combretum molle. *Afr. J. Biotechnol.* **2018**, *17*, 255–262.
63. Ek-Ramos, M.J.; Gomez-Flores, R.; Orozco-Flores, A.A.; Rodríguez-Padilla, C.; González-Ochoa, G.; Tamez-Guerra, P. Bioactive products from plant-endophytic Gram-positive bacteria. *Front. Microbiol.* **2019**, *10*, 463. [CrossRef]
64. Fadiji, A.E.; Babalola, O.O. Elucidating mechanisms of endophytes used in plant protection and other bioactivities with multifunctional prospects. *Front. Bioeng. Biotechnol.* **2020**, *8*, 467. [CrossRef]
65. Fitri, S.N.A.; Gofar, N. Increasing of rice yield by using growth promoting endophytic bacteria from Swamp land. *J. Trop. Soils* **2010**, *15*, 271–276. [CrossRef]

66. Peng, G.; Yuan, Q.; Li, H.; Zhang, W.; Tan, Z. *Rhizobium oryzae* sp. nov., isolated from the wild rice *Oryza alta*. *Int. J. Syst. Evol. Microbiol.* **2008**, *58*, 2158–5163. [CrossRef] [PubMed]
67. Pedraza, R.O.; Bellone, C.H.; de Bellone, C.; Boa Sorte, S.; Teixeira, P.M.F.; Dos, S.K.R. Azospirillum inoculation and nitrogen fertilization effect on grain yield and on the diversity of endophytic bacteria in the phyllosphere of rice rainfed crop. *Eur. J. Soil Biol.* **2009**, *45*, 36–43. [CrossRef]
68. Yanni, Y.G.; Dazzo, F.B. Enhancement of rice production using endophytic strains of *Rhizobium leguminosarum* bv. trifolii in extensive field inoculation trials within the Egypt Nile delta. *Plant Soil* **2010**, *336*, 129–142. [CrossRef]
69. Verma, S.K.; Kingsley, K.L.; Bergen, M.S.; Kowalski, K.P.; White, J.F. Fungal disease prevention in seedlings of rice (*Oryza Sativa*) and other grasses by growth-promoting seed-associated endophytic bacteria from invasive *Phragmites australis*. *Microorganisms* **2018**, *6*, 21. [CrossRef]
70. Doni, F.; Suhaimi, N.S.M.; Mispan, M.S.; Fathurrahman, F.; Marzuki, B.M.; Kusmoro, J.; Uphoff, N. Microbial contributions for rice Production: From conventional crop management to the use of 'Omics' Technologies. *Int. J. Mol. Sci.* **2022**, *23*, 737. [CrossRef]
71. Tie, J.; Qiao, Y.; Jin, N.; Gao, X.; Liu, Y.; Lyu, J.; Zhang, G.; Hu, L.; Yu, J. Yield and rhizosphere soil environment of greenhouse Zucchini in response to different planting and breeding waste composts. *Microorganisms* **2023**, *11*, 1026. [CrossRef]
72. Sullins, K.N.; Dillard, S.L.; Held, D.W.; Carroll, E.P. Utility of plant growth-promoting Rhizobacteria for sustainable production of Bermudagrass forage. *Microorganisms* **2023**, *11*, 863. [CrossRef]
73. Adachi, S.; Yamamoto, T.; Nakae, T.; Yamashita, M.; Uchida, M.; Karimata, R.; Ichihara, N.; Soda, K.; Ochiai, T.; Ao, R.; et al. Genetic architecture of leaf photosynthesis in rice revealed by different types of reciprocal mapping populations. *J. Exp. Bot.* **2019**, *70*, 5131–5144. [CrossRef] [PubMed]

microorganisms

Article

Antifungal Activity of Phyllospheric Bacteria Isolated from *Coffea arabica* against *Hemileia vastatrix*

Katty Ogata-Gutiérrez, Carolina Chumpitaz-Segovia, Jesus Lirio-Paredes and Doris Zúñiga-Dávila *

Laboratorio de Ecología Microbiana y Biotecnología, Departmento de Biología, Facultad de Ciencias, Universidad Nacional Agraria La Molina, Av. La Molina s/n, Lima 15024, Peru; kogata@lamolina.edu.pe (K.O.-G.); jlirio1@icloud.com (J.L.-P.)
* Correspondence: dzuniga@lamolina.edu.pe

Abstract: Peru is one of the leading countries that produce and export specialty coffees, favorably positioned in the international markets for its physical and organoleptic cup qualities. In recent years, yellow coffee rust caused by the phytopathogenic fungus *Hemileia vastatrix* stands out as one of the main phytosanitary diseases that affect coffee culture yields. Many studies have demonstrated bacteria antagonistic activity against a number of phytopathogen fungi. In this context, the aim of this work was to select and characterize phyllospheric bacteria isolated from *Coffea arabica* with antagonistic features against coffee rust to obtain biocontrollers. For that purpose, a total of 82 phyllospheric bacteria were isolated from two coffee leaf rust-susceptible varieties, *typica* and *caturra roja*, and one tolerant variety, *catimor*. Of all the isolates, 15% were endophytic and 85% were epiphytes. Among all the isolates, 14 were capable of inhibiting the mycelial radial growth of *Mycena citricolor*, and *Colletotrichum* sp. 16S rRNA gene sequence-based analysis showed that 9 isolates were related to *Achromobacter insuavis*, 2 were related to *Luteibacter anthropi* and 1 was related to *Rodococcus ceridiohylli*, *Achromobacter marplatensis* and *Pseudomonas parafulva*. A total of 7 representative bacteria of each group were selected based on their antagonistic activity and tested in germination inhibition assays of coffee rust uredinospores. The CRRFLT7 and TRFLT8 isolates showed a high inhibition percentage of urediniospores germination (81% and 82%, respectively), similar to that obtained with the chemical control (91%). An experimental field assay showed a good performance of both strains against rust damage too, making them a promising alternative for coffee leaf rust biocontrol.

Keywords: bacteria; phyllosphere; biocontroller; coffee rust; biotroph

Citation: Ogata-Gutiérrez, K.; Chumpitaz-Segovia, C.; Lirio-Paredes, J.; Zúñiga-Dávila, D. Antifungal Activity of Phyllospheric Bacteria Isolated from *Coffea arabica* against *Hemileia vastatrix*. *Microorganisms* **2024**, *12*, 582. https://doi.org/10.3390/microorganisms12030582

Academic Editors: Vishnu D. Rajput, Anuj Ranjan and Abhishek Chauhan

Received: 11 December 2023
Revised: 10 January 2024
Accepted: 17 January 2024
Published: 14 March 2024

1. Introduction

Coffee is one of the most important crops worldwide and is considered to be one of the main commodities for many developing countries [1]. In Peru, coffee is one of the main agricultural products exported and represents 86% of Peru's total traditional exports. Coffee registered a 3.3% increase in its exportation value compared to that in 2019, due to an improvement in the average price during the last 4 months of the year [2]. Furthermore, Peru, along with Ethiopia, is currently the leading global exporter of organic coffee beans in the world [3] after Mexico and is the main exporter to The United States. Peruvian specialty coffees are recognized for having one of the best physical cup qualities and are becoming famous in many countries, mainly in The United States, Germany and Belgium [4]. Peru has about 425,000 ha of cultivated coffee, representing 10.2% of the national agricultural area with crops. There are currently 236,000 producers involved in the production of coffee, which means that close to 1 million families depend on its cultivation, and there are more than 660,000 ha of coffee cultivated [5]. However, many plant diseases could limit coffee production, causing high economic losses. Coffee leaf rust is one of the most important and aggressive diseases that devastate coffee plants in short periods. Between 2012 and 2013, this biotrophic fungus, *Hemileia vastatrix*, affected about 400,000 ha of coffee cultures, which means a loss of about 414 million soles [6], causing a great impact on coffee farmers whose economy depends mainly on its exportation. The control of this plague could be handled with chemical fungicides and with an adequate management of plant fertilization. However, the indiscriminate use of chemicals and fungicides in crops is an increasing problem since international markets have strict regulations concerning the presence of these chemicals' residue in food and commodities of plant origins. Agriculture is one of the main sources of water and soil pollution [7], causing negative environmental and health [8] effects too. In this context, agriculture based in a sustainable and friendly environment becomes necessary. Microbial inoculants are a natural and organic alternative to chemicals and could be used as biofertilizers and biocontrol agents. Biological control is considered a potent tool for reducing damage caused by certain pathogens; although its performance is not expected to be as effective as chemical fungicides, numerous authors report its efficacy against some phytopathogens. In this sense, it can provide an interesting alternative to fight against coffee diseases sustainably. Bacteria are reported to be the most abundant group of microorganisms that colonize leaves, with culturable counts, growing in a range between 10^6 and 10^{11} per gram of leaf [9]. Because rust is a pathogen that primarily infects coffee leaves, causing plant defoliation, determining the microorganisms that make up the phyllosphere becomes important in plant biocontrol. The phyllosphere is defined as the total aerial area of a plant, mainly leaves; this can serve as a habitat for different microorganisms [10]. Phyllosphere bacteria in this context are the most abundant and may have the potential to suppress some important coffee diseases, including that caused by the phytopathogenic fungi *H. vastatrix*. Bacteria have different mechanisms to lead plant biocontrol, including the production of antagonistic metabolites or certain degradative enzymes, competition and the induction of host resistance, among others. Based on this information, this work pretends to select and characterize coffea phyllospheric bacteria with antagonistic ability against *H. vastatrix* to obtain potential biocontrollers of this biotrophic fungi.

2. Materials and Methods

2.1. Isolation of Bacteria

Bacteria were isolated from leaf samples of *Coffea arabica* of three varieties: two susceptible (*typica* and *caturra roja*) ones that lack genetic resistance to yellow rust, and one tolerant to rust (*catimor*) because it is a hybrid of *timor* (*C. canephora*) and *caturra* (*C. arabica*) varieties. *typica* and *caturra roja* Samples were collected from the Instituto Regional de Desarrollo of the jungle (IRD-Selva), Genova-UNALM, which is located in Chanchamayo-Junin (11°05′42.6298″ S, 75°21′9.4349″ W). Ten leaves of five randomly selected coffee plants were selected for each variety, packed in hermetically sealed bags and delivered into a cooler to the laboratory Laboratorio de Ecología Microbiana y Biotecnología (LEMyB)—UNALM in Lima, Peru for the forward analysis.

2.2. Determination of Phyllospheric Bacterial Populations

The quantification of the endophytic bacteria was carried out according to the methodology proposed by [11]. Leaves were previously disinfected with 70% ethanol and washed with 1% sodium hypochlorite, both for 2 min. Then, they were rinsed with sterile distilled water and macerated in a mortar with a pestle, using 1X PBS buffer in a proportion of 1:20 (*w*/*v*). Serial dilutions of the maceration were made using 0.85% NaCl. Yeast extract mannitol medium (YEM) agar (mannitol 10 g/L, yeast extract 0.5 g/L, K_2HPO_4 0.5 g/L, $MgSO_4 \cdot 7H_2O$ 0.1 g/L, NaCl 0.2 g/L, agar 15 g/L) was used for bacterial growth, quantification, and isolation. Plates were incubated for 15 days at 28 °C after being processed. Meanwhile, the quantification of total epiphytic aerobic bacteria was made using the whole sampled leaves and by washing them with sterile distilled water to clean the excess dust. Subsequently, each leaf was rinsed in a sterile flask with 1X PBS buffer in a proportion of 1:20 (*w*/*v*) and was shaken for 3 min. The spread plate technique in YEM agar was used to obtain epiphytic bacteria for each leaf suspension and its dilutions. The plates were incubated at 28 °C for 15 days to be evaluated. Additionally, both of the PBS suspensions obtained from the endophyte's and the epiphyte's methodologies were treated to obtain spore-forming bacteria too, using the methodology proposed by [12]. For this purpose, dilutions were treated in a water bath at 80 °C for 20 min. Subsequently, 1 mL of each dilution was served in Petri dishes with Tryptone glucose extract (TGE) medium (triptone 5 g/L, meat extract 3 g/L, D-glucose 1 g/L, agar 15 g/L), using the spread plate technique. The plates were incubated at 28 °C for 24 h. The quantification of the bacteria population was reported as colony forming units (CFU)/mL of bacterial culture media. Colonies that showed morphological differences from each other in all the media used were selected and isolated.

2.3. *In Vitro Antifungal Activity Assay against M. citricolor and Colletotrichum sp.*

Antagonistic in vitro assays of the strains against *M. citricolor* and *Colletotrichum* sp. were performed using the dual plate method [13]. A mycelial disk of each fresh fungi culture was placed in the central area of the Potato Dextrose Agar (PDA) (potato extract 4 g/L, Dextrose 20 g, agar 15 g/L) plate and grown at 28 °C for 5 days. In total, 3 µL of each bacterial culture (10^8 CFU/mL) grown in Nutrient Broth was placed at an equal distance of 3 cm from the center of the Petri dish where the fungus was growing. The experiment was Completely Randomized Design (CRD) and carried out in triplicate. Strains that inhibit mycelial growth were considered positive. The plates were incubated in the dark at 28 °C and evaluated daily thereafter for 10 days post-inoculation. The percentage of mycelial growth inhibition was calculated using the following formula: FGI (%) = $(R - r)/R) \times 100$, where R represents the radius of the fungus without any treatment, and r represents the radius of the fungus growing with the evaluated strain (Idris et al., 2007 [13]).

2.4. *DNA Extraction and 16s Ribosomal RNA Gene-Based Phylogenetic Analysis*

Fourteen isolates were selected based on their antifungal activity against *M. citricolor* and *Colletotrichum* sp. The genomic DNA of bacterial cultures was extracted using the AxyPrep Bacterial Genomic DNA Miniprep Kit (Axygen Scientific, Union City, CA, USA), according to the manufacturer's instructions. 16S ribosomal RNA gene amplification (1500 bp) was performed using the primers fD1 and rD1 [14].

The PCR reaction mix (25 µL) contained: 1X reaction buffer (Fermentas, Waltham, MA, USA), 2 µL (50 ng) of extracted genomic DNA, 1.5 mM $MgCl_2$, 5 pmol of each oligonucleotide, 200 µM dNTPs (Fermentas, Waltham, MA, USA) and Taq polymerase (1U/µL, Fermentas, Waltham, MA, USA). The PCR temperature cycling conditions were as follows: initial denaturation at 94 °C for 3 min, followed by 35 cycles of denaturation at 94 °C for 45 s, annealing at 57 °C for 1 min and elongation at 72 °C for 2 min. The last cycle was followed by a final extension at 72 °C for 5 min. The amplification products were examined using a 1% agarose gel and purified with the AxyPrep TMR PCR Cleanup Kit (Axygen Scientific, Union City, CA, USA) according to the manufacturer's instructions and subsequently sequenced by a commercial service (Macrogen Inc., Seoul, Republic of Korea). The obtained sequences were examined and edited using the BioEdit sequence alignment program [15] and identified through BLASTn (Basic Local Alignment Tool for nucleotides) from the public database of The National Center for Biotechnology Information (NCBI) (http://www.ncbi.nlm.nih.gov (accessed on 28 November 2023)). Multiple alignments were compared using the Clustal X2 software [16]. Phylogenetic analyses were performed using the neighbor joining (NJ) method [17], with Mega6 software, applying 1000 bootstrap test subsets with genetic distances computed using the Tamura Nei model [18]. This analysis was performed with the isolates that demonstrate an inhibitory effect against the phytopathogenic fungi.

2.5. Inhibitory Effects of Phyllospheric Bacterial Isolates on the In Vitro Germination of H. vastatrix Urediniospores

The germination inhibition capacity of selected bacteria against coffee rust urediniospores in vitro [19,20] was tested. Bacterial isolates from the phyllosphere were cultivated in SGM growth medium (Na_2HPO_4 0.5 g/L, Na_2MoO_4 0.6 g/L, KH_2PO_4 0.3 g/L, NaCl 0.1 g/L, $MgSO_4 \cdot 7H_2O$ 0.2 g/L, $CaCO_3$ 0.022 g/L, iron citrate 3.8 mM 5 mL/L, microelements 1 mL/L, yeast extract 0.5 g/L and sugar 5 g/L, pH 7.3 ± 0.1) for 72 h at 150 rpm and 28 °C. From one side, each bacterial concentration was adjusted with a 0.85% saline solution until 10^8 CFU/mL [21,22]. On the other side, coffee rust urediniospores obtained from plants showing disease symptoms were collected with a sterile scalpel. The urediniospores were resuspended in sterile distilled water supplemented with 0.001% tween 80, and the concentration was adjusted to 10^5 urediniospores/mL. In total, 10 µL of bacterial culture and 10 µL of the urediniospores suspension were placed on a concave slide inside a humidity chamber system and incubated in the dark at 23 ± 0.2 °C for 48 h. The commercial fungicide propiconazole was used as a positive control against the *H. vastatrix*. For this assay, the propiconazole concentration was 0.3 g/L, according to the manufacturer's recommendation. The experiment was repeated twice, and each treatment consisted of three replicates.

2.6. Foliar Application of Selected Isolates in Coffee Plants var. Caturra Roja

A field experiment was carried out in June at IRD (Instituto de Desarrollo Regional)-jungle Genova, near Santa Rosa annex (750 m altitude, 11°5′40.634″ S, 75°21′ 50.424″ W). Around 40 six-month-old, inoculated coffee seedlings var. *caturra roja* were transplanted from the nursery to the field. The experiment was configured in a 1.5 m row-spacing configuration, and the plants were planted at 1 × 1 m spacing. To avoid the edge effect, additional coffee plants were transplanted around the perimeter. Selected bacteria inocula (CRRFLT7 and TRFLT8) were tested independently. Propiconazole was used as the chemical control and water was used as the negative control. The bacteria inocula of CRRFLT7 and TRFLT8 isolates were prepared using the SGM broth with a concentration of 10^8 CFU/mL. Each treatment was applied by spraying them on the foliar and the abaxial zone every fifteen days (Figure S1). The damage level of the disease on the different treatments was recorded twice, 7 and 8 months after the transplant, when symptoms appeared. Natural infection with *H. vastatrix* in each treatment was assessed to determine plant damage. The disease incidence percentage (DI%) [23] and disease severity expressed as the pathogen disease index (PDI) were the parameters evaluated, calculated as the following formula:

$$DI\% = \frac{\text{Number of infected leaves}}{\text{Total number of leaves}} \times 100 \quad (1)$$

$$PDI = \frac{\text{Sum of all numerical grade}}{\text{Total number of leaves counted} \times \text{Maximum grade}} \times 100 \quad (2)$$

PDI was calculated using the 0–5 scale of [22] with the following scores: 0 = no symptom; 1 = 0.1–5.0% leaf area affected (l.a.a.); 2 = 5.1–15.0% l.a.a.; 3 = 15.1–30.0% l.a.a.; 4 = 30.1–50% l.a.a.; 5 = 50.1–100% l.a.a. All data were analyzed by an LSD ANOVA multiple range test with ten replicates per treatment.

3. Results

3.1. Microbial Population and Isolation of Phyllospheric Bacteria

The population of endophytic and epiphytic bacteria (Table 1) was quantified as the number of colony-forming units (CFU) per gram of dry weight leaf samples obtained from the coffee plants var. *catimor*, *caturra roja* and *typica*. The abundance of the populations varies according to the group of bacteria studied (Table 1). The varieties *catimor* and *typica* showed endophytic bacteria in the leaf samples, compared to the *caturra roja* variety, in which no endophyte was recovered. In addition, leaves of the symptomatic *typica* variety showed the highest population of endophytes (21 × 10^5 CFU·g^{-1} leaf). On the other hand, the aerobic epiphytic bacteria population varies between 10^3 and 10^9 CFU·g^{-1} leaf, the symptomatic samples of *catimor* and *typica* being those that showed a higher population compared to the asymptomatic ones (Table 1). It was observed that symptomatic samples of the *typica* variety showed the highest population in both endophytic and epiphytic aerobic bacteria. Furthermore, the presence of sporogenic aerobic epiphytic bacteria was different in each variety, showing a higher population (23 × 10^2 CFU/mL) in asymptomatic samples of the *catimor* variety compared to the other samples (Table 1). Based on the cultivable bioprospecting, a total of 82 bacterial strains were isolated from three coffee plant varieties. From these isolates, 11 were endophytes (13%) and 71 were epiphytes (87%). Within this last group, 51% and 49% belonged to the aerobic and sporogenic bacteria, respectively. The bacterial population size depends on the leaf zone, plant variety, leaf health and group of bacteria studied.

Table 1. Population size of total cultivable epiphytic and endophytic bacteria in coffee leaves of coffee yellow rust symptomatic and asymptomatic plants.

SAMPLE	Endophytes	Epiphytes	
	Aerobic Bacteria (CFU/g Leaf)	Aerobic Bacteria (CFU/g Leaf)	Sporogenic Bacteria (CFU/g Leaf)
CS	13 × 10^3	13 × 10^4	23 × 10^2
CRS	0	40 × 10^3	50
CRR	0	14 × 10^5	0
TS	2 × 10	16 × 10^4	0
TR	21 × 10^5	52 × 10^9	28 × 10

CS: *catimor* variety, asymptomatics leaves. CRS: *caturra roja* variety, asymptomatic leaves. CRR: *caturra roja* variety, symptomatic leaves. TS: *typica* variety, asymptomatic leaves. TR: *typica* variety, symptomatic leaves.

3.2. Selection of Fungal Antagonists Using Coffee Phytopathogenic Fungi

According to the results, the CSEDT7 isolate was the only endophyte that exhibited an in vitro antagonistic activity against *M. citricolor* and *Colletotrichum* sp., showing 42% and 20% inhibition of radial growth for each fungi, respectively (Table 2). In the same way, thirteen aerobic epiphytic isolates were capable of significantly inhibiting the development of both fungi. It was observed that all the isolates showed the highest antagonistic activity against *M. citricolor* with an inhibition percentage of more than 33%, CSFLT4 and CSFLT5 showing the best results (44%). When the isolates were confronted against *Colletotrichum* sp. instead, their inhibition capacity dropped. It was observed that the endophytic bacterium CSEDT7 presented an inhibition percentage close to 20% (Table 2).

3.3. Phylogenetic Analysis of Selected Isolates

According to the phylogenetic analysis based on the primers homology to conserved regions of the 16S rRNA gene, the corresponding fragments of the expected nucleotidic dimensions were obtained. When compared with similar fragments on the GenBank database by BLAST, 10 of the 14 isolates of the phyllosphere were found to be related to *Achromobacter insuavis* LMG 26845^T, CSEDT7 was related to *Luteibacter anthropi*, CCUG 25036^T, CSFLT6 was related to *Luteibacter yeojuensis* R2A16-10^T, CRRFLT5 was related to *Rhodococcus cercidiphylli* YIM 65003^T and the CRRFLT7 isolate was related to *Pseudomonas parafulva* AJ 2129^T (Table 2). The NJ phylogenetic tree was performed using all the isolates with antagonistic activity against the phytopathogenic fungi tested. This analysis showed that the isolates were clustered in four branches, as shown by the significant bootstrap values. The isolates CSFLT4, TSFLT4, TRFLT8, CRRFLT8, CSFLT5, TSFLT2, TSFLT10, CRRFLT6, TSFLT8 and TSFLT3 were clustered with *A. insuavis*, the CRRFLT7 isolate was grouped with the *P. parafulva* group, CSEDT7 and CSFLT6 were clustered with *L. antrophi* and CRRFLT5 was clustered with *Rhodococcus cercidiphylli* (Figure 1). The sequences determined in this study were deposited in the GenBank database.

Table 2. Antagonistic activity of bacterial isolates against *M. citricolor* and *Colletrotrichum* spp.

TREATMENTS	Origin	In Vitro Antagonistic Activity Inhibition (%)		Phylogenetic Analysis		
		M. citricolor	*Colletrotrichum* sp.	Closely Related Taxa Identified by 16s RNAr GenBank	Similarity (%)	Accession Number
CSEDT7	endophyte	42.2 de	19.8 i	*Luteibacter anthropi* CCUG 25036$^{(T)}$	99.6	FM212561
CSFLT6	epiphyte	35.6 bc	1.1 ab	*Luteibacter anthropi* CCUG 25036$^{(T)}$	98.8	FM212561
CSFLT4	epiphyte	44.4 e	9.8 g	*Achromobacter insuavis* LMG 26845$^{(T)}$	99.6	HF586506
TSFLT2	epiphyte	37.8 bcd	3.8 cd	*Achromobacter insuavis* LMG 26845$^{(T)}$	100	HF586506
TSFLT10	epiphyte	33.3 b	3.0 bc	*Achromobacter insuavis* LMG 26845$^{(T)}$	99.9	HF586506
TSFLT8	epiphyte	33.3 b	3.8 cd	*Achromobacter insuavis* LMG 26845$^{(T)}$	100	HF586506
TSFLT4	epiphyte	33.33 b	13.0 h	*Achromobacter insuavis* LMG 26845$^{(T)}$	99.9	HF586506
CRRFLT6	epiphyte	35.6 bc	3.0 bc	*Achromobacter insuavis* LMG 26845$^{(T)}$	99.7	HF586506
TSFLT3	epiphyte	35.6 bc	5.7 def	*Achromobacter insuavis* LMG 26845$^{(T)}$	99.9	HF586506
CSFLT5	epiphyte	44.4 e	2.4 bc	*Achromobacter insuavis* LMG 26845$^{(T)}$	100	HF586506
CRRFLT8	epiphyte	38.9 cd	7.7 f	*Achromobacter insuavis* LMG 26845$^{(T)}$	99.9	HF586506
CRRFLT5	epiphyte	38.9 cd	12.6 h	*Rhodococcus cercidiphylli* YIM 65003$^{(T)}$	98.7	EU325542
CRRFLT7	epiphyte	38.9 cd	6.2 ef	*Pseudomonas parafulva* AJ 2129$^{(T)}$	99.5	AB060132
TRFLT8	epiphyte	36.7 bc	5.1 de	*Achromobacter marplatensis* B2$^{(T)}$	99.6	EU150134
control	--------	0.00 a	0.00 a	------------------------	--------	--------------

Different letters indicate differences between treatments (ANOVA LSD $p < 0.05$). $^{(T)}$ The type of that microorganism.

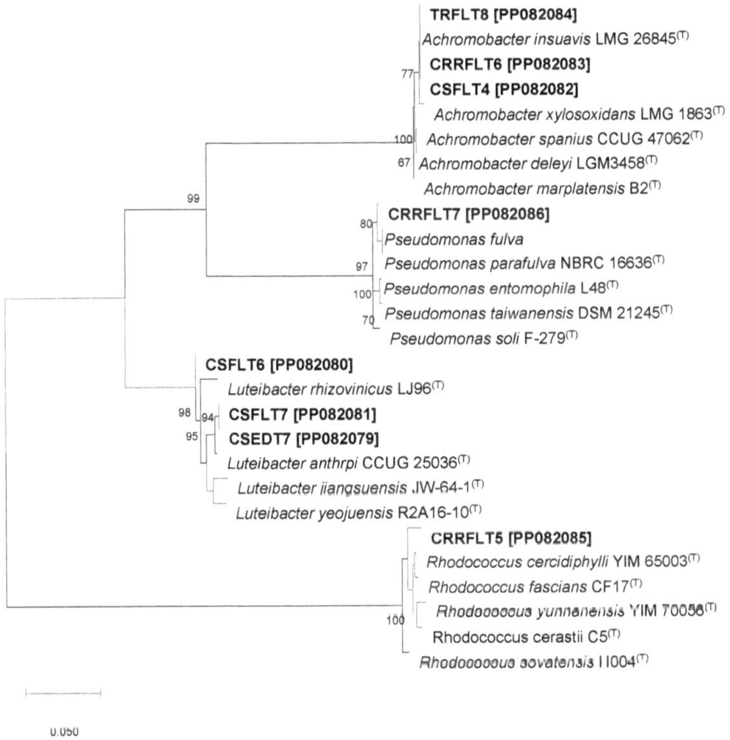

Figure 1. Phylogenetic tree of sequences of the 16S rRNA gene. The phylogenetic reconstruction method of neighbor joining and the distances were calculated in accordance with the Tamura Nei model. The values at the branch points indicate bootstrap support (1000 pseudoreplicates; only values of 50% or above are shown). $^{(T)}$ The type of that microorganism.

3.4. Inhibitory Effect of the Isolates on the Urediniospores Germination Percentage

The isolates CSEDT7, CSFLT6, CSFLT4, CRRFLT5, CRRFLT7, TRFLT8 and TSFLT2 caused a significant inhibition of urediniospores germination, with percentages between 43 to 86% compared to the uninoculated control, which showed the highest number of germination tubes (Table 3). Within these isolates, TRFLT8 (*A. insuavis*) and CRRFLT7 (*P. parafulva*) showed the best performance, with an inhibition percentage of 86 and 82% (Figure 2), respectively. The behavior observed in these isolates was similar to that shown when the chemical control was applied (91%).

Table 3. Effect of bacterial isolates on the germination of *H. vastatrix* urediniospores.

Treatment	Closed Related Taxa (16s RNAr Gene)	Uredinopores Germination (%)
CSEDT7	*Luteibacter anthropi*	13.87 ab † (65.32) ††
CSFLT6	*Luteibacter rhizovicinus*	14.14 ab (64.64)
CSFLT4	*Achromobacter xylosoxidans*	22.59 b (43.49)
CRRFLT5	*Rhodococcus cercidiphylli*	9.2 a (76.99)
CRRFLT7	*Pseudomonas parafulva*	7.38 a (81.54)
TRFLT8	*Achromobacter insuavis*	5.68 a (85.81)
TSFLT2	*Achromobacter insuavis*	14.28 ab (64.29)
Control químico	-	3.58 a (91.05)
Control-	-	39.98 c

† Different letters indicate differences between treatments (ANOVA LSD $p < 0.05$). †† Numbers in parentheses show the % of inhibition.

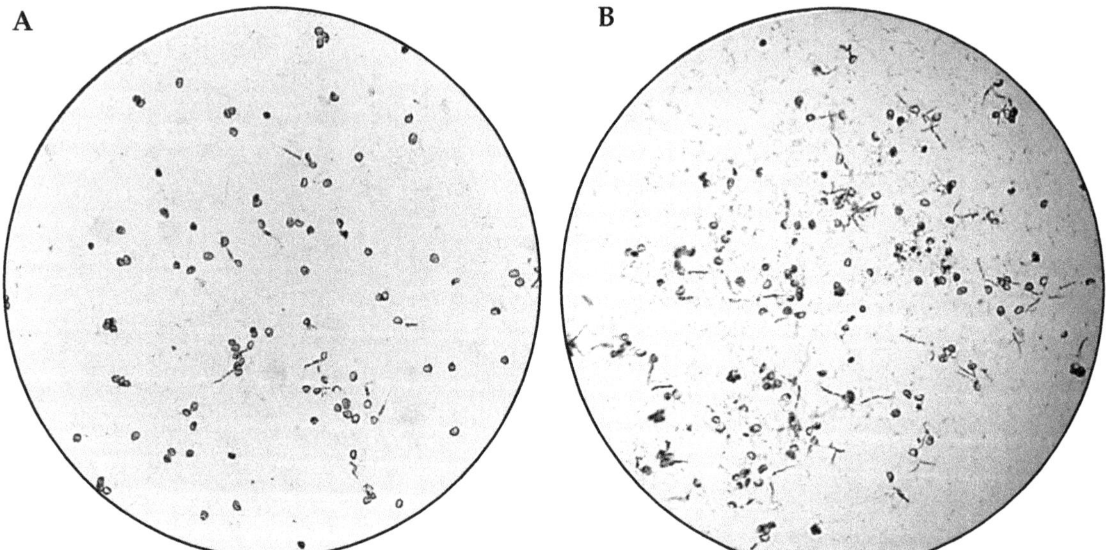

Figure 2. Uredinospores germination of *H. vaxtatrix* (microscopic view at 40× magnification). (**A**) CRRFLT7 treatment, (**B**) negative control.

3.5. Evaluation of Two Isolates as Coffee Rust Biocontrol under Field Conditions

Regarding the number of leaves, the treatments did not show significant differences between them or compared to the control eight months after the transplant. However, there was a positive trend of CRRFLT7 and the chemical control to maintain a relatively high number of leaves compared to the other treatments (Figure 3C). On the other hand, symptoms of coffee rust were observed using the DI and PDI parameters. While DI shows the percentage of infected leaves per plant, PDI shows the severity percentage of disease damage per plant too. According to the statistical analysis, significant differences (p-value < 0.05) in both DI and PDI parameters were observed in plants inoculated with the two isolates and the chemical control compared to the uninoculated plants 8 months after the transplant (Figure 3A,B). The plants inoculated with TRFLT8 (7.7%) and propiconazole (17.9%) showed the same behavior when the disease incidence percentage was evaluated, while the plants inoculated with CRRFLT7 showed a DI of 40.1%. However, the uninoculated control presented an incidence percentage of 77.3%, far outperforming the proven treatments. The severity of the disease resulted in a PDI of 4.4, 15 and 4.6% when propiconazole, CRRFLT7 and TRFLT8 were applied, respectively, compared with the uninoculated control, which showed the greatest severity of damage (29.4%).

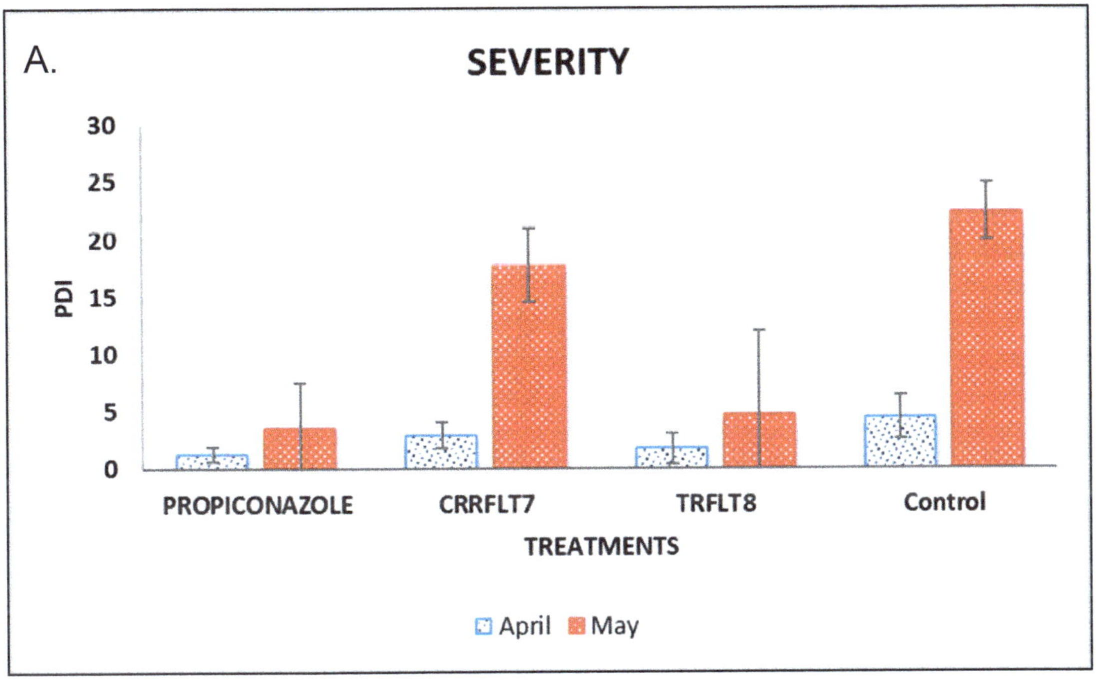

Figure 3. *Cont.*

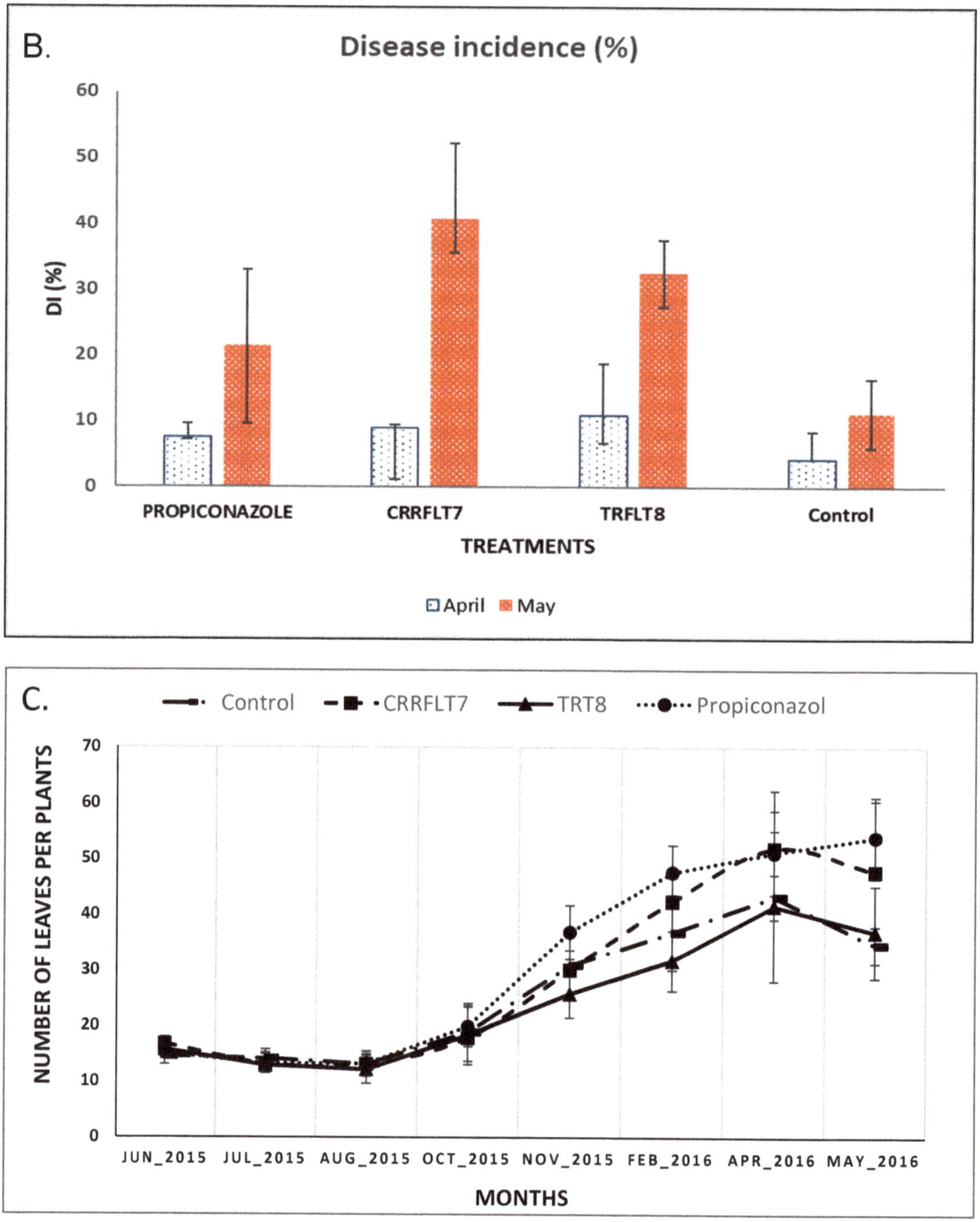

Figure 3. Effect of bacterial inoculation as a biocontroller against coffee rust on field assays. (**A**) Pathogen Disease Index (Severity); (**B**) Disease index (DI%); (**C**) Leaves number per plant for eight months.

4. Discussion

This study examines the population of endophytic and epiphytic phyllospheric bacteria in three coffee plant varieties: *typica, caturra roja* and *catimor,* which are the most extended varieties in this region of Peru. The *typica* and *caturra* varieties are reported to be popular in many countries [24,25] because of their cup quality attributes. Traditionally, coffee production in Peru is only based on certain varieties such as *typica* and *caturra roja.* Less frequently, *pache, mundo novo* and *bourbon* are used, and more recently, *catimor* has been used, which has been widely reported for its tolerance against *H. vastatrix* [26]. The use of the *catimor* variety has not been expanded because its low cup quality is known compared to the susceptible varieties [27]. Despite this, some authors have reported that the quality of *catimor* is not significantly different from that of *typica* or *caturra,* and the quality is not directly related to the variety but also to the climate, altitude, precipitation, harvesting and post-harvesting conditions, among others [28–30]. Moreover, some authors have reported that *catimor* varieties showed similarities in organoleptic and sensory features compared to *typica, caturra* or *bourbon* [29,31]. The results showed in this work, based on a culture-dependent method, revealed that there were fewer endophytic than epiphytic bacteria; these results were similar to those found by [11], who isolated bacteria from coffee phyllosphere. This could be explained by the fact that only a few bacteria are capable of colonizing internal spaces of leaves that are difficult to access. In terms of variety, symptomatic plants of *typica* showed more epiphytic and endophytic bacteria compared to the other varieties. Moreover, in all the varieties, symptomatic plants have shown more bacterial populations than the asymptomatic ones. Similar results were found by [11] in symptomatic and asymptomatic coffee plants infected with *Xyllela fastidiosa*. However, when bacterial populations were found in asymptomatic and symptomatic plants infected with *M. citricolor*, opposite results were observed. Therefore, foliar bacterial populations may vary depending on the host plant and the phytopathogen that colonizes them [32]. From all the isolates, thirteen epiphytes and only one endophyte, showed antagonistic capabilities against two phytopathogen fungi isolated from coffee. In both cases, it was observed that all the tested bacteria have an inhibition activity against *M. citricolor* better than that against *Colletotrichum* sp. The work [33] reported green ZnO nanobiohybrids against these two phytopathogens. The results showed that the percentage of inhibition depends on the concentration of the nanoparticle more than the fungus itself. Bacteria are capable of producing a wide variety of metabolic compounds with antifungal activity, so the effectiveness is linked to the compound that is produced and not to the fungus it faces. Fourteen potential biocontroller strains from the coffee phyllosphere were identified. Molecular identification using the 16S rRNA gene revealed that the isolates belonged to the *Achromobacter, Luteibacter, Rhodococcus* and *Pseudomonas* genera. Related studies on olive's phyllosphere have identified bacteria from the Proteobacteria, Firmicutes and Actinobacteria phyla [34]. Other studies with different crops have found that *Methalobacterium, Pseudomonas* and *Sphingomonas* genera were predominant [35]. When all fourteen microorganisms were tested in in vitro inhibition assays against *H. vastatrix*, only three were able to inhibit rust germination effectively, similar to the chemical control. There are many chemicals used for controlling coffee rust, whose efficacy depends mainly on the timing of the application, severity, weather conditions and agronomic crop management, among other factors. Some of the chemical compounds that are used against *H. vastatrix* are dithiocarbamates, mancozeb, copper compounds, strobilurins and triazoles. The in vitro assay used propiconazole as the chemical control, which belongs to the triazole group. This is a systemic fungicide that acts by inhibiting the biosynthesis of ergosterol [36], a key component in the formation of the fungal cell membrane affecting spore germination. The biocontrol exerted by some bacteria against phytopathogenic fungi can be of different types. It can be direct when there is nutritional competition or the production of certain metabolites that inhibit the growth or development of the phytopathogenic fungus. Alternatively, it can be indirect when a substance produced by the bacteria triggers a response in the plant that affects the fungus. *Pseudomonas* spp. were known to produce a wide range of secondary metabolites that indirectly benefit plant growth by inhibiting some phytopathogenic fungi. One mechanism that is widely reported is the ability of *Pseudomonas* spp. to produce siderophores to obtain Fe^{+3} from the rhizosphere, preventing phytopathogens from accessing this nutrient [37]. Moreover, it has been reported that bacterial siderophores affected spore germination and mycelial growth in some phytopathogenic fungi [38]. *Achromobacter xylosoxidans* has been reported as an endophytic bacterium capable of inhibiting the growth and germination of *Fusarium* spp. Its biocontrol ability has been demonstrated in bean plants, where a decrease in the severity of root rot damage was observed [39]. Other species of *Achromobacter* spp. have been reported as plant growth-promoting bacteria and as a biocontrol agent against some phytopathogenic fungi through the production of hydrolytic enzymes or the activation of the plant's defense by inducing its systemic resistance (SIR). Among these species, *A. insolitus* has shown the capacity to produce cellulases and reduce *Pythium aphanidermatum* growth in in vitro assays [40].

On the field assay, a positive effect of the CRRFLT7 strain and propiconazole on the number of coffee leaves was observed, eight months after the first evaluation. The leaves number is relevant data because *H. vastatrix* attacks coffee leaves, causing defoliation and ultimately leading to the death of the plant. In this context, the effect of the bacteria identified as *Pseudomonas* spp. showed similar behavior to the chemical agent, positioning it as a strain with potential activity to control coffee leaf rust in its early stages. Furthermore, when evaluating the disease index and the percentage of the pathogen disease index (severity), both strains maintained low values, similar to the chemical control. However, there was a better effect observed with the TRFLT8 strain that belongs to the genus *Achromobacter* spp. There was no experimental evidence that confirmed that *P. parafulva* and *A. insuavis* are potentially dangerous pathogens for humans [41,42]. Despite both species being obtained from plants' rhizospheres, some authors report that *A. insuavis* has also been isolated from chronically infected patients [43]. *Pseudomonas* spp. and *Bacillus* spp. are reported as potential biocontrollers against coffee rust in field trials when applied at the onset of the disease in plants with 5% incidence on leaves [12]. In this work, healthy plants were initially used and became naturally infected, allowing for a better observation of the real effect of bacteria during the phenological development of coffee plants. A decrease in damage was observed in both the applied inoculants and propiconazole compared to the uninoculated control. This indicates that the tested inoculants are potential biocontrollers of coffee leaf rust. They should be used within an integrated crop management approach. Currently, there are various chemical products for controlling coffee leaf rust, which are effective at the onset of the disease. However, there are very few biological products that are marketed to address this problem.

5. Conclusions

The use of these bacteria has great economic potential, as the organic coffee market has the best prices in the international market and is subject to fewer fluctuations than the conventional coffee market. The data presented in this study identify two potential biocontrollers not only against leaf rust but also against other phytopathogenic fungi that affect coffee production.

Supplementary Materials: The following supporting information can be downloaded at: https://www.mdpi.com/article/10.3390/microorganisms12030582/s1, Figure S1: Foliar spray inoculation on coffee field assay.

Author Contributions: Conceptualization, review and editing, D.Z.-D.; Methodology, C.C.-S. and J.L.-P.; Investigation, writing and editing, K.O.-G. All authors have read and agreed to the published version of the manuscript.

Funding: This work was supported by CONV-145-2013-FONDECYT and 007-2016-INIA-PNIA/UPMSI/IE projects, Fundación para el Desarrollo Agrario FDA/111/177-UNALM.

Data Availability Statement: Data available on request due to restrictions.

Acknowledgments: The authors would like to thank Jhoselyn Rodriguez-Alvarado for the isolation and characterization of the phytopathogenic fungi *Colletotrichum* sp., Miriam Memenza for the yellow rust isolation and Lee-Anne Maningas for improving the use of English in the manuscript.

Conflicts of Interest: The authors declare no conflicts of interest.

References

1. Ponte, S. The 'Latte Revolution' Regulation, Markets and Consumption in the Global Coffee Chain. *World Dev.* **2002**, *30*, 1099–1122. [CrossRef]
2. JNC (Junta Nacional del Café). Noticias. 2019. Available online: https://juntadelcaf.org.pe/produccion-y-exportaciones-de-cafe-apuntan-a-ser-mejores-durante-el-2021/ (accessed on 31 May 2021).
3. USAID (United States Agency for International Development). Sustainable Development. 2023. Available online: https://www.usaid.gov/peru/sustainable-development (accessed on 5 October 2023).
4. UNDP (United Nations Development Programme). Perking up the Coffee Business. 2017. Available online: https://stories.undp.org/perking-up-the-coffee-business (accessed on 11 June 2018).
5. Inbound Peru. Peru among World's Top Ten Arabica Coffee Producers. 2023. Available online: https://inboundperu.com/2023/08/28/peru-among-worlds-top-ten-arabica-coffee-producers/13080/ (accessed on 1 December 2023).
6. JNC (Junta Nacional del Café). Cafetalero. 2018. Available online: https://juntadelcafe.org.pe/wp-content/uploads/2019/07/cafetalero-ultimo-1.pdf (accessed on 5 October 2023).
7. Martinho, V.J.P.D. Exploring the Topics of Soil Pollution and Agricultural Economics: Highlighting Good Practices. *Agriculture* **2020**, *10*, 24. [CrossRef]
8. Nicolopoulou-Stamati, P.; Maipas, S.; Kotampasi, C.; Stamatis, P.; Hens, L. Chemical Pesticides and Human Health: The Urgent Need for a New Concept in Agriculture. *Front. Public Health* **2016**, *4*, 148. [CrossRef] [PubMed]
9. Vorhol Thompson, I.P.; Bailey, M.J.; Fenlon, J.; Fermor, T.R.; Lilley, A.K.; Lynch, J.M.; McCormack, P.J.; McQuilken, M.P. Quantitative and qualitative seasonal change in the microbial community from the phyllosphere of sugar beet (*Beta vulgaris*). *Plant Soil* **1993**, *150*, 177–191. [CrossRef]
10. Vorholt, J.A. Microbial life in the phyllosphere. *Nat. Rev.* **2012**, *10*, 828–840. [CrossRef] [PubMed]
11. Salas-Bastos, A. Aislamiento e Identificación de Endófitos y Epífitos en Hojas de Café Colectadas en Dos Zonas de Costa Rica y su Posible Empleo Como Biocontroladores de *Mycena citricolor*. Bachelor's Thesis, Escuela de Biología, Instituto Tecnológico de Costa Rica, Cartago, Costa Rica, 2010; p. 113.
12. Haddad, F.; Maffia, L.A.; Mizubuti, E.S.G.; Teixeira, H. Biological control of coffee rust by antagonistic bacteria under field conditions in Brazil. *Biol. Control* **2009**, *49*, 114–119. [CrossRef]
13. Idris, H.; Labuschagne, N.; Korsten, L. Screening rhizobacteria for biological control of Fusarium root and crown rot of sorghum in Ethiopia. *Biol. Control* **2007**, *40*, 97–106. [CrossRef]
14. Weisburg, W.G.; Barns, S.M.; Pelletier, D.A.; Lane, D.J. 16S ribosomal DNA amplification for phylogenetic study. *J. Bacteriol.* **1991**, *173*, 697–703. [CrossRef]
15. Hall, T.A. BioEdit: A user-friendly biological sequence alignment editor and analysis program for Windows 95/98/NT. *Nucleic Acids Symp. Ser.* **1999**, *41*, 95–98.
16. Larkin, M.A.; Blackshields, G.; Brown, N.P.; Chenna, R.; McGettiga, P.A.; McWilliam, H.; Valentin, F.; Wallace, I.M.; Wilm, A.; Lopez, R.; et al. Clustal W and Clustal X version 2.0. *Bioinformatics* **2007**, *223*, 2947–2948. [CrossRef]
17. Saitou, N.; Nei, M. The neighbor-joining method: A new method for reconstructing phylogenetic trees. *Mol. Biol. Evol.* **1987**, *4*, 406–425. [PubMed]
18. Tamura, K.; Stecher, G.; Peterson, D.; Filipski, A.; Kumar, S. MEGA6: Molecular Evolutionary Genetics Analysis version 6.0. *Mol. Biol. Evol.* **2013**, *30*, 2725–2729. [CrossRef] [PubMed]
19. Franco, H.; Alves, H.; Soares, I.; Vieira, F.; Bettiol, W. Bioprospecting endophytic bacteria for biological control of coffee leaf rust. *Sci. Agric.* **2006**, *63*, 32–39.
20. Haddad, F.; Saraiva, R., Mizubuti, E.; Romeiro, R.; Maffia, L. Antifungal compounds as a mechanism to control *Hemileia vastatrix* by antagonistic bacteria. *Trop. Plant Pathol.* **2013**, *38*, 398–405. [CrossRef]
21. Freitas, G.d.S.; Moreira, A.; Prudencio, M.F.; Hungria, M.; Soares Filho, C.V. Foliar Spray Inoculation with Plant Growth Promoting Bacteria Associated with Nitrogen Dosees in *Megathyrsus maximus* cv. BRS Zuri. *Agronomy* **2023**, *13*, 1040. [CrossRef]
22. Fu, H.Z.; Marian, M.; Enommoto, T.; Hieno, A.; Ina, H.; Suga, H.; Shimizu, M. Biocontrol of Tomato Bacterial Wilt by Foliar Spray Application of a Novel Strain of Endophytic *Bacillus* sp. *Microbes Environ.* **2020**, *35*, ME200789. [CrossRef]
23. Gashaw, G.; Alemu, T.; Tesfaye, K. Evaluation of disease incidence and severity and yield loss of finger millet varieties and mycelial growth inhibition of *Pyricularia grisea* isolates using biological antagonists and fungicides in vitro condition. *J. Appl. Biosci.* **2014**, *73*, 5883–5901.
24. Durairaj, C.G.; Karthikeyan Ganapathy, N.; Karuppuchamy, P. Predisposition effect of *Liriomyza trifolii* damage to *Alternaria* leaf spot disease in tomato. *Karnataka J. Agric. Sci.* **2010**, *23*, 161–162.
25. González-Vázquez, A.; Alcántara-Sánchez, D. Calidad organoléptica de variedades tradicionales élite de *Coffea arabica* L. en Chiapas, México. *Cienc. Lat. Rev. Cient. Mult.* **2022**, *6*, 5218–5231. [CrossRef]

26. Juárez-González, T.; Maldonado-Astudillo, Y.I.; González-Mateos, R.; Ramírez-Sucre, M.O.; Álvarez-Fitz, P.; Salazar, R. Caracterización fisicoquímica y sensorial de café de la montaña de Guerrero. *Rev. Mex. Cienc. Agrícolas* **2021**, *12*, 1057–1069. [CrossRef]
27. Alvarado, L.; Vértiz, R.; Jimenez, J.; Borjas, B.; Castro, V.; Julca-Otiniano, A. Caracterización agronómica de 95 accesiones en el banco de germoplasma de café en Chanchamayo (Perú), año 2016. *Rev. ECI Peru* **2017**, *14*, 41–49.
28. World Coffee Research. Arabica Coffee Varieties. Available online: https://varieties.worldcoffeeresearch.org/es/varieties (accessed on 1 May 2019).
29. Van Der Vossen, H. The cup quality of disease-resistant cultivars of Arabica coffee (*Coffea arabica*). *Expl. Agric.* **2009**, *45*, 323–332. [CrossRef]
30. Julca-Otiniano, A.; Alarcón-Águila, G.; Alvarado-Huamán, L.; Borjas-Ventura, R.; Castro-Cepero, V. Comportamiento de tres cultivares de café (Catimor, Colombia y Costa Rica 95) en el valle de el Perené, Junín, Perú. Chil. *J. Agric. Anim.* **2018**, *34*, 205–215. [CrossRef]
31. Ahmed, S.; Brinkley, S.; Smith, E.; Sela, A.; Theisen, M.; Thibodeau, C.; Warne, T.; Anderson, E.; Van Dusen, N.; Giuliano, P.; et al. Climate Change and Coffee Quality: Systematic Review on the Effects of Environmental and Management Variation on Secondary Metabolites and Sensory Attributes of *Coffea arabica* and *Coffea canephora*. *Front. Plant Sci.* **2021**, *12*, 708013. [CrossRef]
32. Marroquín Miranda de Hernández, L.A. Evaluación del Perfil de Taza y de las Características Químicas de Cultivares de Café Derivados de la Variedad Catimor Resistentes a la Roya del Cafeto (*Hemileia vastatrix*), Comparados con las Variedades Utilizadas Comúnmente, en las Verapaces. Master's Thesis, Facultad de Ingeniería, Universidad del Valle de Guatemala, Guatemala City, Guatemala, 2014.
33. Arciniegas-Grijalba, P.A.; Patiño-Portela, M.C.; Mosquera-Sánchez, L.; Guerra Sierra, B.E.; Muñoz-Florez, J.E.; Erazo-Castillo, L.A.; Rodríguez-Páez, J.E. ZnO-based nanofungicides: Synthesis, characterization and their effect on the coffee fungi *Mycena citricolor* and *Colletotrichum* sp. *Mater. Sci. Eng. C* **2019**, *98*, 808–825. [CrossRef]
34. Mina, D.; Pereira, J.A.; Lino-Neto, T. Epiphytic and Endophytic Bacteria on Olive Tree Phyllosphere: Exploring Tissue and Cultivar Effect. *Microb. Ecol.* **2020**, *80*, 145–157. [CrossRef]
35. Huang, S.; Zha, X.; Fu, G. Affecting Factors of Plant Phyllosphere Microbial Community and Their Responses to Climatic Warming—A Review. *Plants* **2023**, *12*, 2891. [CrossRef]
36. Sasirekhamani Ebenezer, P.; Vijayan, V.; Nirmal Nevedhana, K.B. The consequences of inhibition of ergosterol biosynthesis in *Sclerotinia sclerotiorum* (Lib) de bary by propiconazole. *Int. J. Curr. Res.* **2013**, *5*, 2022–2027.
37. Ibrahim, E.; Nasser, R.; Hafeez, R.; Ogunyemi, S.O.; Abdallah, Y.; Khattak, A.A.; Shou, L.; Zhang, Y.; Ahmed, T.; Atef Hatamleh, A.; et al. Biocontrol Efficacy of Endophyte *Pseudomonas poae* to Alleviate Fusarium Seedling Blight by Refining the Morpho-Physiological Attributes of Wheat. *Plants* **2023**, *12*, 2277. [CrossRef]
38. Khan, A.; Singh, P.; Kumar, R.; Das, S.; Singh, R.K.; Mina, U.; Agrawal, G.K.; Rakwal, R.; Sarkar, A.; Srivastava, A. Antifungal Activity of Siderophore Isolated from *Escherichia coli* against *Aspergillus nidulans* via Iron-Mediated Oxidative Stress. *Front. Microbiol.* **2021**, *3*, 729032. [CrossRef]
39. Mohamadpoor, M.; Amini, J.; Ashengroph, M.; Azizi, A. Evaluation of biocontrol potential of *Achromobacer xylosoxidans* strain CTA8689 against common bean root rot. *Physiol. Mol. Plant. Pathol.* **2022**, *117*, 101769. [CrossRef]
40. Yuliar, R.; Nuramida, N.; Salmah, A. Beneficial effect of *Achromobacter insolitus* MB20 and manures in reducing *Pythium aphanidermatum* disease in cucumber. *IOP Conf. Ser. Earth Environ. Sci.* **2021**, *948*, 012053. [CrossRef]
41. Kumari, K.; Rawat, V.; Shadan, A.; Sharma, P.K.; Deb, S.; Singh, R.P. In-depth genome and pan-genome analysis of a metal-resistant bacterium *Pseudomonas parafulva* OS-1. *Front. Microbiol.* **2023**, *14*, 1140249. [CrossRef]
42. Almeida, O.A.C.; De Araujo, N.O.; Mulato, A.T.N.; Persinoti, G.F.; Sforça, M.L.; Calderan-Rodrigues, M.J.; Oliveira, J.V.d.C. Bacterial volatile organic compounds (VOCs) promote growth and induce metabolic changes in rice. *Front. Plant Sci.* **2023**, *13*, 1056082. [CrossRef]
43. Veschetti, L.; Sandri, A.; Patuzzo, C.; Melotti, P.; Malerba, G.; Lleo, M.M. Genomic characterization of Achromobacter species isolates from chronic and occasional lung infection in cystic fibrosis patients. *Microb. Genom.* **2021**, *7*, 000606. [CrossRef]

microorganisms

MDPI

Article

Biofertilizer Based on Biochar and Metal-Tolerant Plant Growth Promoting Rhizobacteria Alleviates Copper Impact on Morphophysiological Traits in *Brassica napus* L.

Adarsh Kumar [1,*], Galina Borisova [1], Maria Maleva [1], Tripti [1], Grigory Shiryaev [1], Anastasia Tugbaeva [1], Artem Sobenin [2] and Irina Kiseleva [1,*]

[1] Institute of Natural Sciences and Mathematics, Ural Federal University, 620002 Yekaterinburg, Russia
[2] Institute of Mining of the Ural Branch of RAS, 620075 Yekaterinburg, Russia
* Correspondence: adarsh.biorem@gmail.com (A.K.); irina.kiseleva@urfu.ru (I.K.)

Abstract: Metal tolerant plant growth-promoting (PGP) rhizobacteria are promising for enhancing plant productivity under copper (Cu) stress. Present pot scale experiment was conducted on *Brassica napus* L. to check the efficiency of rhizobacteria isolated from the rhizosphere of *Tussilago farfara* L. growing on Cu-contaminated soils. Out of fifty Cu tolerant strains, three isolates which showed multiple PGP traits such as indole-3-acetic acid (IAA) synthesis, phosphate (PS) solubilization, siderophore and ammonia production were identified preliminarily by morphological and physiological characteristics followed by 16S rRNA gene sequencing. The best *Bacillus altitudinis* strain TF16a which showed IAA: 15.5 mg L^{-1}, PS: 215 mg L^{-1}, siderophore halo zone ratio of 3.0 with high ammonia production was selected to prepare a biochar-based biofertilizer (BF). Seedling test showed maximum growth of *B. napus* shoot and root in presence of 5% of BF and this concentration was selected for further experiment. The pot experiment included four treatments: control (soil), 100Cu (100 mg Cu kg^{-1} soil), 5%BF (*v/v*), and 5%BF+100Cu, which were carried out for 30 days, after which the morphological, physiological, and biochemical parameters of *B. napus* were studied. The Cu treatment caused its accumulation in shoot and root up to 16.9 and 30.4 mg kg^{-1} DW, respectively, and increased malondialdehyde (MDA) content by 20%. Application of BF with copper led to the decrease in the Cu accumulation by 20% for shoot and 28% for root while MDA content was the same as in the control. Both treatments of BF with and without Cu increased chlorophyll *a* and *b* content by 1.3 times on average as well as non-enzymatic antioxidants such as soluble phenolic compounds (1.3 times) and free proline (1.6 times). Moreover, BF + Cu led to the increase in the biomass of shoot and root by 30 and 60%, respectively, while there was no significant effect on the growth characteristics of plants after the addition of BF without Cu. The study elucidates that BF based on *B. altitudinis* strain TF16a and biochar can be a promising bioformulation which could increase rapeseed growth under the moderate Cu concentration in soil.

Keywords: *Bacillus altitudinis*; rapeseed; bioformulation; heavy metal; morphological parameters; lipid peroxidation; photosynthetic pigments; non-enzymatic antioxidants

Citation: Kumar, A.; Borisova, G.; Maleva, M.; Tripti; Shiryaev, G.; Tugbaeva, A.; Sobenin, A.; Kiseleva, I. Biofertilizer Based on Biochar and Metal-Tolerant Plant Growth Promoting Rhizobacteria Alleviates Copper Impact on Morphophysiological Traits in *Brassica napus* L. *Microorganisms* **2022**, *10*, 2164. https://doi.org/10.3390/microorganisms10112164

Academic Editors: Anuj Ranjan, Vishnu D. Rajput and Abhishek Chauhan

Received: 30 September 2022
Accepted: 27 October 2022
Published: 31 October 2022

1. Introduction

Extensive agriculture is being practiced all around the world to fulfil the human demand from thousands of years. In recent times, numerous synthetic and chemical additives were used for cultivation of crops to enhance their productivity. Although, such additives have tremendously increased the productivity of crops, they also caused an increase in contaminants in the soil [1,2]. Heavy metals (HMs) are the widely distributed chemical pollutants which persist in the environment [3–5]. Copper is one of the HMs which is of great concern. Presence of Cu in the chemical fertilizers, mainly insoluble-phosphorous, as well as different pesticides increased its concentration in the cultivated crops and thus created an alarming situation [6–8]. According to Alengebawy et al. [7], the copper concentration in P-fertilizers varies around the world from 7 to 225 mg Cu kg^{-1} and is about 13 mg Cu kg^{-1} in European Union, whereas N-fertilizers contain 2–1450 and 1.9 mg Cu kg^{-1}, respectively. Copper is an essential metal required by both plants and humans; however, the prolonged fertilizer application leads to its accumulation in agricultural soils and can cause severe health issues [9–12]. In case of plants, an increase in Cu concentration induces oxidative stress through enhanced production of various reactive oxygen species (ROS), resulting in damage of membrane lipids, enzymes, DNA, RNA, decrease in pigment and nutrient content, and plant growth [13,14]. Continuous use of these chemicals is altering the agricultural lands into unusable. Thus, there is a great need of an alternative technologies for sustainable enhanced agricultural production [2,4,7].

In past few decades, plant growth-promoting rhizobacteria (PGPR) gained significant importance in crop yield production. They are capable to improve plant nutrition by phosphate solubilizing (PS) and ammonia (AM) production; promote plant growth by producing indole-3-acetic acid (IAA); protect plants from pathogenic attacks by synthesizing hydrogen cyanide (HCN); and immobilize HMs by producing siderophores (SP) [2,3,15]. Additionally, they support soil biota by mitigating different abiotic stress factors such as HMs, drought, and salinity. However, PGPR cannot improve the physical properties of soil and thus there is a need for their amalgamation with an ecofriendly carrier to enhance the shelf-life and efficiency.

Numerous solid carrier substrates such as coal, alginate, peat, compost, perlite, talc etc., were used to support the PGPR [4,16]. Biochar is also one of such substrates which is generated by thermochemical conversion of biomass in limited oxygen supply. The processing of biomass to biochar involves drying, grinding, pyrolysis, and separation [2]. Biochar is a carbonaceous material which possess multitudinous pores which can retain water, therefore can improve moisture content (MC) when applied to the soil. Its high specific surface area, water holding capacity (WHC) along with presence of carbon, hydrogen, nitrogen, macro-, and micronutrients with several nonpolar or polar substances, which have a strong affinity to inorganic ions such as HMs ions, nitrate, and phosphate makes it the most suitable carrier substrate for amalgamation with PGPR to prepare a stable and efficient biofertilizer [2].

Brassica napus L. (rapeseed) is an oil yielding fast growing plant with large biomass, used as vegetable. It can sequester the majority of metals in the underground part, thus participates in production of clean above-ground biomass and yield [17].

In the present study, we hypothesize that biofertilizer (BF) prepared from wood biochar combined with metal-tolerant PGPR can improve the biometric growth parameters and alleviate the copper impact on the morphophysiological traits of *B. napus*. The specific objectives of the study include: (a) isolation, identification, and characterization of metal tolerance and plant growth-promoting (PGP) traits (IAA, PS, SP, AM, and HCN) of rhizobacteria from Cu-contaminated site, (b) determination of biochar + PGPR based BF efficiency using Petri plate assay, and (c) evaluation of the BF influence on biometric growth parameters (shoot and root length and biomass) as well as physiological and biochemical characteristics (the content of lipid peroxidation products, photosynthetic pigments, soluble phenolic compounds, and free proline) of rapeseed plant with and without of copper.

2. Materials and Methods

2.1. Sampling and Preparation

To isolate the metal-tolerant PGPR, one of the most dominant plant species *Tussilago farfara* L. growing on Cu smelter influenced site (55°29′44.8″ N; 60°14′50.5″ E) near to Karabash town (Chelyabinsk region, Russia) was selected. The rhizospheric soil was collected in sterile zip-lock bags, and transferred to the laboratory. One-third of the soil sample was stored at 4 °C for bacterial isolation and the rest were air dried, sieved, and oven dried at 70 °C for physicochemical characterization according to the methods reported by Maiti [18].

2.2. Isolation, Morphological Characterization, and Genetic Identification

To isolate the rhizospheric bacteria, 10 g of soil sample (on dry weight basis) was transferred in 250 mL of Erlenmeyer flask and mixed with 90 mL of phosphate buffer (pH 6.5), shaken at 180 rpm, 2 h at 28 °C.

The soil suspension was serially diluted to 10^{-1}–10^{-7} with sterile distilled water. Luria-Bertani (LB) agar plates supplemented with Cu (10 g of tryptone, 5 g of yeast extract, 10 g of NaCl, 17 g of agar, and 100 mg of Cu per L, pH 7.0 ± 0.2) were inoculated with 100 µL of serially diluted samples and incubated at 28 °C for next three days in a microbiological incubator to obtain the separate colonies. Morphologically different colonies were picked carefully and further grown on LB-media to obtain pure colonies. The isolates were preliminarily identified by their morphology (color, texture, shape, growth, height, size, pigmentation and optical property), Gram staining, and verified by Bergey's manual [19].

Briefly, genetic identification was based on sequencing the 16S rRNA gene. It was amplified by PCR using genomic DNA as a template and bacterial primers, 27f (50-AGAGTTTGATCMTGGCTCAG-30) and 1492r (50-TACGGYTACCTTGTTACGACTT-30). The automated sequencing was performed as described by Kumar et al. [15]. The obtained sequences were tallied to nucleotide sequences in GenBank using the BLASTn program. The Mega 11 software was used for phylogenetic analysis. The 16S rRNA sequences were aligned using ClustalW. The phylogenetic tree was constructed using the Neighbor-Joining Method, genetic distances were generated using the Tamura-Nei model. The numbers at the branches are bootstrap confidence percentages from 1000 bootstrapped trees.

2.3. Minimum Inhibitory Concentration, PGP Attributes, Drought Tolerance and Antibiotic Sensitivity Test

The minimum inhibitory concentration (MIC) was determined using the methods reported by Kumar et al. [15]. Briefly, the isolates were subjected to increasing metal concentrations (for Cu: 250–1750 mg L^{-1}; Cd: 50–750 mg L^{-1}; Cr: 100–1000 mg L^{-1}; Pb: 500–3500 mg L^{-1} and Ni: 250–2250 mg L^{-1}), allowed to grow for 5 days at 28 °C in agar plate until their growth arrests at particular concentration. Precipitation of metals was prevented by adding EDTA (ethylenediaminetetraacetic acid). Polyethylene glycol (PEG) 6000 was used to test the level of drought tolerance of isolated strains by growing them in tryptone yeast-extract and glucose (TYEG) media for 3 days at 28 °C supplemented with PEG 6000 (5% to 25%) and compared with the control [15]. PGP attributes such as IAA and SP were estimated by the method reported by Brick et al. [20] and Schwyn and Neilands [21], respectively, whereas P-solubilization and AM production were checked as described earlier [4]. Antibiotic sensitivity test was performed according to Kumar et al. [15].

2.4. Carrier Characterization, Biofertilizer Preparation, and Survival Test of Strain TF16a

The biochar produced from birch wood, obtained from domestic manufacturer (Russia) was sieved through 300 mesh size followed by drying for 2 days at 70 °C, and double sterilized at 110 °C for 20 min [2]; and tested for its physicochemical properties. The pH and electrical conductivity (ES) of biochar were determined in water slurry (1:5, *w/v*). The MC and WHC were determined by standard methods [22].

For BF preparation, the strain of metal-tolerant and drought-resistant bacteria (*Bacillus altitudinis* strain TF16a) was grown overnight in 250 mL LB broth (10 g of tryptone, 5 g of yeast extract, 10 g of NaCl per L, pH 7.0 ± 0.2) for 2 days shaking at 150 rpm at 28 °C on orbital shaker-incubator (Biosan ES-20/60, Riga, Latvia). The bacterial culture was centrifuged at 5000× *g* rpm, 10 min and the pellet was washed twice with phosphate buffer (pH 6.5). The obtained pellet was diluted to 99 mL (~10^8 cfu mL^{-1}) and added with 1 mL glycerol. The cell suspension was aseptically added to 250 g of biochar, and left overnight in laminar over flow until the moisture level reached 25–35%. The prepared biofertilizer was sealed in sterile bags and stored at room temperature for use in seedling and pot scale experiments.

The BF was tested to check the shelf-life by comparing its cfu count at the beginning and the end of the three-month time period. Five grams of BF was suspended in 45 mL sterile buffer, shaken at 160 rpm, 45 min at 28 °C, serially diluted, put on LB plates, and cfu were counted and recorded after 3 days of transfer.

2.5. Seedling Growth Test and Plant Development Assay

The healthy seeds of *B. napus* were soaked in sterile Millipore water for 24 h at room temperature. Four treatments: 0%BF (Millipore water, control), 2.5%BF, 5%BF, and 7.5%BF (BF:Millipore water, *w/v*) were prepared and 20 fully imbibed seeds were transferred on Petri plates fitted with circular sterile filter paper (five replicates for each treatment). The seeds were grown for 7 days in a plant growth chamber under 14:10 (day:night) photoperiod, illumination 150 ± 20 µM m^{-2} s^{-1} provided by phytolamps (ULI-P10-18W/SPFR IP40) at 23 ± 2 °C. The percentage of seed germination was calculated as the number of seeds germinated to the total number of seeds sown × 100. The length and biomass of shoot and root were recorded.

The pot experiment was carried out under controlled conditions in September 2020 and repeated in September–October 2021, and the results were averaged. Three *B. napus* plants were grown in 150 mL plastic pots (nine replicates for each treatment) and placed in growth chamber for 30 days under the conditions as described above. Four treatments: control (soil without Cu and BF), 100Cu (100 mg Cu kg^{-1} soil), 5%BF (BF:soil, *v/v*), and 5%BF+100Cu were prepared. The seeds of *B. napus* were sown in double sterilized (130 °C for 15 min) garden soil which showed pH 6.5 ± 0.2, EC 0.4 ± 0.02 dS m^{-1}, and organic carbon 1.75 ± 0.2%. The content of Cu in the soil was low and did not exceed 20.5 ± 2.5 mg kg^{-1}. The following growth characteristics of plants were studied: the rate of seed germination [23], length of shoot and root, dry biomass of aboveground and underground organs, and leaf area [15]. Plant organs were properly washed, dried, and used for Cu determination. Wet ashing of plant samples was carried out by digesting 100 mg sample using 70% nitric acid on a hot plate until the full dissolution was achieved. The Cu content in *B. napus* organs was determined by the atomic absorption spectrometer AA240FS (Varian Australia Pty Ltd., Mulgrave, Victoria, Australia).

2.6. Physiological and Biochemical Parameters

Physiological and biochemical parameters of *B. napus* leaves were determined spectrophotometrically using multimode plate reader Infinite M 200 PRO (Tecan, Grödig, Austria). The content of lipid peroxidation products was measured based on the level of malondialdehyde (MDA) in the reaction with thiobarbituric acid by the absorbtion at 532 and 600 nm [24]. Extraction of photosynthetic pigments was carried out in 80% acetone. The optical density of extracts was measured at 470, 647, and 663 nm and the content of pigments was calculated according to Lichtenthaler [25]. The amount of soluble phenolic compounds (including flavonoids) in plant leaves was determined in 80%-ethanol extracts (24-h extraction in darkness) as described earlier [26]. The total phenolic content was measured in the reaction with the Folin–Ciocalteu reagent at 725 nm. Gallic acid (Sigma-Aldrich Chemie GmbH, Taufkirchen, Germany) was used as a standard [27]. The amount of flavonoids was measured at 412 nm after reacting with 10% aluminum chloride, using rutin (Sigma-Aldrich Chemie GmbH, Taufkirchen, Germany) as a standard [28]. The content of free proline was determined after extraction in boiling water (95–100 °C). Staining was performed with a ninhydrin solution with the addition of glacial acetic acid in an equivalent ratio according to a modified method at a wavelength of 520 nm [26].

2.7. Statistical Analyses

Statistical processing of the results was carried out using STATISTICA 10.0 and Excel 16.0 software. After checking the normality by Shapiro–Wilk test and the homogeneity of variance by Levene's test, the differences between the studied treatments were determined with the nonparametric Kruskal–Wallis H test and Mann–Whitney U test at $p < 0.05$. The relationship between different parameters was estimated by Spearman's rank correlation coefficient. The figures and the tables show the arithmetic mean values (means) and their standard deviations (SD), significant differences between the treatments are indicated by different alphabetical letters.

3. Results and Discussion

3.1. Chemical Properties and Heavy Metal Content in Rhizospheric Soil and Biochar

The rhizospheric soil of *T. farfara* was found slightly acidic in nature (5.40 ± 0.30) with low electrical conductivity (0.3 ± 0.09 dS m^{-1}) and organic carbon content (0.35 ± 0.03%). The soil was found rich in nickel content (550 ± 20 mg kg^{-1}) and had a high magnesium:calcium ratio (4.85 ± 0.25) which reflects its serpentine nature. Because of the continuous operation of century old Cu smelter and geogenic origin, the soil showed maximum concentration for Cu (3058 ± 102 mg kg^{-1}) followed by manganese (991 ± 32 mg kg^{-1}), lead (869 ± 24 mg kg^{-1}), zinc (786 ± 21 mg kg^{-1}), and chromium (566 ± 18 mg kg^{-1}). The biochar used in experiments had circumneutral pH (6.9 ± 0.2) with EC (0.3 ± 0.02 dS m^{-1}), MC (14.0 ± 0.4%), and WHC (81.5 ± 2.5%).

3.2. Bacterial Identification and Characterization

A total of fifty morphologically different isolates, which can tolerate 100 mg Cu L^{-1}, were obtained from rhizospheric soil adhered to the roots of *T. farfara*. Out of fifty, only twelve isolates, which were able to tolerate Cu concentration above 750 mg L^{-1} and Ni above 1250 mg L^{-1}, were further tested for PGP attributes. These isolates were capable to produce IAA and solubilize phosphates, whereas only five produced siderophores. Ammonia production was observed by seven isolates, whereas only one isolate demonstrated HCN production. Three isolates which showed at least four out of five PGP attributes were selected for identification. The 16S rRNA gene sequencing showed >99% similarity of these strains to the reported NCBI database *Bacillus* sp. (strain TF16a), *Arthrobacter* sp. (strain TF16b), and *Pseudomonas* sp. (strain TF16c). *Bacillus, Arthrobacter,* and *Pseudomonas* species were shown to play an important role in plant growth and development in metal contaminated sites and have been mostly described in the literature [2,15,29].

3.3. Minimum Inhibitory Concentration, PGP Attributes, and Drought Resistance of Selected Strains

The characteristics of three selected strains are presented in Table 1. High Cu tolerance was exhibited by all three strains and ranged between 750 and 1750 mg L^{-1}, whereas for Ni it was 1250–2250 mg L^{-1}. They were also able to tolerate a high concentration of Pb, Cr, and Cd.

The rhizosphere of metal-tolerant plant possess different beneficial microorganisms that play a vital role in plant growth. They can produce IAA which promotes root elongation; chelate Fe and other metals by producing siderophore which help in nutrient uptake; solubilize phosphates by excretion of organic acids; and produce ammonia for the enhanced N-uptake [3,16]. For PGP traits, IAA production was found maximal for strain TF16b, whereas P-solubilization for TF16a (Table 1). The halo zone revealed by strain TF16a was maximal among three strains which evidenced its high siderophore producing ability. Ammonia was produced by both TF16a and TF16b strains, whereas none of the three strains showed positive result for hydrogen cyanide (HCN) production.

Table 1. Minimum inhibitory concentration (MIC), plant growth-promoting (PGP) attributes, and drought tolerance of three isolated bacterial strains.

Strain	*Bacillus* sp. TF16a	*Arthrobacter* sp. TF16b	*Pseudomonas* sp. TF16c
	MIC, mg L^{-1}		
Copper	1000	1750	750
Lead	3500	6000	2500
Chromium	2000	2000	1000
Cadmium	750	1000	500
Nickel	2250	2250	1250
	PGP attributes		
Indole-3-acetic acid production, mg L^{-1}	[1] 15.48 ± 0.56 a	20.81 ± 2.85 a	4.84 ± 0.45 b
Solubilized phosphate, mg L^{-1}	215.26 ± 9.45 a	49.21 ± 4.50 c	130.00 ± 6.55 b
Siderophore production *	3.00	1.75	1.71
Ammonia production	+++	+++	-
Hydrogen cyanide production	-	-	-
	Drought tolerance, % (MPa)		
PEG6000 tolerance	25% (−0.75)	25% (−0.75)	20% (−0.49)

[1] Data is presented as means ± SD (n = 4). Different alphabetical letters indicate a significant difference at $p < 0.05$.
* Ratio halo zone diameter to colony diameter; +++ high, - negative.

Drought stress induced by PEG 6000 from −0.05 to −0.73 MPa showed that both TF16a and TF16b strains were capable to tolerate moderate-to-high stress up to −0.73 MPa (25% of PEG 6000). Antibiotic sensitivity test (ampicillin: 10 μg, kanamycin: 30 μg, chloramphenicol: 30 μg, penicillin: 6 μg, tetracycline: 30 μg and streptomycin: 30 μg) revealed resistance to all studied antibiotics, except for streptomycin for TF16a strain. Microorganisms living in rhizosphere of metal stressed soils are often capable to resist drought and antibiotics [3,15]. Among three strains, TF16a revealed the highest siderophore production that was important for chelation of HMs [2,4], thus it was selected for BF preparation by inoculating biochar. Shelf-life study of BF showed no significant decrease in the cfu count with only 3% decrease in MC and no significant change in pH after 90 days of storage.

The high sequence similarity and phylogeny based on ClustalW indicates that strain TF16a belongs to *Bacillus altitudinis* (Figure 1). The 16S rRNA sequence was submitted into NCBI with the accession number OK103906. The phylogenetic trees showing the relationship of partially sequences 16S rRNA gene of TF16b and TF16c strains are presented in Figures S1 and S2.

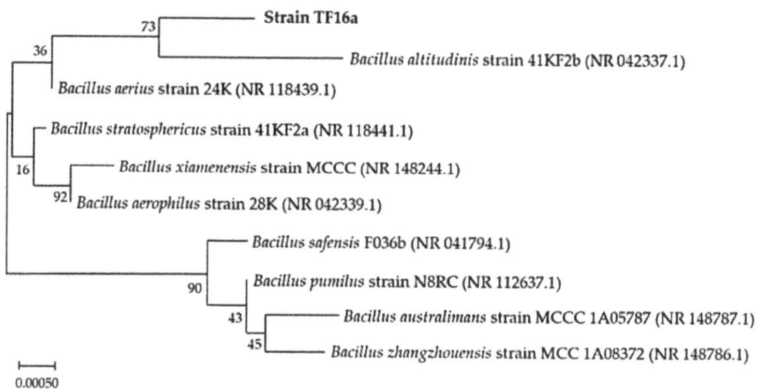

Figure 1. Phylogenetic tree based on partial sequences of the 16S rRNA gene of *Bacillus altitudinis* strain TF16a and identified bacteria from the NCBI database.

3.4. Biometric Growth Parameters and Copper Accumulation

3.4.1. Seedling Growth Test

Seedling growth test was performed in sterile Petri plates using different proportion of BF i.e., 0%, 2.5%, 5%, and 7.5%. Application of BF significantly increased the percentage of seed germination and was found maximal for 5%BF (95%) followed by 2.5%BF (90%) = 7.5%BF (90%) > control (70%) after 7 days of seedling growth. In our previous experiment on *B. napus*, different concentrations of woody biochar (from 2.5% to 10%) without PGPR did not affect the seed germination. However, application of biochar at all studied concentrations reduced the root length, but did not considerably change the shoot length and seedling biomass [30]. In the present study the maximal seedling shoot and root length (Figure 2a) and fresh biomass (Figure 2b) were found with 5%BF which could be due to the presence of PGPR strain TF16a together with biochar. We suppose that drought resistance of TF16a provided chance to plants to grow efficiently even in limited water conditions. Water deficit could be further compensated by biochar through retaining the moisture in its porous structure for a longer time [4].

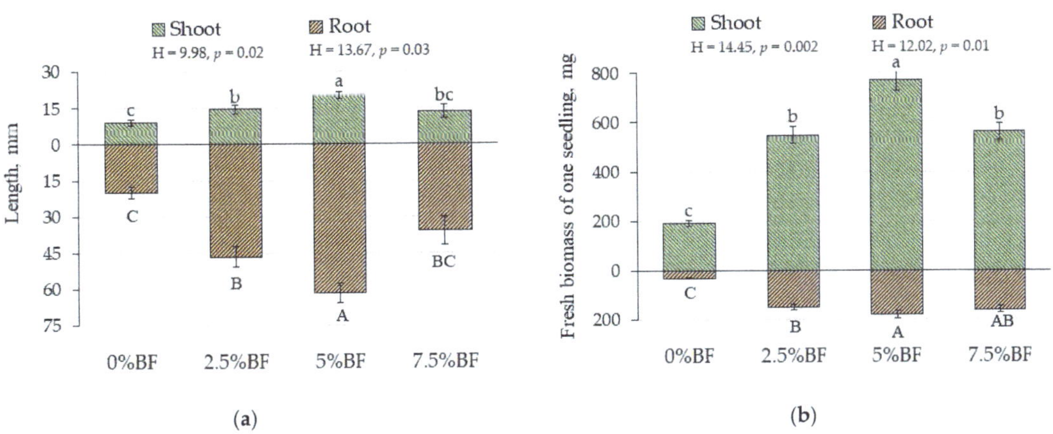

Figure 2. *B. napus* seedling growth parameters: (**a**) length, and (**b**) fresh biomass of shoot and root (means ± SD; $n = 5$). Different alphabetical letters indicate a significant difference at $p < 0.05$.

3.4.2. Pot Scale Experiment

The application of biofertilizers can affect the morphological parameters of plants [4,15]. Seedling growth test showed maximum growth with 5%BF application; thus, it was selected further for pot scale experiment (Figure 2). In our study, the shoot length of *B. napus* did not change at all treatments (Table 2). At the same time, the root length increased 1.5-fold with application of BF both single and combined with Cu (Table 2). It was also noted [15,31] that the combined use of BF with Cu, Zn, and/or Cd could increase the length of plant shoot and root.

Earlier it was reported that application of 5% biochar increased shoot length of *B. napus* [30] and *Phacelia tanacetifolia* [32], but did not affect (in the case of phacelia) or decrease (rapeseed) root length. We suppose that in the present study the opposite effect of BF based on biochar and PGP-active rhizobacteria was associated with the ability of PGPR to synthesize IAA, and thus stimulate *B. napus* root growth.

Application of 100 mg kg^{-1} of Cu led to the significant increase in the root length which suggests limited bioavailability of Cu in the soil and its transfer into the roots. Although there was no difference in shoot length, a great significant difference was observed in fresh and dry biomass of shoot and root after application of 5%BF+100Cu compared to all three treatments including control (Table 2, Figure 3a). When BF was added together with Cu, the dry weight of the shoot and root increased by 30 and 60%, respectively, and leaf area was greater than in any other treatment (Table 2). The variability of morphological parameters was not high, the coefficient of variation did not exceed 14%.

Table 2. Morphological parameters of *B. napus* grown in soil with the addition of Cu and BF.

Parameter	Control	100Cu	5%BF	5%BF+100Cu
Shoot length, cm	[1] 11.63 ± 0.74 a / 6.38	13.46 ± 0.93 a / 6.92	10.76 ± 0.50 a / 4.67	10.38 ± 0.53 a / 5.09
Root length, cm	6.07 ± 0.34 c / 5.60	11.33 ± 0.93 a / 8.16	9.25 ± 0.51 b / 5.53	9.42 ± 0.42 b / 3.40
Shoot FW, mg	350.25 ± 44.50 b / 12.70	338.38 ± 47.59 b / 14.06	298.50 ± 37.81 b / 12.67	423.25 ± 20.67 a / 13.82
Root FW, mg	20.87 ± 1.52 c / 7.25	25.01 ± 2.10 b / 8.15	26.16 ± 0.98 b / 3.74	35.32 ± 2.29 a / 6.47
Shoot DW, mg	58.01 ± 6.11 b / 10.54	57.91 ± 7.29 b / 12.56	50.825 ± 6.09 b / 11.97	75.96 ± 2.21 a / 2.90
Root DW, mg	3.63 ± 0.29 b / 8.07	3.67 ± 0.35 b / 9.62	3.69 ± 0.14 b / 3.86	5.83 ± 0.37 a / 6.29
Leaf area, cm^2	4.23 ± 0.38 bc / 8.85	4.64 ± 0.31 b / 6.67	3.98 ±0.14 c / 3.47	5.70 ± 0.31 a / 5.43

[1] Data are presented in the numerator as means ± SD (n = 9); in the denominator as a coefficient of variation (CV), %. Different alphabetical letters indicate a significant difference at $p < 0.05$. FW: fresh weight; DW: dry weight; BF: biofertilizer.

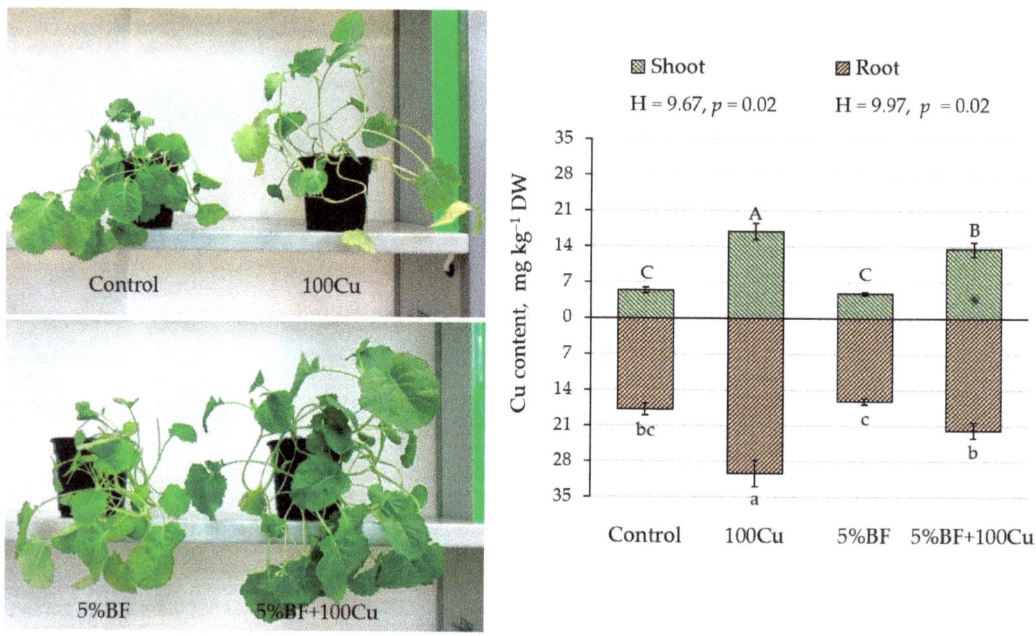

(a) (b)

Figure 3. (**a**) *B. napus* plants after 30-day pot experiment, (**b**) copper accumulation in shoot and root after addition of Cu and BF (means ± SD; n = 9). Different alphabetical letters indicate a significant difference at $p < 0.05$.

Glick [32] and Tripti et al. [4] reported that application of metal and drought tolerant PGPR plays a vital role in shoot and root growth by alleviating the environmental abiotic stress. It can be assumed that since copper is an essential element, in the conditions of low Cu supply the application of BF could provide stimulating effect on plant growth. When copper was added separately, its concentration was too high, while BF could reduce its toxic effect due to exudates secreted by rhizobacteria [9].

In the present study, the addition of copper to the soil increased its accumulation in the shoot by 3.1 times and root by 1.7 times compared with the control (Figure 3b). At the same time, when copper was added together with 5%BF, its accumulation was lower by 1.3 times compared with single Cu treatment. The decrease in copper accumulation when combined with BF can be explained by metal sorption by bacteria and biochar as previously was reported [9,13,33]. Copper was mainly accumulated in the underground part of plants, while its concentration in the shoot was much lower [13,34]. Aust et al. [35] reported that effect of PGPR on the accumulation of metals in *B. napus* depended on a particular bacterial strain and bacterial genus due to different mechanisms.

3.5. Plant Physiological and Biochemical Parameters

It is known that an excess of HMs can cause oxidative stress in plants due to the generation of ROS [13,14]. Since copper is a redox active metal, it is involved in the direct generation of ROS, such as superoxide radical, hydrogen peroxide, and hydroxyl radical [36]. These active molecules are involved in free radical chain reactions with membrane lipids and proteins [37]. The action of 100Cu led to a significant increase in the content of MDA in the leaves of *B. napus* by 20% (Figure 4). At the same time, after the application of 5%BF single and combined with Cu the content of MDA did not differ from the control. Similar results were observed in Cu-enriched pot scale experiment where the PGPR decreased the MDA content along with Cu accumulation in shoot and root [13]. It is possible that the use of BF contributed to maintaining the stability of membrane structures either due to partial absorption of the copper ions by biochar and/or by chelating ability of PGPR [13,34], or to the activation of the antioxidant defense system [14].

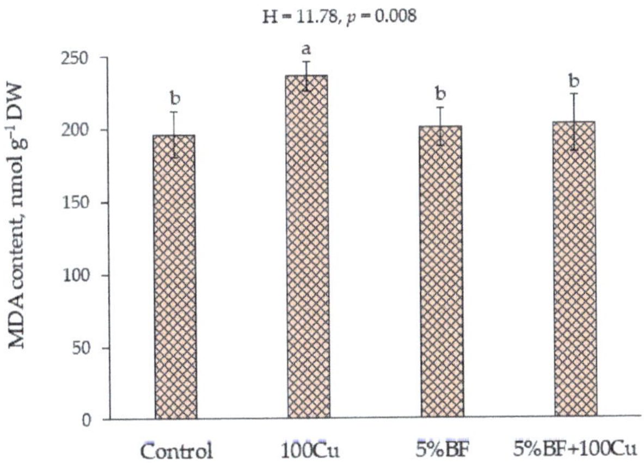

Figure 4. Lipid peroxidation product (MDA) content in the leaves of *B. napus* grown in soil with the addition of Cu and BF (means ± SD; n = 9). Different alphabetical letters indicate significant difference at $p < 0.05$.

The pigment complex is the key characteristic of photosynthetic apparatus in plants under stress. When copper was applied with and without 5%BF, the content of Chl *a* increased 1.3 times, the similar trend was found for Chl *b* (Figure 5). High positive correlation was found between Cu concentration in *B. napus* shoot and chlorophyll content in leaves (r_s = 0.69 for Chl *a* and 0.81 for Chl *b*, at $p < 0.05$) (Table S1). In case of separate application of BF the content of photosynthetic pigments did not change. Copper is an important essential element and plays a significant role in plant physiology, which well explains the rise of chlorophylls content when applied [9,38]. In case of carotenoids, there were no substantial differences in their content.

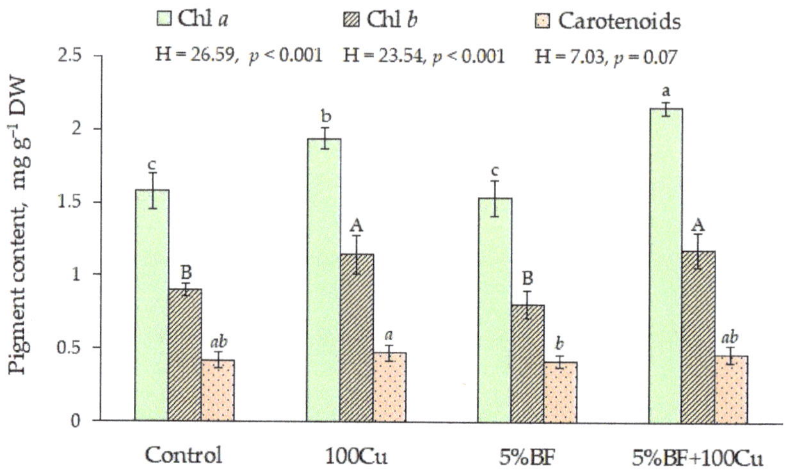

Figure 5. The content of photosynthetic pigments in the leaves of *B. napus* grown in the soil with the addition of Cu and BF (means ± SD; $n = 9$). Different alphabetical letters indicate significant difference at $p < 0.05$.

To protect against ROS, plants use both antioxidant defense enzymes and non-enzymatic antioxidants such as phenolic compounds, proline, ascorbic acid, glutathione, tocopherol, etc. Under abiotic stress conditions, non-enzymatic antioxidants directly quench ROS due to the presence of certain functional groups [14].

Phenolic compounds are an important group of non-enzymatic antioxidants. Their antioxidant effect is realized in several ways: neutralization of ROS (superoxide anion and hydroxyl radicals), chelation of HMs using hydroxyl and carboxyl functional groups, as well as nucleophilic aromatic rings [39,40]. When 5%BF was added with and without Cu, the content of soluble phenolics in the leaves of *B. napus* increased on average by 25%, while in copper-treated soil without BF increased to the greatest extent by 40% (Figure 6a). This confirms their important role in the antioxidant defense of plants. Furthermore, the 5%BF+100Cu application modulated the antioxidant capability in plant which was also reported by other authors in the presence of PGPR [13]. The amount of flavonoids changed similarly to the total content of soluble phenolics: the higher amounts were found in the plants grown in Cu-treated soil (with or without BF). The proportion of flavonoids in the total content of soluble phenolics varied from 31 to 36% (Figure 6a).

One of the most important non-enzymatic antioxidants involved in the protection of plant cells from ROS is free proline. Accumulation of proline in plant cells is considered to be a non-specific defense reaction to various stress factors, including HMs [41,42]. In case of 5%BF+100Cu treatment, the content of proline significantly increased (1.7 times relative to the control) (Figure 6b). With separate addition of copper, its content also has risen by 1.4 times. It is known that PGPR can increase the accumulation of proline in cells under stress conditions and thereby increase plant resistance [43]. However, in our experiment, the increase of proline content was observed only in the presence of Cu (single and, especially, combined with BF).

Spearman's rank correlation analysis revealed high positive correlation between Cu concentration in shoot of *B. napus* and total soluble phenolics ($r_s = 0.77$ at $p < 0.05$, Table S1), including flavonoids ($r_s = 0.71$), and free proline ($r_s = 0.67$) in leaves, which indicates a significant role of these non-enzymatic antioxidants in plant protective reactions.

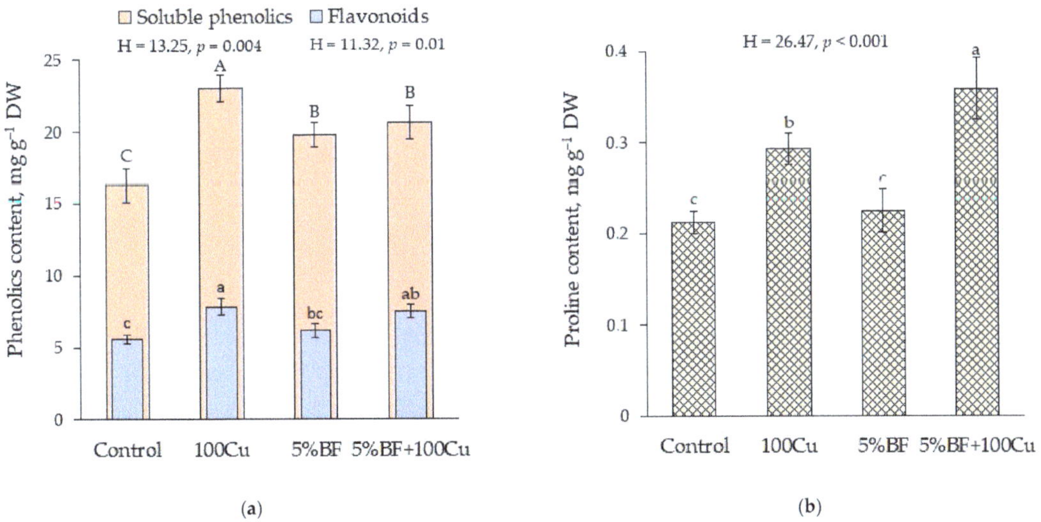

Figure 6. The content of (**a**) phenolic compounds and (**b**) free proline in the leaves of *B. napus* grown in the soil with the addition of Cu and BF (means ± SD; $n = 9$). Different alphabetical letters indicate significant difference at $p < 0.05$.

4. Conclusions

Bacillus altitudinis strain TF16a isolated from the rhizospheric soil of copper smelter influenced site showed a high tolerance to Cu and was able to exhibit multiple plant growth-promoting attributes (indole-3-acetic acid, siderophore, ammonia production, and phosphate solubilization). Application of 5% biofertilizer (BF) based on *B. altitudinis* strain TF16a and wood biochar increased the growth of *Brassica napus* in seedling growth test. Pot scale experiment with application of 5%BF, with and without Cu (100 mg kg^{-1}) after 30 days of *B. napus* vegetation demonstrated that Cu treatment caused accumulation of this metal in shoot and root and increased the content of malondialdehyde (MDA) in leaves as a marker of stress. Application of BF with Cu led to the decrease in the Cu accumulation in comparison with single copper addition in *B. napus* shoot and root, while MDA content remained the same as in the control. Moreover, combined treatment enhanced the growth parameters: shoot and root length and biomass, as compared to other studied treatments. The results evidenced that application of 5%BF+100Cu not only improved the growth of the rapeseed plants but also alleviated the Cu impact by synthesizing non-enzymatic antioxidants (soluble phenolic compounds and free proline) that was confirmed by high positive correlation with Cu content in *B. napus* shoot. The study suggests that BF based on *B. altitudinis* strain TF16a and biochar can be a promising bioformulation which could increase rapeseed growth under moderate Cu concentration in soil.

Supplementary Materials: Supporting information can be downloaded at: https://www.mdpi.com/article/10.3390/microorganisms10112164/s1. Table S1: Spearman's rank correlation between Cu concentration in shoot of *B. napus* and studied physiological and biochemical parameters. Figure S1: Phylogenetic tree based on partial sequences of the 16S rRNA gene of *Arthrobacter* sp. strain TF16b with other related sequences and identified bacteria from the NCBI database. Figure S2: Phylogenetic tree based on partial sequences of the 16S rRNA gene of *Pseudomonas* sp. strain TF16c with other related sequences and identified bacteria from the NCBI database.

Author Contributions: Conceptualization, A.K., G.B., M.M., T. and I.K.; data curation, G.B., M.M. and A.K.; formal analysis, A.K., T., G.B., M.M. and I.K.; funding acquisition, A.K. and M.M.; investigation, A.K., G.B., M.M., T., G.S., A.T. and A.S.; methodology, A.K., T., G.B. and M.M.; project administration, A.K., G.B., M.M. and I.K.; resources, M.M., G.B. and I.K.; software, A.K., T., M.M. and G.S.; validation,

A.K. and T.; visualization, A.K., G.B., M.M., A.T. and T.; writing—original draft, A.K., G.B., M.M., G.S., T. and I.K.; writing—review and editing, A.K., G.B., M.M., T. and I.K. All authors have read and agreed to the published version of the manuscript.

Funding: The work was supported and funded by Russian Science Foundation (Project No. 21-76-00011), https://rscf.ru/project/21-76-00011 for soil sampling and preparation; isolation, morphological characterization and genetic identification of rhizospheric bacteria; carrier characterization and biofertilizer preparation; and partly (assessment of plant morphological and physiological traits) by the Ministry of Science and Higher Education of the Russian Federation within the framework of State Assignment FEUZ-2021-0014.

Institutional Review Board Statement: Not applicable.

Informed Consent Statement: Not applicable.

Data Availability Statement: Data is contained within the article.

Acknowledgments: The authors acknowledge the work support from UrFU, Yekaterinburg. The authors also acknowledge the multiple anonymous reviewers for their constructive comments.

Conflicts of Interest: The authors declare no conflict of interest.

References

1. Ajmal, A.W.; Yasmin, H.; Hassan, M.N.; Khan, N.; Jan, B.L.; Mumtaz, S. Heavy Metal-Resistant Plant Growth-Promoting *Citrobacter werkmanii* Strain WWN1 and *Enterobacter cloacae* Strain JWM6 Enhance Wheat (*Triticum aestivum* L.) Growth by Modulating Physiological Attributes and Some Key Antioxidants Under Multi-Metal Stress. *Front. Microbiol.* **2022**, *13*, 815704. [CrossRef] [PubMed]
2. Tripti; Kumar, A.; Kumar, V.; Bruno, L.B.; Rajkumar, M. Synergism of Industrial and Agricultural Waste as a Suitable Carrier Material for Developing Potential Biofertilizer for Sustainable Agricultural Production of Eggplant. *Horticulturae* **2022**, *8*, 444. [CrossRef]
3. Ma, Y.; Oliveira, R.S.; Nai, F.; Rajkumar, M.; Luo, Y.; Rocha, I.; Freitas, H. The hyperaccumulator *Sedum plumbizincicola* harbors metal-resistant endophytic bacteria that improve its phytoextraction capacity in multi-metal contaminated soil. *J. Environ. Manag.* **2015**, *156*, 62–69. [CrossRef]
4. Tripti; Kumar, A.; Usmani, Z.; Kumar, V. Biochar and flyash inoculated with plant growth promoting rhizobacteria act as potential biofertilizer for luxuriant growth and yield of tomato plant. *J. Environ. Manag.* **2017**, *190*, 20–27. [CrossRef] [PubMed]
5. Kumar, A.; Tripti; Raj, D.; Maiti, S.K.; Maleva, M.; Borisova, G. Soil Pollution and Plant Efficiency Indices for Phytoremediation of Heavy Metal(loid)s: Two-Decade Study (2002–2021). *Metals* **2022**, *12*, 1330. [CrossRef]
6. Li, X.; Zhang, J.; Gong, Y.; Liu, Q.; Yang, S.; Ma, J.; Zhao, L.; Hou, H. Status of copper accumulation in agricultural soils across China (1985–2016). *Chemosphere* **2020**, *244*, 125516. [CrossRef]
7. Alengebawy, A.; Abdelkhalek, S.T.; Qureshi, S.R.; Wang, M.Q. Heavy Metals and Pesticides Toxicity in Agricultural Soil and Plants: Ecological Risks and Human Health Implications. *Toxics* **2021**, *9*, 42. [CrossRef]
8. Gülser, F.; Erdoğan, E. The effects of heavy metal pollution on enzyme activities and basal soil respiration of roadside soils. *Environ. Monit. Assess.* **2008**, *145*, 127–133. [CrossRef]
9. Yruela, I. Copper in plants. *Braz. J. Plant Physiol.* **2005**, *17*, 145–156. [CrossRef]
10. Arora, N.K.; Chauhan, R. Heavy metal toxicity and sustainable interventions for their decontamination. *Environ. Sustain.* **2021**, *4*, 1–3. [CrossRef]
11. Vardhan, K.H.; Kumar, P.S.; Panda, R.C. A review on heavy metal pollution, toxicity and remedial measures: Current trends and future perspectives. *J. Mol. Liq.* **2019**, *290*, 111197. [CrossRef]
12. Wang, X.; Liu, W.; Li, Z.; Teng, Y.; Christie, P.; Luo, Y. Effects of long-term fertilizer applications on peanut yield and quality and plant and soil heavy metal accumulation. *Pedosphere* **2020**, *30*, 555–562. [CrossRef]
13. Latef, A.A.A.H.; Zaid, A.; Abo-Baker, A.B.A.E.; Salem, W.; Abu Alhmad, M.F. Mitigation of Copper Stress in Maize by Inoculation with *Paenibacillus polymyxa* and *Bacillus circulans*. *Plants* **2020**, *9*, 1513. [CrossRef] [PubMed]
14. Hasanuzzaman, M.; Bhuyan, M.H.M.B.; Zulfiqar, F.; Raza, A.; Mohsin, S.M.; Mahmud, J.A.; Fujita, M.; Fotopoulos, V. Reactive Oxygen Species and Antioxidant Defense in Plants under Abiotic Stress: Revisiting the Crucial Role of a Universal Defense Regulator. *Antioxidants* **2020**, *9*, 681. [CrossRef]
15. Kumar, A.; Tripti; Voropaeva, O.; Maleva, M.; Panikovskaya, K.; Borisova, G.; Rajkumar, M.; Bruno, L.B. Bioaugmentation with copper tolerant endophyte *Pseudomonas lurida* strain EOO26 for improved plant growth and copper phytoremediation by *Helianthus annuus*. *Chemosphere* **2021**, *266*, 128983. [CrossRef]
16. Maheshwari, D.K.; Dubey, R.C.; Agarwal, M.; Dheeman, S.; Aeron, A.; Bajpai, V.K. Carrier based formulations of biocoenotic consortia of disease suppressive *Pseudomonas aeruginosa* KRP1 and *Bacillus licheniformis* KRB1. *Ecol. Eng.* **2015**, *81*, 272–277. [CrossRef]

17. Mourato, M.P.; Moreira, I.N.; Leitão, I.; Pinto, F.R.; Sales, J.R.; Louro Martins, L. Effect of Heavy Metals in Plants of the Genus *Brassica*. *Int. J. Mol. Sci.* **2015**, *16*, 17975–17998. [CrossRef] [PubMed]
18. Maiti, S.K. *Ecorestoration of the Coalmine Degraded Lands*; Springer: New York, NY, USA, 2013; p. 361.
19. Bergey, D.H.; Holt, J.G.; Krieg, P.; Sneath, P.H. *Bergey's Manual of Determinative Bacteriology*; Williams and Wilkins: Baltimore, MD, USA, 1994.
20. Bric, J.M.; Bostock, R.M.; Silverstone, S.E. Rapid In Situ Assay for Indoleacetic Acid Production by Bacteria Immobilized on a Nitrocellulose Membrane. *Appl. Environ. Microbiol.* **1991**, *57*, 535–538. [CrossRef]
21. Schwyn, B.; Neilands, J.B. Universal chemical assay for the detection and determination of siderophores. *Anal. Biochem.* **1987**, *160*, 47–56. [CrossRef]
22. Feng, L.; Roughley, R.J.; Copeland, L. Morphological Changes of Rhizobia in Peat Cultures. *Appl. Environ. Microbiol.* **2002**, *68*, 1064–1070. [CrossRef]
23. Wijewardana, C.; Reddy, K.R.; Krutz, L.J.; Gao, W.; Bellaloui, N. Drought stress has transgenerational effects on soybean seed germination and seedling vigor. *PLoS ONE* **2019**, *14*, e0214977. [CrossRef] [PubMed]
24. Heath, R.L.; Packer, L. Photoperoxidation in isolated chloroplasts: I. Kinetics and stoichiometry of fatty acid peroxidation. *Arch. Biochem. Biophys.* **1968**, *125*, 189–198. [CrossRef]
25. Lichtenthaler, H.K. Chlorophylls and carotenoids: Pigments of photosynthetic membranes. *Method. Enzymol.* **1987**, *148*, 350–382. [CrossRef]
26. Maleva, M.; Borisova, G.; Filimonova, E.; Lukina, N.; Chukina, N.; Ermoshin, A.; Tugbaeva, A.; Voropaeva, O. Adaptive Redox Reactions Promote Naturalization of Rare Orchid *Epipactis atrorubens* on Serpentine Dumps Post Asbestos Mining. *Horticulturae* **2022**, *8*, 603. [CrossRef]
27. Singleton, V.L.; Orthofer, R.; Lamuela-Raventos, R.M. Analysis of total phenols and other oxidation substrates and antioxidants by means of Folin–Ciocalteu reagent. *Methods Enzymol.* **1999**, *299*, 152–178. [CrossRef]
28. Chang, C.C.; Yang, M.H.; Wen, H.M.; Chern, J.C. Estimation of Total Flavonoid Content in Propolis by Two Complementary Colorimetric Methods. *J. Food Drug Anal.* **2002**, *10*, 178–182. [CrossRef]
29. Zhang, Y.F.; He, L.Y.; Chen, Z.J.; Wang, Q.Y.; Qian, M.; Sheng, X.F. Characterization of ACC deaminase-producing endophytic bacteria isolated from copper-tolerant plants and their potential in promoting the growth and copper accumulation of *Brassica napus*. *Chemosphere* **2011**, *83*, 57–62. [CrossRef] [PubMed]
30. Shiryaev, G.I.; Maleva, M.G.; Borisova, G.G.; Kumar, A. Parameters of rape seed growth when adding biochar in different concentration. In *The State of the Environment, Environmental Problems and Ways to Solve Them, Proceedings of the All-Russian Scientific-practical Conference, Ust-Ilimsk, Russia, 20–21 December 2021*; BSU Publishing House: Irkutsk, Russia, 2022; pp. 95–101. (In Russian)
31. Dabrowska, G.; Hrynkiewicz, K.; Trejgell, A.; Baum, C. The effect of plant growth-promoting rhizobacteria on the phytoextraction of Cd and Zn by *Brassica napus* L. *Int. J. Phytorem.* **2017**, *19*, 597–604. [CrossRef]
32. Borisova, G.; Maleva, M.; Atambire, A.; Davydova, D.; Tripti. Wood Biochar as an Amendment for Enhanced Growth of Phacelia tanacetifolia. *AIP Conf. Proceed.* **2021**, *2388*, 040007. [CrossRef]
33. Glick, B.R. Bacteria with ACC deaminase can promote plant growth and help to feed the world. *Microbiol. Res.* **2014**, *169*, 30–39. [CrossRef]
34. Abbaszadeh-Dahaji, P.; Omidvari, M.; Ghorbanpour, M. *Plant-Microbe Interaction: An Approach to Sustainable Agriculture. Increasing Phytoremediation Efficiency of Heavy Metal-Contaminated Soil Using PGPR for Sustainable Agriculture*; Choudhary, D.K., Varma, A., Tuteja, N., Eds.; Springer: Singapore, 2016; pp. 187–204.
35. Mwamba, T.M.; Ali, S.; Ali, B.; Lwalaba, J.L.; Liu, H.; Farooq, M.A.; Shou, J.; Zhou, W. Interactive effects of cadmium and copper on metal accumulation, oxidative stress, and mineral composition in *Brassica napus*. *Int. J. Environ. Sci. Technol.* **2016**, *13*, 2163–2174. [CrossRef]
36. Aust, S.D.; Morehouse, L.A.; Thomas, C.E. Role of metals in oxygen radical reactions. *J. Free Radi. Biol. Med.* **1985**, *1*, 3–25 [CrossRef]
37. Kanoun-Boulé, M.; Vicente, J.A.F.; Nabais, C.; Prasad, M.N.V.; Freitas, H. Ecophysiological tolerance of duckweeds exposed to copper. *Aquat. Toxicol.* **2009**, *91*, 1–9. [CrossRef]
38. El-Meihy, R.M.; Abou-Aly, H.E.; Youssef, A.M.; Tewfike, T.A.; El-Alkshar, E.A. Efficiency of heavy metals-tolerant plant growth promoting bacteria for alleviating heavy metals toxicity on sorghum. *Environ. Exp. Bot.* **2019**, *162*, 295–301. [CrossRef]
39. Kısa, D.; Elmastaş, M.; Öztürk, L.; Kayır, Ö. Responses of the phenolic compounds of *Zea mays* under heavy metal stress. *Appl. Biol. Chem.* **2016**, *59*, 813–820. [CrossRef]
40. Foti, M.C. Antioxidant properties of phenols. *J. Pharm. Pharmacol.* **2007**, *59*, 1673–1685. [CrossRef] [PubMed]
41. Aslam, M.; Saeed, M.S.; Sattar, S.; Sajad, S.; Sajjad, M.; Adnan, M.; Iqbal, M.; Sharif, M.T. Specific Role of Proline Against Heavy Metals Toxicity in Plants. *Ind. J. Pure App. Biosci.* **2017**, *5*, 27–34. [CrossRef]
42. Kaur, G.; Asthir, B. Proline: A key player in plant abiotic stress tolerance. *Biol. Plant.* **2015**, *59*, 609–619. [CrossRef]
43. Ansary, M.H.; Rahmani, H.A.; Ardakani, M.R.; Paknejad, F.; Habibi, D.; Mafakheri, S. Effect of *Pseudomonas fluorescent* on Proline and Phytohormonal Status of Maize (*Zea mays* L.) under Water Deficit Stress. *Annal. Biol. Res.* **2012**, *3*, 1054–1062.

microorganisms

Article

Role of *Rahnella aquatilis* AZO16M2 in Phosphate Solubilization and Ex Vitro Acclimatization of *Musa acuminata* var. Valery

Daniela Landa-Acuña [1], Marcia Toro [1,2,*], Ricardo Santos-Mendoza [1] and Doris Zúñiga-Dávila [1,*]

1 Laboratorio de Ecología Microbiana y Biotecnología, Departamento de Biología, Facultad de Ciencias, Universidad Nacional Agraria La Molina, Lima 15024, Peru; danielanda17@gmail.com (D.L.-A.); rsantosme@gmail.com (R.S.-M.)
2 Centro de Ecología Aplicada, Instituto de Zoología y Ecología Tropical, Facultad de Ciencias, Universidad Central de Venezuela, Caracas 1041-A, Venezuela
* Correspondence: mtoro@lamolina.edu.pe or marcia.toro@ucv.ve (M.T.); dzuniga@lamolina.edu.pe (D.Z.-D.)

Abstract: *Rahnella aquatilis* AZO16M2, was characterized for its phosphate solubilization capacity to improve the establishment and survival of *Musa acuminata* var. Valery seedlings under ex-acclimation. Three phosphorus sources (Rock Phosphate (RF), $Ca_3(PO_4)_2$ and K_2HPO_4) and two types of substrate (sand:vermiculite (1:1) and Premix N°8) were selected. The factorial analysis of variance ($p < 0.05$) showed that *R. aquatilis* AZO16M2 (OQ256130) solubilizes $Ca_3(PO_4)_2$ in solid medium, with a Solubilization Index (SI) of 3.77 at 28 °C (pH 6.8). In liquid medium, it was observed that *R. aquatilis* produced 29.6 mg/L soluble P (pH 4.4), and synthesized organic acids (oxalic, D-gluconic, 2-ketogluconic and malic), Indole Acetic Acid (IAA) (33.90 ppm) and siderophores (+). Additionally, acid and alkaline phosphatases (2.59 and 2.56 μg pNP/mL/min) were detected. The presence of the pyrroloquinoline-quinone (PQQ) cofactor gene was confirmed. After inoculating AZO16M2 to *M. acuminata* in sand:vermiculite with RF, the chlorophyll content was 42.38 SPAD (Soil Plant Analysis Development). Aerial fresh weight (AFW), aerial dry weight (ADW) and root dry weight (RDW) were superior to the control by 64.15%, 60.53% and 43.48%, respectively. In Premix N°8 with RF and *R. aquatilis*, 8.91% longer roots were obtained, with 35.58% and 18.76% more AFW and RFW compared with the control as well as 94.45 SPAD. With $Ca_3(PO_4)_2$, values exceeded the control by 14.15% RFW, with 45.45 SPAD. *Rahnella aquatilis* AZO16M2 favored the ex-climatization of *M. acuminata* through improving seedling establishment and survival.

Keywords: pyrroloquinoline-quinone (PQQ) gene; 16S gene; tricalcium phosphate; organic acids; rock phosphate; plant-growth-promoting rhizobacteria (PGPR)

Citation: Landa-Acuña, D.; Toro, M.; Santos-Mendoza, R.; Zúñiga-Dávila, D. Role of *Rahnella aquatilis* AZO16M2 in Phosphate Solubilization and Ex Vitro Acclimatization of *Musa acuminata* var. Valery. *Microorganisms* **2023**, *11*, 1596. https://doi.org/10.3390/microorganisms11061596

Academic Editors: Vishnu D. Rajput, Anuj Ranjan and Abhishek Chauhan

Received: 1 May 2023
Revised: 4 June 2023
Accepted: 9 June 2023
Published: 16 June 2023

1. Introduction

Phosphorus (P) is an essential macronutrient required by all plants [1,2]. It is considered, after nitrogen, as the most critical element for agricultural production, of vital importance in plant growth and development [2,3] because of its role in important biomolecules such as nucleic acids, phospholipids and nucleotides [4,5] as well as photosynthesis and various metabolic processes [6]. Widely distributed in nature, it is affected by soil type, pH, vegetation type, microbial activity, and fertilizer inputs [7]. The forms of P present in the soil can be classified into organic and inorganic phosphorus; however, lack of accessibility is common, because it is found in highly insoluble forms in soils [4], whose availability depends on a progressive insufficiency of its natural sources, relative edaphic scarcity, high retention by the soil matrix, lack of natural replenishment and low mobility compared to that of other minerals [8].

Although many soils have a relatively large reserve of total P (about 0.05% on average), only 0.1% of the total is available to plants [9]. Still, this situation is not entirely improved through the application of P fertilizers, as the soluble phosphorus present in fertilizers easily and rapidly precipitates in insoluble forms with Ca^{2+}, Fe^{3+}, Al^{3+}, Co^{2+} or Zn^{2+} cations, or is adsorbed to calcium carbonates, aluminum oxide, iron oxide and aluminum silicate, depending on the particular soil properties [10,11], such that much of the added P loses effectiveness due to a series of abiotic and biotic biochemical processes such as fixation, sorption and immobilization, in addition to representing an environmental risk due to leaching, runoff and erosion [12].

Meanwhile, plants assimilate phosphorus in soluble inorganic forms, such as monobasic ($H_2PO_4^{-1}$) and dibasic (HPO_4^{-2}) ions [13–15]. In view of this, research recognizes the benefits of the application of bacterial strains to agricultural crops, which favor the mobilization of P in the soil, in addition to offering an alternative for improving crop productivity [16]. In this context, some microorganisms play a very important role in the biochemistry of the phosphorus cycle in natural and agricultural ecosystems, due to their ability to transform insoluble P into forms that are assimilable by plants ($H_2PO_4^{-1}$ and HPO_4^{-2}), either through acidification, chelation, exchange reactions or formation of polymeric substances [17], which facilitates the mobilization of nutrients contained in fertilizers [18], in addition to favoring solubilization or mineralization, which influences three dynamic phenomena of the P cycle: sorption–desorption, dissolution–precipitation and mineralization–immobilization [16,19].

Phosphate solubilizing bacteria (PSB) are free-living microorganisms, which may or may not be associated with plant roots and exert a plant-growth-promoting effect. They also play an important role in plant nutrition through increasing available P [20]. Among the plant-growth-promoting effects (PGPR) of phosphate solubilizing bacteria (PSB), the production of auxins, acetic acid-3-indole-3-acetic acid is considered [21,22]. In addition to the ability to mobilize organic and inorganic phosphates, they are considered the most important in plant nutrition [23,24]. Therefore, studies focused on the isolation, characterization, and evaluation of phosphate solubilizing capacity [25] have been carried out with the aim of understanding the solubilizing mechanism and improving crop yield [26,27]. Such mechanisms are associated with a pH decrease and the release of organic acids [28]; on the other hand, they can solubilize P through chelation activities [29] as well as secretion of enzymes such as phosphatase, phytase and C-P lyase that favor the mineralization of organic phosphorus [30].

Banana is a plant that, during the vegetative and reproductive stages, requires an important input of fertilizers [31]. Although it requires only small amounts of P, it is one of the most used nutrients in fertilization [32]. Among the most used forms of phosphorus in banana crops are monoammonium phosphate (10–12% N and 48–61% P_2O_5) and diammonium phosphate (18% N and 46%P_2O_5); other sources of phosphorus are the following: triple superphosphate (20% P) and single superphosphate (12–18% P_2O_5, 18–21% Ca and 11–12% S). Generally, two applications per year are sufficient [33]. Likewise, the presence and dynamics of P in the soil depend on various factors. That is why in certain countries it has become a crop that is highly dependent on inorganic chemical inputs that are easily assimilated by plants, which creates a risk to the environment due to contamination of water resources, leaching, the susceptibility of crops to diseases [34] and human, plant, soil and environmental health [35]. In addition, the banana plant is susceptible to different types of biotic and abiotic stresses; in such sense, phosphate solubilizing plant-growth-promoting bacteria can exert a beneficial effect, increase nutrient bioavailability and enhance beneficial symbiosis with the host [36].

Agriculture in Peru is one of the most important economic activities in the country. Therefore, for farmers, obtaining better yields represents a challenge in production. This productivity, among other conditions, is limited by the availability of phosphorus in the soil, which leads to the excessive use of fertilizers. The mechanism used by certain plant-growth-promoting bacteria in phosphate solubilization represents a valuable tool that can improve the availability of P in soils. *Rahnella aquatilis* AZO16M2, could improve phosphorus availability under different pH and temperature conditions, through different mechanisms such as the production of organic acids, phosphatases and siderophores, as well as influencing the establishment and survival of *M. acuminata*, "Banana" seedlings, during the first growth phase under ex-acclimation conditions.

2. Materials and Methods

2.1. Semi-Quantitative Estimation of Phosphate Solubilization Capacity: Effects of Temperature and Initial pH of the Medium

From a pure culture of *R. aquatilis* AZO16M2 previously isolated from potato rhizosphere [37], prepared in nutrient broth at 28 °C for 24 h, 10 µL were inoculated at three equidistant points in Petri dishes with NBRIP solid medium, supplemented with $Ca_3(PO_4)_2$, $AlPO_4$ and $FePO_4$ as the sole source of phosphorus. The pH of the medium was previously adjusted to pH 4.5, 5.5, 5.5, 5.5, 6.8, 8 and 9 according to the phosphorus source tested. They were then incubated at temperatures of 4 °C, 6 °C, 28 °C, 35 °C and 40 °C for a period of 18 days. The 0.5% bromocresol green indicator added to the NBRIP medium helped to show the decrease in pH as well as the translucent zones around the inoculation points. The diameter of the solubilization halo (translucent zone around the colony) was measured daily. The effect of temperature and pH was performed on the determination of the solubilization index (SI), according to the formula SI = Halo diameter (mm)/Colony diameter (mm) [38,39] and the determination of the Relative Solubilization Efficiency (RSE), which indicates the range of strain performance in the medium expressed as a percentage in relation to the diameter of the inoculation point: RSE = solubilization halo diameter/colony diameter × 100 [40].

The data obtained were subjected to an analysis of variance ($p < 0.05$) under a complete randomized design (CRD) with factorial arrangement. The factors were type of phosphorus source, temperature with four levels (4 °C, 6 °C, 28 °C, 35 °C) and pH with six levels (4.5, 5.5, 5.5, 5.5, 6.8, 8 and 9). In case of significance, the treatments were compared with Duncan to determine differences between phosphorus source type, pH and temperature.

2.2. Quantitative Estimation of Phosphate Solubilizing Capacity

The inoculum was prepared in nutrient broth and incubated at 28 °C for 24 h. Then, 1 mL (1×10^7 CFU/mL) was inoculated in 150 mL of NBRIP liquid medium with $Ca_3(PO_4)_2$, $AlPO_4$ or FeO_4 as the sole phosphorus source and incubated at 28 °C at 150 rpm. The incubation time was a function of the phosphate solubilizing activity and growth rate of each strain. The quantification of soluble phosphorus was calculated according to the Phosphomolybdenum Blue technique. For this, 5 mL of the bacterial culture was taken every 2 h, centrifuged at 10,000× *g* rpm for 10 min and filtered through 0.22 µm cellulose acetate membranes. The filtrate was then subjected to photometric determination at 690 nm, in triplicate according to the standard method of Murphy and Riley [41]. At the same time, viable cells were counted using the plate incorporation technique and pH measurement.

2.3. Determination of Siderophores

Pure cultures in LB medium with 24 h of incubation were used. An amount of 10 µL of the culture was added on plates containing Chrome Azurol Sulfonate Agar (CAS) according to the method of Schwyn and Neilands [42] as detailed in Louden et al. [43], as well as following the detailed preparation of Louden et al. [44]. The plates were incubated for three to five days at 28 °C. The presence of siderophores in the medium was assessed based on the appearance of a color change in the medium surrounding the inoculation point [44,45]. The assay was performed 2 times and in triplicate each time. An additional pure culture was used as a negative control.

2.4. Quantification of Acid and Alkaline Phosphatases

Measurement of acid and alkaline phosphatases was performed under the protocol of [46]. It was carried out from the supernatant of NBRIP liquid culture medium with $Ca_3(PO_4)_2$ inoculated with the pure strains and incubated at 28 °C for 24 h. Once centrifuged at 150 rpm, 100 μL of p-nitrophenyl phosphate (pNPP) solution (0.05 M), 50 μL of $MgCl_2$ (0.02 M) and 0.1 M universal buffer at pH 6.5 or pH 11, depending on whether for acid or alkaline phosphatase, were added to 50 μL of the supernatant. The samples were then incubated for 1 h at 37 °C, after which the reaction was stopped with 500 μL NaOH (2 N), and its absorbance was measured at 410 nm. The amount of phosphatase obtained corresponds to the amount of p-nitrophenol (pNP) released per minute and per ml (μg pNP/mL/min).

2.5. Quantification of Indole Acetic Acid (IAA)

For this purpose, LB culture broths supplemented with L-tryptophan (100 mg/L) inoculated with the pure strains were used. The broths were incubated at 28 ± 2 °C in the dark for 48 h. The cultures were then centrifuged at 9500 rpm for 15 min. The supernatants were used for IAA determination via Salkovski's method [47,48].

2.6. Presence of the PQQ Gene Involved in Phosphate Solubilization

Based on the results obtained via high-performance liquid chromatography, which evidences the production of organic acids, we proceeded to verify the presence of the gene encoding the cofactor pyrroquinoline-quinine (PQQ) associated with glucose dehydrogenase (GDH), present in phosphate solubilizing bacteria [23,49]. For this purpose, to verify the identity of the selected strains, genomic DNA was extracted from pure strains, whose 16S rRNA gene sequencing was analyzed with the universal primers rD1 (5′-TAAGGAGGACGTGATCCAGCC-3′) and 800r (5′-TACCAGGGTATCTAATCC-3′). Likewise, primers were constructed for the recognition of PQQ genes and amplification via the polymerase chain reaction (PCR). The amplified sequences were compared with the GenBank database (https://www.ncbi.nlm.nih.gov/genbank/, accessed on 12 January 2023) through NCBI (http://www.ncbi.nlm.nih.gov/, accessed on 12 January 2023) and the use of the BLAST program (Basic Local Alignment Search Tool, https://blast.ncbi.nlm.nih.gov/Blast.cgi, accessed on 12 January 2023); for verification of alignments, the BioEdit program version 7.0 was used. The evolutionary analysis was performed with the MEGA-X program version 10.0.5. The phylogenetic tree and evolutionary analysis was performed with the program MEGA-X version 10.0.5. The sequences of species of the genus *Rahnella* used as outgroup sequences were obtained from the GenBank database.

2.7. Provenance of Clones of Musa Acuminata var. Valery in Nursery

The clones of *M. acuminata* were supplied by the companies Avo Hass Perú S.A.C. and Biotechnology Center S.A.C. and came from in vitro multiplication in temporary immersion systems of 21 days of cultivation. All clones were cultivated and manipulated in sterile conditions. They were then subjected to transplant preparation.

2.8. Transplanting, Irrigation and Fertilization

The preparation of *M. acuminata* seedlings from multiplication in temporary immersion systems consisted of initial washing with sterile water, with subsequent removal of incipient roots and necrotic leaves to clear the corm of the seedling to allow direct contact with the substrate and the bacterial inoculum; the whole procedure was carried out in a laminar flow chamber. Then, the seedlings were transplanted in the substrates according to each treatment (sand:vermiculite or Premix N°8), followed by the application of the phosphorous source as background fertilization and finally the bacterial inoculum was placed at the base of the corm of each seedling. They were then covered and fixed with remaining portions of the substrate and placed in the climate chamber until they were ready to be taken to the greenhouse weeks later.

Macronutrients and micronutrients were added using Hoagland and Arnon [50] nutrient solution devoid of P, starting from the second week after planting, according to the needs of the crop: approximately 100 mL per day per seedling. After three weeks, all treatments were placed in the greenhouse.

2.9. Statistical Analysis and Experimental Design

For the statistical analysis, the Statistical Package for Social Sciences (SPSS) program of the IBM company version 26 was used. The data obtained in each experiment were subjected to an analysis of variance under a Complete Randomized Design (CRD) with a factorial arrangement, according to the factors tested. The factors tested were substrate factor with two levels (substrate sand:vermiculite, 1:1 [*v*/*v*] and formulated substrate Premix® N°8), phosphorus source factor with two levels ($Ca_3(PO_4)_2$ and rock phosphate, RP) for the treatments that considered the association with bacteria and three levels for the controls (rock phosphate, $Ca_3(PO_4)_2$, K_2HPO_4) and bacteria factor with the levels established for that moment. An additional absolute control (B0 = No bacteria) and a control without phosphorus source (P0) were added for the tested substrate types. For each treatment, 5 replicates were considered, with a total of 100 experimental units. In all the experimental units with phosphorus source, the equivalent of 31 g of P was applied according to the sources $Ca_3(PO_4)_2$ and rock phosphate; the amounts were calculated according to their molecular weight and were applied at a single moment in the sowing of the clones. In case of significance, the treatments were compared using Duncan's test to determine the differences between the bacteria, substrate type and phosphorus source tested. The probability of alpha error of less than 5% was considered significant.

2.10. Parameters Evaluated

The parameters evaluated were associated with the growth variables of the seedlings such as plant height, root length, root fresh weight, aerial fresh weight, root dry weight, aerial dry weight, number of leaves, total leaf area and chlorophyll content (SPAD). The measurement of the response variables was carried out in five replicates of each treatment. Leaf area was estimated via a non-destructive method, using a regression equation on an initial sample of 100 leaves belonging to additional units of the experiment, which relates the maximum leaf length and width. According to Rodriguez et al. [51] and Khan et al. [52], leaf chlorophyll content was estimated with the SPAD 502 Plus Chlorophyll Meter. Since the amount of light captured by the SPAD 502 cell is inversely proportional to the amount of light used by the chlorophyll, the equipment processes the signal, and the absorbance is quantified in dimensional values ranging from 0 to 199, so that the SPAD units will be the same according to the green tone of the leaves [53].

3. Results

3.1. Effect of Temperature and Initial pH in Solid Medium

The measurement of solubilization halos is part of the preliminary evaluation of the phosphate solubilization capacity [40,54] in solid medium (NBRIP) (Figure 1) through the measurement of the Solubilization Index (SI) and the Relative Solubilization Efficiency (RSE). *R. aquatilis* AZO16M2 only produced solubilization halos in the presence of $Ca_3(PO_4)_2$, but not in $AlPO_4$ and $FePO_4$. *R. aquatilis* AZO16M2 generated a maximum SI of 3.77 (28 °C) at pH 6.8. With greater stability at temperatures between 6 °C and 28 °C, at pH 6.8, IS was close to 3.67 and 3.77. Likewise, it was able to solubilize $Ca_3(PO_4)_2$ in a wide pH range from 4.5 to 9.5, where halo generation was generated (Figure 2). As for the ERS, the maximum efficiencies obtained were 266.67 (6 °C) and 276.85 (28 °C) at pH 6.8 (Figure 3).

3.2. Quantification of Available Phosphorus and pH Evolution

Quantification of available P was carried out in NBRIP liquid medium with $Ca_3(PO_4)_2$, $AlPO_4$ and $FePO_4$ as the only P source, accompanied by pH measurement and evaluation of population growth of *R. aquatilis* AZO16M2 (Table 1). Cell concentration was associated with the decrease in pH for the phosphorus sources tested (Table 1).

Minimum soluble P values were quantified in the presence of $AlPO_4$ and $FePO_4$ compared to the medium with $Ca_3(PO_4)_2$. Meanwhile, the maximum cell concentration was detected in the presence of $Ca_3(PO_4)_2$, with a slight increase in the presence of $FePO_4$ and a gradual decrease without reaching maximum peaks in the presence of $AlPO_4$ (Table 1).

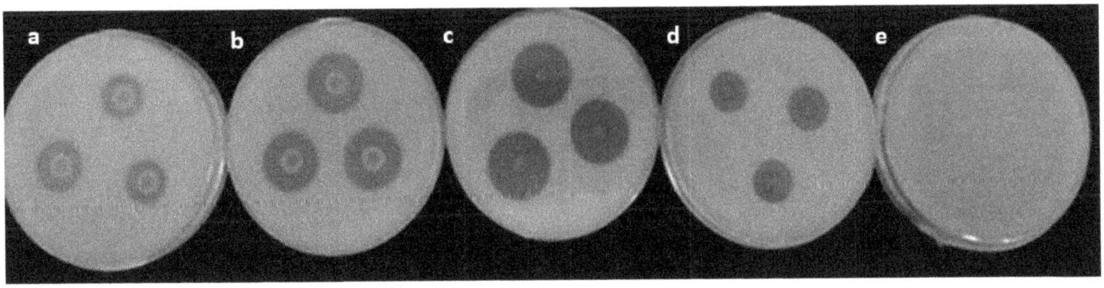

Figure 1. Solubilization halos of *R. aquatilis* (AZO16M2) in NBRIP with $Ca_3(PO_4)_2$ at different temperatures at pH 6.8. From left to right: temperatures tested (**a**) 4 °C, (**b**) 6 °C, (**c**) 28 °C, (**d**) 37 °C and (**e**) control without bacteria.

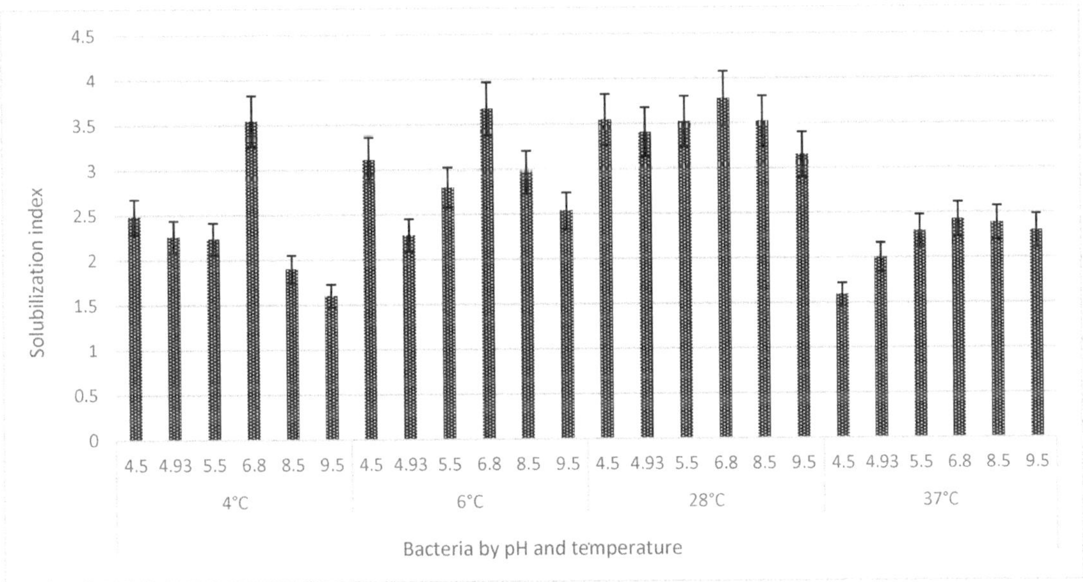

Figure 2. Solubilization index in $Ca_3(PO_4)_2$ medium under different temperature and pH conditions. The bars correspond to the average of four replicates per treatment and the bars to the standard deviation.

Table 1. Quantification of available phosphorus in liquid medium with different phosphate sources.

	NBRIP–$AlPO_4$			NBRIP–$FePO_4$			NBRIP–$Ca_3(PO_4)_2$		
Strain	**Population (ufc/mL)**	**pH**	**mg/L P**	**Population (ufc/mL)**	**pH**	**mg/L P**	**Population (ufc/mL)**	**pH**	**mg/L P**
AZO16M2	33×10^3	3.48	2.4	33×10^6	3.27	1.1	47×10^{10}	4.4	29.6

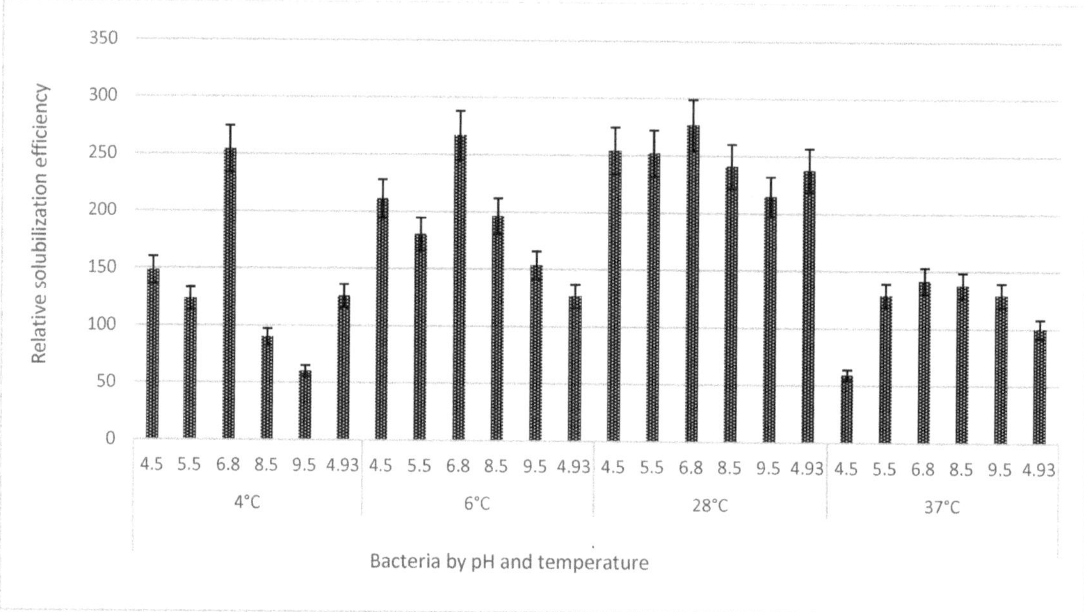

Figure 3. Relative solubilization efficiency in $Ca_3(PO_4)_2$ medium under different temperature and pH conditions. The bars correspond to the average of four replicates per treatment and the bars to the standard deviation.

3.3. Synthesis of Siderophores, Phosphatases, AIA and Organic Acids

The synthesis capacity of siderophores of *R. aquatilis* was determined through observing the color change from blue to brown around the inoculation point in CAS medium (Figure 4). Likewise, *R. aquatilis* AZO16M2 showed synthesis capacity of organic acids (oxalic acid, D-gluconic acid and 2-ketogluconic acid). Alkaline and acid phosphatases were also demonstrated (Table 2).

Figure 4. Presence of clear (yellow) halo in CAS medium produced by *R. aquatilis* AZO16M2.

Table 2. Synthesis of AIA, siderophores and alkaline and acid phosphatases of *R. aquatilis* AZO16M2.

Strain	AIA (ppm)	Siderophores	Alkaline Phosphatase (μg pNP/mL/min)	Acid Phosphatase (μg pNP/mL/min)
AZO16M2	33.90	(+)	2.56	2.59

The concentration of organic acids synthesized by *R. aquatilis* AZO16M2 varied according to the pH and the temperatures tested (Table 3). Maximum concentrations of oxalic acid were detected at 28 °C, at pH 6.8 and 8.5, while 2-ketogluconic acid registered higher concentrations at 6 °C, 28 °C and 35 °C, at pH 6.8 at each temperature, unlike D-gluconic acid, with presence at 4 °C and 6 °C, at pH 5.5 (Table 3).

Table 3. Content of organic acids synthesized by *R. aquatilis* AZO16M2 at different temperatures and pH.

Temperature (C°)	pH	Oxalic Acid [1] (μg/mL)		SD [2]	2-Keto-Gluconic [1] Acid (μg/mL)		SD [2]	D-Gluconic [1] Acid (μg/mL)		SD [2]
4	5.50	6.5373	ab	0.00802	4.3873	c	0.26034	13.8660	i	0.03012
	6.80	6.5313	ab	0.01850	1.5193	b	0.03900	13.8683	i	0.12920
	8.50	6.5397	ab	0.00902	0.0000	a	0.00000	13.7850	hi	0.01229
6	5.50	6.5320	ab	0.00985	6.0627	d	0.03573	13.7517	hi	0.06621
	6.80	6.5277	ab	0.00929	8.6717	e	0.10340	13.6753	h	0.01986
	8.50	6.5377	ab	0.00723	0.0000	a	0.00000	11.4910	g	0.02081
28	5.50	6.5680	b	0.00458	5.5910	cd	0.08266	9.3400	e	0.02443
	6.80	7.0137	d	0.01250	8.3220	e	0.08800	7.3273	a	0.00902
	8.50	6.6770	c	0.03995	5.6997	cd	0.29612	7.5240	b	0.12450
35	5.50	6.5380	ab	0.01652	6.3663	d	0.03623	10.0033	f	0.00723
	6.80	6.4887	a	0.00902	8.0493	e	2.67808	8.9277	c	0.02532
	8.50	6.5790	b	0.08055	5.9113	d	0.60680	9.2110	d	0.12010

[1] Concentrations of organic acids with distinct letters are significantly different, according to Duncan's test ($\alpha < 0.05$). Confidence limit 95%. [2] Standard deviation.

3.4. Presence of the PQQ Gene Involved in Phosphate Solubilization

As part of the study, we proceeded to corroborate the identity of the AZO16M2 bacterium with the use of the 16S rRNA gene. The percentage of similarity with the species described in the NCBI database was obtained and entered with the accession code designation OQ256130. The analysis of the PQQ gene of *R. aquatilis* AZO16M2 was carried out using the neighbor-joining tree test with the statistical method of maximum likelihood, under the Tamura three-parameter substitution model (Gamma distribution 2.82). The consistency of the phylogenetic tree is sustained at 500 Bootstrap replications. Phylogenetic analysis sought to relate *R. aquatilis* AZO16M2 to other *Rahnella* species (outgroups) based on the PQQ gene sequence. The analysis grouped AZO16M2 into a single clade with 75% certainty with the sequences of *Rahnella aquiatilis* CIP, *R. aquatilis* HX2 and *R. aquatilis* KM25 (Figure 5). Likewise, it can be observed that it has a similarity of 72% with other strains of *R. aquatilis* from the neighboring clade. Analysis of the PQQ sequence establishes a phylogenetic relationship between members of the same species that can synthesize organic acids such as gluconic acid.

3.5. Response of *M. acuminata var. Valery* in Association with Phosphate-Solubilizing PGPR Bacteria in a Sand:Vermiculite Substrate

When sand and vermiculite were used as substrate in a 1:1 (*v*/*v*) ratio, a significant interaction was observed between the factor type of bacteria and phosphorus source in the following variables evaluated: For a variable number of leaves, the effect of the bacteria factor was statistically equal between AZO16M2 and the controls, not surpassing the soluble phosphorus source (K_2HPO_4) applied. Regarding the phosphorus source factor, a better response was found in the application of RP compared to the application of $Ca_3(PO_4)_2$, for the variables chlorophyll content, aerial fresh weight, aerial dry weight and root dry weight, surpassing the values obtained by the control (Table 4). On the other hand, no significant differences were observed in the treatments with $Ca_3(PO_4)_2$ as the only phosphorus source.

The influence of *R. aquatilis* on nutrient uptake by *M. acuminata* var. Valery, under ex vitro conditions, is substantially associated with the type of substrate and phosphorus source for some nutrients. Thus, a marked superiority in K% and Mg% is observed with respect to the control with RP, as well as K% with $Ca_3(PO_4)_2$ in sand:vermiculite substrate. On the other hand, nutrient uptake was not substantial to the type of substrate and phosphorus source for N% and P% with a slight increase in Ca%, K% and Mg% in the presence of $Ca_3(PO_4)_2$ and rock phosphate, while Ca% did not show changes with respect to the controls (Table 5).

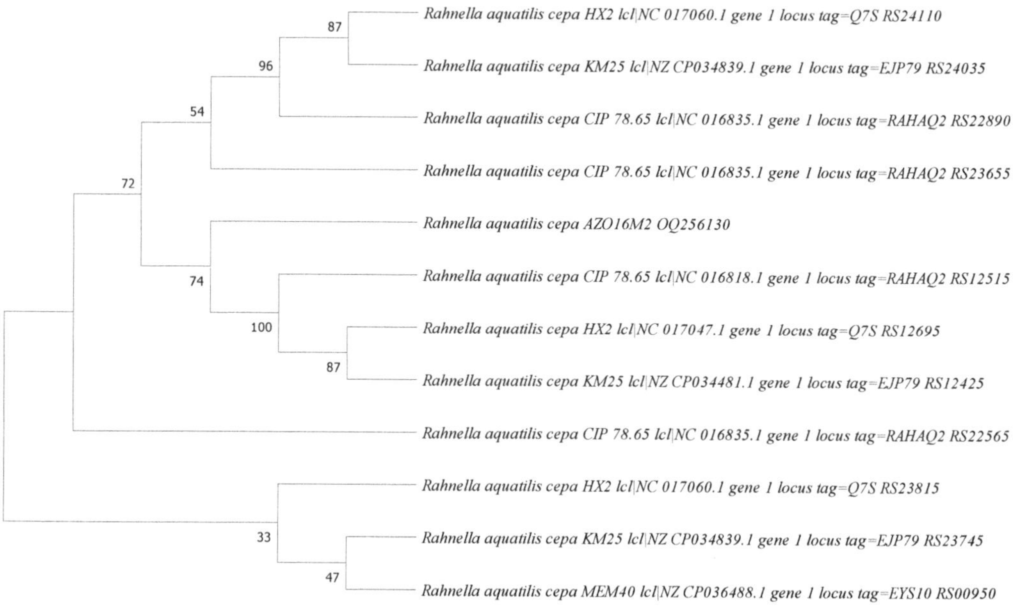

Figure 5. Phylogenetic tree built according to the maximum likelihood statistical method that highlights the relative position of *R. aquatilis* (AZO16M2) in relation to other *Rahnella* species, based on the pyrroquinoline-quinone (PQQ) gene sequence. The sequences of the *Rahnella* species were chosen according to the presence of the PQQ gene as an outgroup.

Table 4. Factorial means of the response of *M. acuminate* var. Valery in association with phosphate-solubilizing PGPR bacteria, *R. aquatilis* AZO16M2, in sand:vermiculite substrate.

Strain	Phosphorus Source	Number of Leaves [1] (cm)		Plant Length [1] (cm)		Chlorophyll Content [1] (SPAD)		Aerial Fresh Weight [1] (g)		Aerial Dry Weight [1] (g)		Root Dry Weight [1] (g)	
AZO16M2	$Ca_3(PO_4)_2$	7.8	b	13.9	b	55.28	a	8.89	b	0.83	b	1.33	b
	rock phosphate	6.6	a	13.4	a	42.38	a	3.18	a	0.38	a	0.23	a
[2] B0	$Ca_3(PO_4)_2$	8.2	b	16.6	b	54.00	a	12.06	b	1.15	b	1.46	b
	K_2HPO_4	8.4	c	19.7	c	48.67	a	16.49	cb	1.38	cb	1.45	c
	[3] P0	5.6	a	8.4	a	44.59	a	1.04	a	0.13	a	0.18	a
	rock phosphate	5.2	a	8.4	a	41.65	a	1.14	a	0.15	a	0.13	a

[1] Values with distinct letters are significantly different, according to Duncan's test ($\alpha < 0.05$). Confidence limit 95%. [2] Treatment without bacteria. [3] Treatment without phosphorus source.

Table 5. Macronutrient content of the leaves of *M. acuminata* var. Valery under ex vitro conditions in association with phosphate-solubilizing PGPR bacteria, *R. aquatilis* AZO16M2, in sand:vermiculite substrate.

Strain	Phosphorus Source	N (%)	P (%)	K (%)	Ca (%)	Mg (%)
Control [1] B0	K_2HPO_4	2.69	0.42	6.00	0.70	0.69
Control [1] B0	$Ca_3(PO_4)_2$	3.53	0.21	4.91	0.83	0.85
Control [1] B0	rock phosphate	3.53	0.10	3.90	0.83	0.82
Control [1] B0	[2] P0	2.83	0.12	5.70	0.60	0.81
AZO16M2	$Ca_3(PO_4)_2$	3.16	0.15	5.23	0.73	0.63
AZO16M2	rock phosphate	4.00	0.11	6.20	0.69	1.11

[1] Treatment without bacteria. [2] Treatment without phosphorus source.

3.6. Response of M. acuminata var. Valery in Association with Phosphate-Solubilizing PGPR Bacteria in PREMIX® N°8 Substrate

When using PREMIX® formulated substrate, a significant interaction was found between the factor type of bacteria and phosphorus source in the following characteristics evaluated: root length, aerial fresh weight, root fresh weight and chlorophyll content, when the phosphorus source used was tricalcium phosphate (Table 6).

Table 6. Factorial means of the response of *M. acuminata* var. Valery in association with phosphate-solubilizing PGPR bacteria, *R. aquatilis* AZO16M2, on Premix N°8 substrate.

Strain	Phosphorus Source	Root Length [1] (cm)		Aerial Fresh Weight [1] (g)		Root Fresh Weight [1] (g)		Chlorophyll Content [1] (SPAD)	
AZO16M2	$Ca_3(PO_4)_2$	39.30	b	20.23	b	35.68	bc	45.45	a
	rock phosphate	25.80	b	18.91	b	23.98	bc	94.45	a
[2] B0	$Ca_3(PO_4)_2$	25.56	b	20.12	b	30.63	bc	38.16	a
	K_2HPO_4	21.74	a	8.97	a	8.65	a	35.39	a
	[3] P0	27.20	ab	20.99	b	34.86	c	44.18	a
	rock phosphate	23.50	b	12.15	b	19.48	b	40.04	a

[1] Values with distinct letters are significantly different, according to Duncan's test ($\alpha < 0.05$). Confidence limit 95%. [2] Treatment without bacteria. [3] Treatment without phosphorus source.

Regarding the influence of *R. aquatilis* on the uptake of nutrients by *M. acuminata* var. Valery, in ex vitro conditions, it is not substantial, associated with the type of substrate and phosphorus source. Thus, a slight superiority is observed in N%, P% and K% compared to controls with RP in Premix No.8 substrate; a slight increase in Ca%, K% and Mg% is evident in the presence of $Ca_3(PO_4)_2$ and rock phosphate compared to the controls (Table 7).

Table 7. Macronutrient content of the leaf part of *M. acuminata* var. Valery under ex vitro conditions in association with phosphate-solubilizing PGPR bacteria, *R. aquatilis* AZO16M2, in Premix N°8 substrate.

Strain	Phosphorus Source	N (%)	P (%)	K (%)	Ca (%)	Mg (%)
Control [1] B0	K_2HPO_4	2.16	0.42	4.86	0.73	0.74
Control [1] B0	$Ca_3(PO_4)_2$	2.46	0.42	5.69	0.85	0.52
Control [1] B0	rock phosphate	2.32	0.40	2.75	0.80	0.73
Control [1] B0	[2] P0	2.04	0.40	4.15	2.15	0.50
AZO16M2	$Ca_3(PO_4)_2$	2.27	0.41	5.79	0.60	0.65
AZO16M2	rock phosphate	2.10	0.39	6.75	1.01	0.78

[1] Treatment without bacteria. [2] Treatment without phosphorus source.

The interaction between *R. aquatilis* AZO16M2, the phosphorus source and the type of substrate with respect to the establishment of *M. acuminata* var. Valery seedlings under ex vitro conditions allowed *M. acuminata* var. Valery seedlings to obtain a survival rate of 75%, surpassing the control with $Ca_3(PO_4)_2$ in sand:vermiculite, whose percentage is similar to the control with K_2HPO_4. On the other hand, in Premix N°8 with $Ca_3(PO_4)_2$, there was a higher survival rate of 87%, a value that exceeds the controls (Table 8).

Table 8. Survival percentage of *M. acuminata* var. Valery under ex vitro conditions.

Strain	Phosphorus Source	Survival (%) Sand:Vermiculite	Survival (%) Premix N°8
Control [1] B0	K_2HPO_4	75	50
Control [1] B0	$Ca_3(PO_4)_2$	62.5	37.5
Control [1] B0	rock phosphate	50	50
Control [1] B0	[2] P0	37.5	37
AZO16M2	$Ca_3(PO_4)_2$	75	87
AZO16M2	rock phosphate	25	25

[1] Treatment without bacteria. [2] Treatment without phosphorus source.

4. Discussion

The measurement of solubilization halos is part of the preliminary evaluation of the phosphate solubilization capacity [40,54] (Figure 1), carried out through the determination of IS and ERS [55–57]. The presence of halos around the colonies of *R. aquatilis* AZO16M2 evidenced the acidification of the culture medium as determinant in phosphate solubilization described in bacteria [22,58–60]. The production of organic acids causes the reduction of the pH of the medium [61–64] and is a determining factor in phosphate solubilization (Table 1), as reported in several bacterial species [65,66]. The synthesis of organic acids is related to the presence of the PPQ gene in some bacteria. According to Behera et al. [23], the main mechanism is the production of gluconic acid from the direct oxidation of glucose, a process that is carried out by glucose dehydrogenase and the cofactor pyrroloquinoline quinone (PQQ). Confirmation of the existence of the PQQ gene in *R. aquatilis* AZO16M2 and sequence analysis phylogenetically links it to members of the same genus *Rahnella* that can synthesize gluconic acid. Moreover, its regulation is a function of the conditions given during microbial growth [67], which confirms the variation of IS and ERS according to the pH and temperature conditions imposed, with maximum IS and ERS detected between 6 °C and 28 °C at pH 6.8 (Figures 2 and 3). Likewise, in liquid medium, the type of organic acid synthesized was a function of the variation of these parameters (pH and temperature) (Table 3). On the other hand, several cloning and gene expression studies have reported the importance of the PQQ gene as a necessary cofactor for gluconic acid synthesis [68–71].

Differential production of organic acids among bacterial species, attributed to substrate conditions depending on the phosphorus source used [46,65,72,73], has been reported by several researchers [74–76]. *R. aquatilis* AZO16M2, in liquid medium with different phosphorus sources, yielded minimum soluble P values when using $AlPO_4$ and $FePO_4$ compared to $Ca_3(PO_4)_2$ (29.6 mg/L soluble P) (Table 1). The results obtained from the use of $AlPO_4$ and $FePO_4$ would not be associated with the reduction of the pH of the medium. However, pH reduction in $AlPO_4$ media is caused by the synthesis of organic acids that act as chelating agents, where their ligands or anions form one or more bonds with the Al^{3+} ions of soluble phosphates to form cyclic structures and release phosphates [77]. The results obtained in the present investigation suggest that the drastic reduction in pH is caused by $AlPO_4$ and $FePO_4$ and not by the activity of the bacteria; on the contrary, they impede the correct solubilizing activity of *R. aquatilis* AZO16M2, reflected in the low concentrations of soluble phosphorus obtained during the tests (Table 1). Even so, there are other mechanisms carried out by AZO16M2 such as the synthesis of siderophores (Table 2), which capture phosphorus from Fe–P compounds in response to iron stress [49,65,78], favoring the release of available phosphorus (organic or inorganic) [48], reported in several bacteria [74,75,79,80]. Likewise, *R. aquatilis* AZO16M2 synthesizes acid and alkaline phosphatases (Table 2), enzymes involved in the phosphate mineralization process [81] with effects on plant growth [57]. Additionally, their presence is based on the use of the substrate p-nitrophenyl phosphate [82]. Several studies have reported the presence of phosphatase genes [83] in PGPR capable of synthetizing acid and alkaline phosphatases as main mechanisms in phosphorus release [46,72]. In addition, *R. aquatilis* AZO16M2 has the ability to synthetize IAA, an important phytohormone in plant growth and development [24,55], which promotes root development and better plant establishment [84].

When sand and vermiculite were used as substrate in a 1:1 (v/v) ratio, there was a significant interaction between the factor bacteria (*R. aquatilis* AZO16M2), phosphorus source and substrate, with a better response in the application of rock phosphate on the variables: aerial fresh weight, aerial dry weight and root dry weight (Table 4), superior to the control by 64.15%, 60.53% and 43.48%, considering that the RP contains between 20% and 40% of P_2O_5, a characteristic that makes it more accessible to the plant [85]. Another variable is the chlorophyll content (42.38 SPAD) (Table 4), which corresponds to the foliar Mg% (1.11%) (Table 5). Mg is a fundamental part of the chlorophyll molecule, activator of carbohydrate, fat and protein metabolism, as well as an interventor in the transport of phosphates [86]. In contrast, $Ca_3(PO_4)_2$ treatments did not outperform the established controls, including the K_2HPO_4 control, which is a more accessible source of phosphorus for the plant [85]. This behavior could be due to the intervention of root exudates, a weak cation exchange at the root level or due to the characteristics of the substrate. Singh et al. [87] argue that cation exchange is effected in sandy and young soils, which lead to low absorption of cations, especially calcium and magnesium. This behavior corresponds to the leaf content of macronutrients (Table 5) whose percentage of Mg, Ca, K, N and P did not exceed the controls established in the trial. It is worth mentioning that *R. aquatilis* AZO16M2 synthesizes of AIA favors root proliferation [69,70], promotes its development and influences markedly in plant survival in the presence of $Ca_3(PO_4)_2$ (Table 8) despite having obtained values very close to the controls.

Similarly, a marked interaction was determined between the factors substrate (Premix N°8), bacteria (*R. aquatilis* AZO16M2) and phosphorus source. When rock phosphate was used, 8.91% longer roots were obtained, with 35.58% and 18.76% more aerial and root fresh weight compared to the control as well as a higher chlorophyll content (94.45 SPAD) (Table 6). While in the presence of $Ca_3(PO_4)_2$, only the root fresh weight variable exceeded the control by 14.15%, as did the chlorophyll content (45.45 SPAD) (Table 6). The fresh weight gained in the presence of RP and $Ca_3(PO_4)_2$ is due to the need for low pH to favor its solubilization [2,23], a condition stimulated by the organic nature of Premix N°8, as well as by the capacity of AZO16M2 to synthesize organic acids [68,71]. In addition, the substrate favors the hydration of the seedlings and improves the formation of floccules that allow the conservation of humidity and retention of nutrients, generating an advantageous environment for the establishment of symbiosis between AZO16M2 and the seedlings. All this favors a better response of *M. acuminata* seedlings to the ex vitro adaptation period, overcoming structural, physiological and anatomical anomalies [88] typical of in vitro maintenance, leading to higher seedling survival observed in the presence of $Ca_3(PO_4)_2$ (Table 8). The low survival percentages with rock phosphate would be due to its low solubility, circumstances that may vary according to the igneous source of the rock phosphate where it comes from [89] and the interaction between organic matter concentration and calcium (Ca) linked to mass substitution [90], which would explain the low survival of *M. acuminata* in ex vitro conditions with RP on Premix N°8 substrate. This condition corresponds to the foliar Ca concentration of the treatments with rock phosphate (Table 7).

5. Conclusions

Semiquantitative and quantitative characterization under in vitro conditions, determined that *R. aquatilis* AZO16M2 solubilizes phosphates due to its ability to synthesize siderophores, acid and alkaline phosphatases, and organic acids such as oxalic acid, 2-ketogluconic acid and D-gluconic acid, synthesized on different pH and temperature conditions. In addition, it was confirmed the presence of the PQQ gene in *R. aquatilis* involved in the synthesis of organic acids.

When comparing the ex vitro establishment test of *M. acuminata* in both substrates, it was observed that the increase in dry matter in sand:vermiculite was higher than in the phosphorus source. Thus, AZO16M2 in the presence of RF had a positive impact on chlorophyll content and aerial and root dry weight. On the other hand, the substrate Premix N°8 showed a clear tendency to increase fresh matter, depending on the phosphorus source used, with root length and chlorophyll content being the variables that were most noticeable in the presence of RF, and root dry weight being most noticeable when the source was $Ca_3(PO_4)_2$. This interaction between the bacteria, the substrate and the phosphorus source influenced the survival percentage, which shows a better adaptation of AZO16M2 in Premix N°8 with $Ca_3(PO_4)_2$, with a survival percentage of 87% compared to 37 % in the control.

For the future, it is necessary to deepen the study of microbial consortia with different plant-growth-promoting capacities in association with phosphate-solubilizing bacteria, particularly with *R. aquatilis* AZO16M2, in ex vitro conditions to know their feasibility as co-inoculants and future inoculants for commercial use in open-field banana crops.

Author Contributions: Conceptualization, D.L.-A., M.T. and D.Z.-D.; methodology, D.L.-A., M.T., R.S.-M. and D.Z.-D.; validation, D.L.-A.; formal analysis, D.L.-A., M.T. and D.Z.-D.; investigation, D.L.-A.; resources, D.Z.-D.; writing—preparation of the original draft, D.L.-A.; writing—revising and editing, D.L.-A., M.T. and D.Z.-D.; supervision, M.T. and D.Z.-D.; project administration, D.Z.-D.; funding acquisition, D.Z.-D. All authors have read and agreed to the published version of the manuscript.

Funding: Daniela Milagros Landa-Acuña was funded by scholarship grant No 177-2015-FONDECYT from the National Council of Science, Technology and Technological Innovation of Peru. Project No 009-2017-FONDECYT from the National Council of Science, Technology and Technological Innovation of Peru.

Data Availability Statement: Not applicable.

Acknowledgments: We thank M.T. for the revision of the English language in this manuscript. To Issaack Vásquez Romero, founder of the company AvoHass-Perú, for the donation of the in vitro seedlings of *Musa acuminata* var. Valery. To Katty Ogata for providing the *R. aquatilis* (AZO16M2) strain used in this study.

Conflicts of Interest: The authors declare no conflict of interest.

References

1. Lareen, A.; Burton, F.; Schäfer, P. Plant root-microbe communication in shaping root microbiomes. *Plant Mol. Biol.* **2016**, *90*, 575–587. [CrossRef] [PubMed]
2. Kaur, G.; Sudhakara, M. Influence of P-solubilizing bacteria on crop yield and soil fertility at multilocational sites. *Eur. J. Soil Biol.* **2014**, *61*, 35–40. [CrossRef]
3. Ahmad, M.; Ahmad, M.; El-Naggar, A.H.; Usman, A.R.; Abduljabbar, A.; Vithanage, M.; Elfaki, J.; Abdulelah, A.F.; Al-Wabel, M.I. Aging effects of organic and inorganic fertilizers on phosphorus fractionation in a calcareous sandy loam soil. *Pedosphere* **2018**, *28*, 873–883. [CrossRef]
4. Babá, C.; Nahas, E. Solubilização do fosfato de ferro em meio de cultura. *Pesqueira Agropecu. Bras.* **2008**, *43*, 529–535. [CrossRef]
5. Malhotra, H.; Sharma, S.; Pandey, R. Phosphorus nutrition: Plant growth in response to deficiency and excess. In *Plant Nutrients and Abiotic Stress Tolerance*; Hasanuzzaman, M., Fujita, M., Oku, H., Nahar, K., Hawrylak-Nowak, B., Eds.; Springer: Singapore, 2018; pp. 171–190. [CrossRef]
6. Babá, C.; Nahas, E. Enhanced Solubilization of Iron and Calcium phosphates by *Aspergillus niger* by the Addition of Alcohols. *Braz. Arch. Biol. Technol.* **2013**, *56*, 181–189. [CrossRef]
7. Rooney, D.; Clipson, N. Phosphate addition and plant species alters microbial community structure in acidic upland grassland soil. *Microb. Ecol.* **2009**, *57*, 4–13. [CrossRef]
8. Beltrán, M. La solubilización de fosfatos como estrategia microbiana para promover el crecimiento vegetal. *Corpoica Cienc. Tecnol. Agropecu.* **2014**, *15*, 101–113. [CrossRef]
9. Xiaolong, L.; Lijin, L.; Jinshui, Y.; Baozhen, L.; Hongli, Y. Mechanisms for Solubilization of Various Insoluble Phosphates and Activation of Immobilized Phosphates in Different Soils by an Efficient and Salinity-Tolerant *Aspergillus niger* Strain An2. *Appl. Biochem. Biotechnol.* **2015**, *175*, 2755–2768. [CrossRef]
10. Chang, C.H.; Yang, S.S. Thermo-tolerant phosphate-solubilizing microbes for multi-functional biofertilizer preparation. *Bioresour. Technol.* **2009**, *100*, 1648–1658. [CrossRef]

11. Deng, Y.; Teng, W.; Tong, Y.-P.; Chen, X.-P.; Zou, C.-Q. Phosphorus efficiency mechanisms of two wheat cultivars as affected by a range of phosphorus levels in the field. *Front. Plant Sci.* **2018**, *9*, 1614. [CrossRef]
12. Bindraban, P.S.; Dimkpa, C.O.; Pandey, R. Exploring phosphorus fertilizers and fertilization strategies for improved human and environmental health. *Biol. Fertil. Soils* **2020**, *56*, 299–317. [CrossRef]
13. Banerjee, S.; Palit, R.; Sengupta, C.; Standing, D. Stress induced phosphate solubilization by *Arthrobacter* sp. and *Bacillus* sp. Isolated from tomato rhizosphere. *Aust. J. Crop Sci.* **2010**, *4*, 378–383.
14. Krishnaraj, P.; Goldstein, A. Cloning of a *Serratia marcescens* DNA fragment that induces quinoprotein glucose dehydrogenase-mediated gluconic acid production in *Escherichia coli* in the presence of stationary phase *Serratia marcescens*. *FEMS Microbiol. Lett.* **2001**, *205*, 215–220. [CrossRef] [PubMed]
15. Beever, R.E.; Burns, D.J.W. Phosphate uptake, storage and utilization by fungi. *Adv. Bot. Res.* **1980**, *8*, 127–219. [CrossRef]
16. Bargaz, A.; Elhaissoufi, W.; Khourchi, S.; Benmrid, B.; Borden, K.A.; Rchiad, Z. Benefits of phosphate solubilizing bacteria on belowground crop performance for improved crop acquisition of phosphorus. *Microbiol. Res.* **2021**, *252*, 126842. [CrossRef]
17. Delvasto, P.; Valverde, A.; Ballester, A.; Igual, J.M.; Muñoz, J.A.; González, F.; Blázquez, M.L.; García, C. Characterization of brushite as a re-crystallization product formed during bacterial solubilization of hydroxyapatite in batch cultures. *Soil Biol. Biochem.* **2006**, *38*, 2645–2654. [CrossRef]
18. Hong-Joo, S.; Geun-Tae, P.; Mi-Sun, C.; Moon-Soo, H. Solubilization of insoluble inorganic phosphates by a novel salt- and pH-tolerant *Pantoea agglomerans* R-42 isolated from soybean rhizosphere. *Bioresour. Technol.* **2006**, *97*, 204–210. [CrossRef]
19. Sharma, S.B.; Sayyed, R.Z.; Trivedi, M.H.; Gobi, T.A. Phosphate solubilizing microbes: Sustainable approach for managing phosphorus deficiency in agricultural soils. *SpringerPlus* **2013**, *2*, 587. [CrossRef]
20. Rodríguez, H.; Fraga, R.; Gonzales, T.; Bashan, Y. Genetics of phosphate solubilization and its applications bacteria for im-proving plant growth-promoting. *Plant Soil* **2006**, *287*, 15–21. [CrossRef]
21. Suleman, M.; Id, S.Y.; Rasul, M.; Yahya, M.; Atta, M.; Mirza, M.S. Phosphate solubilizing bacteria with glucose dehydrogenase gene for phosphorus uptake and beneficial effects on wheat. *PLoS ONE* **2018**, *9*, e0204408. [CrossRef]
22. Emami, S.; Alikhani, H.A.; Pourbabaei, A.A.; Etesami, H.; Sarmadian, F.; Motessharezadeh, B. Effect of rhizospheric and endophytic bacteria with multiple plant growth promoting traits on wheat growth. *Environ. Sci. Pollut. Res.* **2019**, *26*, 19804–19813. [CrossRef] [PubMed]
23. Behera, B.; Singdevsachan, S.; Mishra, R.; Dutta, S.; Thatoi, H. Diversity, mechanism and biotechnology of phosphate solubilizing microorganism in mangrove—A review. *Biocatal. Agric. Biotechnol.* **2014**, *3*, 97–110. [CrossRef]
24. Gyaneshwar, P.; Naresh, G.; Parekh, L.J.; Poole, P.S. Role of soil microorganisms in improving P nutrition of plants. *Plant Soil* **2022**, *245*, 83–93. [CrossRef]
25. Hernández-Leal, T.I.; Carrión, G.; Heredia, G. In vitro phosphate solubilization by a strain of *Paecilomyces lilacinus* (Thom) Samson. *Agrociencia* **2011**, *45*, 881–892.
26. Ghosh, R.; Soma, B.; Mukherjee, R.; Mandal, N. Role of phosphate solubilizing *Burkholderia colonization* spp. For successful and growth promotion (Lycopodiaceae) of *Lycopodium cernuum* L. in lateritic belt of Birbhum district of West Bengal, Indian. *Microbiol. Res.* **2016**, *183*, 80–91. [CrossRef]
27. Oliveira, C.A.; Alves, V.M.C.; Marriel, I.E.; Gomes, E.S.; Scotti, M.R.; Carneiro, N.P.; Guimaraes, C.T.; Schaffert, R.E.; Sá, N.M.H. Phosphate solubilizing microorganisms isolated from rhizosphere of maize cultivated in an oxisol of the Brazilian Cerrado Biome. *Soil Biol. Biochem.* **2009**, *41*, 1782–1787. [CrossRef]
28. Liu, C.; Mou, L.; Yi, J.; Wang, J.; Liu, A.; Yu, J. The Eno gene of *Burkholderia cenocepacia* strain 71-2 is involved in phosphate solubilization. *Curr. Microbiol.* **2019**, *76*, 495–502. [CrossRef]
29. Billah, M.; Khan, M.; Bano, A.; Hassan, T.U.; Munir, A.; Gurmani, A.R. Phosphorus and phosphate solubilizing bacteria: Keys for sustainable agriculture. *Geomicrobiol. J.* **2019**, *36*, 904–916. [CrossRef]
30. Othman, R.; Panhwar, Q.A. Phosphate-solubilizing bacteria improves nutrient uptake in aerobic rice. In *Phosphate Solubilizing Microorganisms*; Springer: Cham, Switzerland, 2014; pp. 207–224. [CrossRef]
31. Hoe, T.K.; Sarmidi, M.R.; Alwee, S.S.R.S.; Zakaria, Z.A. Oil palm and banana root colonization potential of locally isolated nitrogen-fixing and phosphate-solubilizing bacteria. In *E3S Web of Conferences*; EDP Sciences: Les Ulis, France, 2020; Volume 202, p. 01006. [CrossRef]
32. Bolfarini, A.C.B.; Putti, F.F.; Souza, J.M.A.; Silva, M.d.S.; Ferreira, R.B.; Leonel, M.; Leonel, S. Yield and nutritional evaluation of the banana hybrid "FHIA-18" as influenced by phosphate fertilization. *J. Plant Nutr.* **2020**, *43*, 1331–1342. [CrossRef]
33. Nyombi, K. Diagnosis and management of nutrient constraints in bananas (*Musa* spp.). In *Fruit Crops*; Elsevier: Amsterdam, The Netherlands, 2020; pp. 651–659. [CrossRef]
34. Lacorte, G.H.; Flamiano, R.S.; Alinsug, M.V.; Tagaloguin, P.M. Screening for Potential Plant Growth-Promoting Rhizobacteria (PGPR) Associated with Five *Musa* spp. Cultivars in SOCCSKSARGEN, Philippines. *J. Eng. Environ. Agric. Res.* **2018**, *1*, 5. [CrossRef]
35. Sharma, R.; Dahiya, A.; Sindhu, S.S. Harnessing Proficient Rhizobacteria to Minimize the Use of Agrochemicals. *Int. J. Curr. Microbiol. Appl. Sci.* **2018**, *7*, 3186–3197. [CrossRef]
36. Mendapara, P.; Khunt, M.D.; Chaudhari, J. Isolation and Characterization of Multi-trait PGPR from Banana (*Musa paradisiaca*) Rhizosphere. *Int. J. Pure Appl. Biosci.* **2020**, *8*, 37–45. [CrossRef]

37. Ogata-Gutiérrez, K.; Alvarado, D.; Chumpitaz-Segovia, C.; Zúñiga-Dávila, D. Characterization of plant growth promoting rhizobacteria isolated from the rhizosphere of Peruvian highlands native crops. *Int. J. Plant Soil Sci.* **2016**, *11*, 1–8. [CrossRef] [PubMed]
38. Pande, A.; Pandey, P.; Mehra, S.; Singh, M.; Kaushik, S. Phenotypic and genotypic characterization of phosphate solubilizing bacteria and their efficiency on the growth of maize. *J. Genet. Eng. Biotechnol.* **2017**, *15*, 379–391. [CrossRef] [PubMed]
39. De Oliveira-Longatti, S.M.; Marra, L.M.; Lima Soares, B.; Bomfeti, C.A.; Da Silva, K.; Avelar Ferreira, P.A.; De Souza Moreira, F.M. Bacteria isolated from soils of the western Amazon and from rehabilitated bauxite-mining areas have potential as plant growth promoters. *World J. Microbiol. Biotechnol.* **2014**, *30*, 1239–1250. [CrossRef]
40. Spagnoletti, F.N.; Tobar, N.E.; Di Pardo, A.F.; Chiocchio, V.M.; Lavado, R.S. Dark septate endophytes present different potential to solubilize calcium, iron and aluminum phosphates. *Appl. Soil Ecol.* **2017**, *111*, 25–32. [CrossRef]
41. Murphy, J.; Riley, J.P. A modified single solution method for the determination of phosphate in natural Waters. *Anal. Chim. Acta* **1992**, *27*, 31–36. [CrossRef]
42. Schwyn, B.; Neilands, J.B. Universal chemical assay for the detection and determination of siderophores. *Anal. Biochem.* **1987**, *160*, 47–56. [CrossRef]
43. Louden, B.C.; Haarmann, D.; Lynne, A.M. Use of blue agar CAS assay for siderophore detection. *J. Microbiol. Biol. Educ.* **2011**, *12*, 51–53. [CrossRef]
44. Alka, S.; Kalyani, D.; Shukla, P.K.; Atul, S.; Rubina, L.; Ramteke, P.W. Comparative analysis of production of hydrogen cyanide with production of siderophore and phosphate solubilization activity in plant growth promoting bacteria. *Vegetos* **2018**, *31*, 130–135. [CrossRef]
45. Verma, V.; Joshi, K.; Mazumdar, B. Study of Siderophore Formation in Nodule-Forming Bacterial Species. *Res. J. Chem. Sci.* **2012**, *2*, 26–29.
46. Anzuay, M.S.; Ruiz, M.G.; Ludueña, L.M.; Angelini, J.G.; Barros, G.; Pastor, N.; Taurian, T. Growth promotion of peanut (*Arachis mays hypogaea* L.) and maize (*Zea mays* L.) plants by single and mixed solubilizing cultures of efficient phosphate bacteria that are tolerant to abiotic stress and pesticides. *Microbiol. Res.* **2017**, *199*, 98–109. [CrossRef]
47. Pandya, N.D.; Desai, P.V.; Jadhav, H.P.; Sayyed, R.Z. Plant growth-promoting potential of *Aspergillus* sp. NPF7, isolated from wheat rhizosphere in South Gujarat, India. *Environ. Sustain.* **2018**, *1*, 245–252. [CrossRef]
48. Nithyapriya, S.; Lalitha, S.; Sayyed, R.Z.; Reddy, M.S.; Dailin, D.J.; El Enshasy, H.A.; Suriani, N.L.; Herlambang, S. Production, purification, and characterization of bacillibactin siderophore of Bacillus subtilis and its application for improvement in plant growth and oil content in sesame. *Sustainability* **2021**, *13*, 5394. [CrossRef]
49. Rawat, P.; Das, S.; Shankhdhar, D.; Shankhdhar, S.C. Phosphate-Solubilizing Microorganisms: Mechanism and Their Role in Phosphate Solubilization and Uptake. *J. Soil Sci. Plant Nutr.* **2020**, *21*, 49–68. [CrossRef]
50. Hoagland, D.C.; Arnon, D.L. The Water Culture Methods for Growing Plants without Soil. University of California. 1950. Agricultural Experimental Station Circular N° 347. Available online: https://sayurankitadotcom.files.wordpress.com/2016/04/the-water-culture-method-for-growing-plants-without-soil.pdf (accessed on 17 April 2020).
51. Rodríguez, M.; Alcántar, G.; Aguilar, A.; Etchevers, J.; Santizó, J. Estimación de la concentración de nitrógeno y clorofila en tomate mediante un medidor portátil de clorofila. *Terra* **1998**, *16*, 135–141.
52. Khan, N.; Bano, A.; Rahman, M.A.; Guo, J.; Kang, Z.; Babar, M. Comparative physiological and metabolic analysis reveals a complex mechanism involved in drought tolerance in chickpea (*Cicer arietinum* L.) induced by PGPR and PGRs. *Sci. Rep.* **2019**, *9*, 2097. [CrossRef]
53. Hurtado, E.; González-Vallejos, F.; Roper, C.; Bastías, E.; Mazuela, P. Propuesta para la determinación del contenido de clorofila en hojas de tomate. *Idesia* **2017**, *35*, 129–130. [CrossRef]
54. Liu, Z.; Li, Y.C.; Zhang, S.; Fu, Y.; Fan, X.; Patel, J.S.; Zhang, M. Characterization of phosphate-solubilizing bacteria isolated from calcareous soils. *Appl. Soil Ecol.* **2015**, *96*, 217–224. [CrossRef]
55. Teng, Z.; Chen, Z.; Zhang, Q.; Yao, Y.; Song, M.; Li, M. Isolation and characterization of phosphate solubilizing bacteria from rhizosphere soils of the Yeyahu Wetland in Beijing, China. *Environ. Sci. Pollut. Res.* **2018**, *26*, 33976–33987. [CrossRef]
56. Chawngthu, L.; Hnamte, R.; Lalfakzuala, R. Isolation and characterization of rhizospheric phosphate solubilizing bacteria from wetland paddy field of Mizoram, India. *Geomicrobiol. J.* **2020**, *37*, 366–375. [CrossRef]
57. Matos, A.D.; Gomes, I.C.; Nietsche, S.; Xavier, A.A.; Gomes, W.S.; Dos Santos, J.A.; Pereira, M.C. Phosphate solubilization by endophytic bacteria isolated from banana trees. *An. Da Acad. Bras. De Cienc.* **2017**, *89*, 2945–2954. [CrossRef] [PubMed]
58. Wang, Y.Y.; Li, P.S.; Zhang, B.X.; Wang, Y.P.; Meng, J.; Gao, Y.F.; He, X.M.; Hu, X.M. Identification of phosphate-solubilizing microorganisms and determination of their phosphate-solubilizing activity and growth-promoting capability. *BioResources* **2020**, *15*, 2560–2578. [CrossRef]
59. Hii, Y.S.; San, C.Y.; Lau, S.W.; Danquah, M.K. Isolation and characterisation of phosphate solubilizing microorganisms from peat. *Biocatal. Agric. Biotechnol.* **2020**, *26*, 101643. [CrossRef]
60. Ludueña, L.M.; Anzuay, M.S.; Angelini, J.G.; Barros, G.; Luna, M.F.; Monge, M.P.; Fabra, A.; Taurian, T. Role of bacterial pyrroloquinoline quinone inphosphate solubilizing ability and in plant growth promotion on strain *Serratia* sp. S119. *Symbiosis* **2017**, *72*, 31–43. [CrossRef]
61. Ingle, K.P.; Padole, D.A. Phosphate solubilizing microbes: An overview. *Int. J. Curr. Microbiol. Appl. Sci.* **2017**, *6*, 844–852. [CrossRef]

62. Maliha, R.; Samina, K.; Najma, A.; Sadia, A.; Farooq, L. Organic acids production and phosphate solubilization by phosphate solubilizing microorganisms under in vitro conditions. *Pak. J. Biol. Sci.* **2004**, *7*, 187–196. [CrossRef]
63. Park, K.H.; Lee, O.M.; Jung, H.I.; Jeong, J.H.; Jeon, Y.D.; Hwang, D.Y.; Lee, C.Y.; Son, H.J. Rapid solubilization of insoluble phosphate by a novel environmental stress—Tolerant *Burkholderia vietnamiensis* M6 isolated from ginseng rhizospheric soil. *Appl. Microbiol. Biotechnol.* **2010**, *86*, 947–955. [CrossRef]
64. Zaidi, A.; Khan, M.S.; Ahemad, M.; Oves, M.; Wani, P.A. Recent Advances in Plant Growth Promotion by Phosphate Solubilizing Microbes. In *Microbial Strategies for Crop Improvement*; Khan, M.S., Zaidi, A., Musarrat, J., Eds.; Springer: Berlin/Heidelberg, Germany, 2009; pp. 23–50. [CrossRef]
65. Li, L.; Chen, R.; Zuo, Z.; Lv, Z.; Yang, Z.; Mao, W.; Liu, Y.; Zhou, Y.; Huang, J.; Song, Z. Evaluation and improvement of phosphate solubilization by an isolated bacterium *Pantoea agglomerans* ZB. *World J. Microbiol. Biotechnol.* **2020**, *36*, 27. [CrossRef]
66. Sharon, J.A.; Hathwaik, L.T.; Glenn, G.M.; Imam, S.H.; Lee, C.C. Isolation of efficient phosphate solubilizing bacteria capable of enhancing tomato plant growth. *J. Soil Sci. Plant Nutr.* **2016**, *16*, 525–536. [CrossRef]
67. An, R.; Moe, L. Regulation of Pyrroloquinoline Quinone-Dependent Glucose Dehydrogenase Activity in the Model Rhizosphere-Dwelling Bacterium Pseudomonas putida KT2440. *Appl. Environ. Microbiol.* **2016**, *82*, 4955–4964. [CrossRef] [PubMed]
68. Rodríguez, H.; Rossolini, G.M.; Gonzalez, T.; Jiping, L.; Glick, B.R. Isolation of a gene from *Burkholderia cepacia* IS-16 encoding a protein that facilitates phosphatase activity. *Curr. Microbiol.* **2000**, *140*, 362–366. [CrossRef]
69. Asghar, H.N.; Zahir, Z.A.; Arshad, M. Screening rhizobacteria for improving the growth, yield, and oil content of canola (*Brassica napus* L.). *Aust. J. Agric. Res.* **2004**, *55*, 187–194. [CrossRef]
70. Khalid, A.; Arshad, M.; Zahir, Z.A. Screening plant growth-promoting rhizobacteria for improving groth and yield of wheat. *J. Appl. Microbiol.* **2004**, *96*, 473–480. [CrossRef] [PubMed]
71. Ahmed, N.; Shahab, S. Phosphate solubilization: Their mechanism genetics and application. *Internet J. Microbiol.* **2009**, *9*, 1–19. [CrossRef]
72. Behera, B.C.; Yadav, H.; Singh, S.K.; Mishra, R.R.; Sethi, B.K.; Dutta, S.K.; Thatoi, H.N. Phosphate solubilization and acid phosphatase activity of *Serratia* sp. isolated from mangrove soil of Mahanadi River delta, Odisha, India. *J. Genet. Eng. Biotechnol.* **2017**, *15*, 169–178. [CrossRef]
73. Vyas, P.; Gulati, A. Organic acid production in vitro and plant growth promotion in maize under controlled environment by phosphate-solubilizing fluorescent Pseudomonas. *BMC Microbiol.* **2009**, *9*, 174. [CrossRef]
74. Bechtaoui, N.; Raklami, A.; Benidire, L.; Tahiri, A.I.; Göttfert, M.; Oufdou, K. Effects of PGPR co-inoculation on growth, phosphorus nutrition and phosphatase/phytase activities of faba bean under different phosphorus availability conditions. *Pol. J. Environ. Stud.* **2020**, *29*, 1557–1565. [CrossRef]
75. Rfaki, A.; Nassiri, L.; Ibijbijen, J. Selection and Characterization of Phosphate Solubilizing Bacteria Isolated from the Rhizosphere of Three Cultivated Legumes in Morocco. *Adv. Trends Biotechnol. Genet.* **2019**, *1*, 119–129. [CrossRef]
76. Chen, Q.; Liu, S. Identification and characterization of the phosphate-solubilizing bacterium *Pantoea* sp. S32 in reclamation soil in Shanxi, China. *Front. Microbiol.* **2019**, *10*, 2171. [CrossRef]
77. Bashan, Y.; Kamnev, A.; de-Bashan, L. Tricalcium phosphate is inappropriate as a universal selection factor for isolating and testing phosphate—Solubilizing bacteria that enhance plant growth: A proposal for an alternative procedure. *Biol. Fertil. Soils* **2013**, *49*, 465–469. [CrossRef]
78. Whitelaw, M.A. Growth promotion of plants inoculated with phosphate-solubilizing fungi. *Adv. Agron.* **2000**, *69*, 99–151. [CrossRef]
79. Buono, N.I.; Ulla, E.L. Efectos de la inoculación con bacterias solubilizadoras de fosfato en tabaco (*Nicotiana tabacum* L.) y pimiento (*Capsicum annuum* L.) en condiciones controladas. *Rev. Agron. Noroeste Argent.* **2016**, *36*, 45–54.
80. Magallon-Servin, P.; Antoun, H.; Taktek, S.; de-Bashan, L.E. Designing a multi-species inoculant of phosphate rock-solubilizing bacteria compatible with arbuscular mycorrhizae for plant growth promotion in low-P soil amended with PR. *Biol. Fertil. Soils* **2020**, *56*, 521–536. [CrossRef]
81. Kumar, R., Shastri, B. Role of phosphate-solubilising microorganisms in sustainable agricultural development. In *Agro-Environmental Sustainability*; Singh, J., Seneviratne, G., Eds.; Springer: Cham, Switzerland, 2017; pp. 271–303. [CrossRef]
82. Glick, B.R. Modulation of plant ethylene levels by the enzyme ACC deaminase. *FEMS Microbiol. Lett.* **2005**, *251*, 1–7. [CrossRef]
83. Yuan, L.; Li, L.; Zheng, F.; Shi, Y.; Xie, X.; Chai, A.; Li, B. The complete genome sequence of *Rahnella aquatilis* ZF7 reveals potential beneficial properties and stress tolerance capabilities. *Arch. Microbiol.* **2020**, *202*, 483–499. [CrossRef]
84. Takahashi, H. Auxin biology in roots. *Plant Root* **2013**, *7*, 49–64. [CrossRef]
85. Nath, D.; Maurya, B.R.; Meena, V.S. Documentation of five potassium-and phosphorus-solubilizing bacteria for their K and P-solubilization ability from various minerals. *Biocatal. Agric. Biotechnol.* **2017**, *10*, 174–181. [CrossRef]
86. López, M.; España, M.; Toro, M. Efficiency of phosphorus absorption in you will cultivate of sorghum of different tolerance from the aluminum toxicity. *Agron. Trop.* **2007**, *57*, 205–218.
87. Singh, M.; Dotaniya, M.L.; Mishra, A.; Dotaniya, C.K.; Regar, K.L.; Lata, M. Role of biofertilizers in conservation agriculture. In *Conservation Agriculture: An Approach to Combat Climate Change in Indian Himalaya*; Bisht, J.K., Meena, V.S., Mishra, P.K., Pattanayak, A., Eds.; Springer: Singapore, 2016; pp. 113–134. [CrossRef]
88. Montes-Cruz, S.; Lalama-Aguirre, J.M.; Echeverría-Félix, J.M.; Salazar-Torres, S.M. Factores bióticos y abióticos que influyen en la aclimatación de las vitroplantas en invernadero. *Dominio De Las Cienc.* **2016**, *2*, 63–89. [CrossRef]

89. Chien, S.N.; Prochnow, L.I.; Mikkelsen, R. Agronomic use of phosphate rock for direct application. *RAE* **2010**, *75*, 25.
90. Castro, L.N.; Melgar, R. Rocas fosfóricas. Minerales para la agricultura en latinoamérica. Universidad Nacional de San Martín, Buenos Aires. *Proyecto CYTED* **2018**, *13*, 1–29.

microorganisms

MDPI

Review

Solid-State Fermentation: Applications and Future Perspectives for Biostimulant and Biopesticides Production

Alessandro Mattedi [1], Enrico Sabbi [1], Beatrice Farda [1], Rihab Djebaili [1], Debasis Mitra [2], Claudia Ercole [1], Paola Cacchio [1], Maddalena Del Gallo [1] and Marika Pellegrini [1,*]

1 Department of Life, Health and Environmental Sciences, University of L'Aquila, Via Vetoio, Coppito, 67100 L'Aquila, Italy; alexmtt89@gmail.com (A.M.); enrico.sabbi@student.univaq.it (E.S.); beatrice.farda@graduated.univaq.it (B.F.); rihab.djebaili@guest.univaq.it (R.D.); claudia.ercole@univaq.it (C.E.); paola.cacchio@univaq.it (P.C.); maddalena.delgallo@univaq.it (M.D.G.)

2 Department of Microbiology, Raiganj University, Raiganj 733134, India; debasismitra3@raiganjuniversity.ac.in

* Correspondence: marika.pellegrini@univaq.it; Tel.: +39-0862-433258

Abstract: With the expansion of the green products market and the worldwide policies and strategies directed toward a green revolution and ecological transition, the demand for innovative approaches is always on the rise. Among the sustainable agricultural approaches, microbial-based products are emerging over time as effective and feasible alternatives to agrochemicals. However, the production, formulation, and commercialization of some products can be challenging. Among the main challenges are the industrial production processes that ensure the quality of the product and its cost on the market. In the context of a circular economy, solid-state fermentation (SSF) might represent a smart approach to obtaining valuable products from waste and by-products. SSF enables the growth of various microorganisms on solid surfaces in the absence or near absence of free-flowing water. It is a valuable and practical method and is used in the food, pharmaceutical, energy, and chemical industries. Nevertheless, the application of this technology in the production of formulations useful in agriculture is still limited. This review summarizes the literature dealing with SSF agricultural applications and the future perspective of its use in sustainable agriculture. The survey showed good potential for SSF to produce biostimulants and biopesticides useful in agriculture.

Keywords: biostimulants; biopesticides; bioactive compounds; industrial scale-up; fermentations; waste recovery; waste bioconversion; circular economy

Citation: Mattedi, A.; Sabbi, E.; Farda, B.; Djebaili, R.; Mitra, D.; Ercole, C.; Cacchio, P.; Del Gallo, M.; Pellegrini, M. Solid-State Fermentation: Applications and Future Perspectives for Biostimulant and Biopesticides Production. *Microorganisms* **2023**, *11*, 1408. https://doi.org/10.3390/microorganisms11061408

Academic Editor: João Miguel F. Rocha

Received: 8 May 2023
Revised: 19 May 2023
Accepted: 24 May 2023
Published: 26 May 2023

1. Introduction

Sustainability perception has changed significantly since the United Nations (UN) adopted the Sustainable Development Goals (SDGs) in 2015 [1]. Through the promotion of the circular economy, the advancement of renewable energy sources, and more sustainable agriculture, global policies and strategies began social and economic fundamental changes to achieve the green revolution and ecological transition [2,3]. One of the relevant challenges in the achievement of a sustainable agrifood system is the increasing demand for biostimulants and biopesticides to limit or substitute the use of synthetic chemicals [4]. Even if many scientific questions remain unanswered, it has become more and more common to use biostimulants; these products have been extensively used in agriculture, horticulture, and forestry, to promote growth, improve nutrient uptake, protect plants from biotic and abiotic stress [5]. Their potential applications are studied to reduce our dependence on conventional fertilizers: despite their importance during the XX century to increasing crop yields for a growing population, their excess and abuse can lead to significant pollution [6]. Beyond the need for nutrients, crops are at constant exposure to hazards from parasites and other organisms that feed on them. In nature, plants defend themselves through a wide and astonishing range of mechanisms and traits [7,8], including mechanical defenses [9], structural traits [10], and particularly chemical compounds that are disgusting or toxic to phytophages and herbivores [11–13] or attract predators and parasitoids [14] or can contrast the attack of pathogens [15,16] or induce systemic acquired resistance which can help against future aggressions [17]. Plants have undergone millions of years of natural selection that defined their characteristics and adaptations, but human agriculture in the last ten millennia reshaped the traits of cultivated plants by intentional manipulation through the means of artificial selection, thus changing their fitness and pressures in what is considered an extraordinary example of plant-animal mutualistic co-evolution [18–20]. In agriculture, the study of the ecological interactions and the evolutionary patterns related to plant exposure to abiotic and biotic stresses is particularly concerning, as they can significantly reduce yields and damage crop quality, which are necessary to feed people. It is not only a matter of the farmer aiming at saving the entire harvest and avoiding any minor losses: modern cultivars are more productive, but also often more vulnerable compared to their wild progenitors because of a complex interaction of factors, including reduced genetic variability and loss of secondary toxic metabolites [21–24]. Plant phytophages and pathogens can and will lead to dire consequences regarding food provisioning and waste [25,26] economic losses [27] risk of food poisoning due to toxins [28]. The current climate emergency is also a Damocles' sword as global warming will further spread certain phytophages and pathogens, while also exerting higher abiotic stresses on crops, thus exacerbating plant diseases and physiopathies. Third-world countries will be the most impacted, but first-world countries face risks to their food security too [29–31]. In the last century, scientific and technological research developed several ways to mitigate or deter these issues. One of the most important tools is agricultural chemistry. Since damaging organisms are vernacularly termed pests, chemical compounds that repel or kill them are termed pesticides. It is not surprising that pesticides have occupied a key role in agriculture in the last decades to assure a stable source of food and income for farmers, but also to fight the spread of diseases, and overall to sustain a constantly growing population with its increasing demands [32]. However, they did not come without possible risk due to their abuse and mismanagement, and in some cases even from their application alone. While the benefits of agrochemistry have been widely recognized, there was also increasing concern for the collateral effects on the environment [33–35] with a particular focus on human health [36] non-target organisms [37], the development of resistance in target organisms [38–40], and economical costs [41]. Despite this, many farmers still rely on pesticides, which can be seen as counterintuitive, and their consumption has increased worldwide as they often are necessary, particularly herbicides followed by fungicides and insecticides [42]. These issues are of no quick and easy solution. A further problem is that countries in the third world suffer the most harm from crop damage, as they are economically vulnerable, and farmers often lack the instruments and the money to adequately face the most severe issues and recover from losses. They are also more vulnerable to the effects of climate change. The Food and Agriculture Organization of the United Nations estimated that up to 40% of global crop production is lost because of pests and that the current climate change scenarios will result in an increase in pest risk and pesticide usage in agricultural ecosystems. [43]. Agricultural research was and currently is pivotal in the struggle against poverty [44]. Yet, farmers in the poorest countries cannot always afford the most effective and efficient tools, are forced to renounce chemical compounds to protect their crops, or to use older formulations that might be more polluting (sometimes banned in first world countries) and faulty, less safe instruments to release pesticides [45]. In the last decades, research has focused on how to mitigate the environmental effects of pesticides, reduce the need for them, and find alternative tools and strategies for their use. The global goal is to achieve a more sustainable agriculture, while not decreasing yields [46]. Therefore, research in plant protection is key to these ambitious objectives. This has been pursued through more severe regulations, increased technical education [47,48] and the development and choice of less impacting agrochemicals that are more selective or less persistent [49–51], the breeding or engineering of resistant crops [15], the application of evolutionary and ecological thought in agriculture to improve crop selection [19], the optimization of farming practices through systems such as integrated and precision agriculture and ecological intensification [52], search of microorganisms for biological control [53], the usage of useful insects for biological control [14]. The importance of research in crop protection is increased considering the impact of climate change on agriculture, and the need to reduce the impact on the climate of agriculture itself. Reducing the need for agrochemicals and increasing the efficiency of farming techniques can help reduce the emissions of carbon dioxide and greenhouse equivalents [54].

Midst new bio-based products, the use of microbial-based inoculants are gaining increasing interest from research, industrial, and commercial fields. Microbial-based inoculants contain microbial entities with the ability to increase nutrient uptake, shield plants from biotic and abiotic stress, and promote growth (e.g., germination, flowering, fruiting) [5]. These microbes fall within plant growth-promoting microbes (PGPM), beneficial bacteria and fungi that sustain the positive effects on plants by colonizing plant roots and benefit their hosts by controlling the synthesis of phytohormones, boosting soil nutrient availability, and enhancing disease resistance [55].

Since the discovery and description of PGPM, scientific research carried out the isolation and characterization of countless potentially useful strains. However, most of these isolates are not commercialized [56]. The commercialization of bio-based products may be severely hampered by improper microbial inoculant formulation that may not consider the costs linked to the industrial production of the product and its input into the market [57,58]. A successful microbe-based formulation is characterized by efficacy, versatility, practicality, delivery, persistence, commercial viability, and congruity with regulatory frameworks [57]. These aspects, which ensure the high quality of the product and its success in the market, are achieved through a valid scale-up from laboratory to industrial production. The fundamental concerns of industrial fermentations, process optimization, and scale-up are targeted at maintaining optimal and uniform reaction conditions, limiting microbial stress exposure, and boosting metabolic accuracy to maximize product yields and assure consistent product quality [59]. A thorough and detailed process characterization, the identification of the most important process parameters influencing product yield and quality, and their establishment as scale-up parameters to be kept constant as much as possible are required to develop suitable strategies for each individual product, process, and facility [59].

Among the processes applied in the production of microbe-based products (e.g., microbial biomass, enzymes, cell metabolites, etc), submerged state fermentation (SmSF)—also known as liquid state fermentation or submerged liquid fermentation—is the most used technology [60]. However, based on the requirements of microbe/microbes, the growth media, resources, and energy inputs (i.e., large amounts of water and costs of agitation and aeration), and equipment, SmSF could be expensive, resulting in a non-economically sustainable process [61]. Moreover, SmSF is sensitive to several factors, prone to contaminations, has a lack of control of the physical and chemical variables of the process, and some enzymatic and metabolite releases might be challenging [61]. Solid-state fermentation (SSF)—in which bacteria and fungi, are grown on a moist, solid, non-soluble organic material in the absence or almost absence of free-flowing water—is used as an alternative to SmSF for several microbial biotechnology processes [62]. Beyond low energy consumption and other practical advantages than SmSF, SSF allows the bioconversion of organic agricultural and industrial wastes, achieving the circular economy goal [63].

Huge quantities of residues are created annually by industries with agricultural backgrounds. If these leftovers aren't properly disposed of, they might pollute the environment and have a negative impact on both human and animal health. Because most agro industrial wastes are untreated and unused, they are often disposed of by burning, dumping, or unintentional landfilling. These untreated wastes increase several greenhouse gases, which contribute to various climate change issues [64,65]. The recent review of Yafetto emphasized the significance of using SSF to valorize diverse agro-industrial wastes to produce goods with advantages for industry, agriculture, and human health [66].

In view of the need for agro-industrial wastes valorization and economic sustainability of microbial-based products improvement, the purpose of this review is to underline the application of SSF in biostimulants and biopesticides production and encourage research to progress knowledge on the subject. A detailed description of the technology and the laboratory- and industrial-scale instruments were provided. To evaluate the suitability of SSF for this purpose we summarised the literature dealing with the topic using several databases. Limitations, advantages, and future perspectives were also presented.

2. Solid-State Fermentation: Process and Applications

SSF is a three-phase heterogeneous process that combines solid, liquid, and gaseous phases to convert a starting substrate to products with added value. SSF has drawn a lot of interest in the last two decades for the development of industrial bioprocesses, because of its economic and environmental sustainability whilst producing more products with a decreased risk of contamination. Several parameters affect SSF and are essential to the process development's technical and financial viability. As with other bioprocesses (including SmSF), these parameters include the selection of the right microbe/consortium and substrate, and finding the best physical, chemical, and biological process parameters (e.g., pH, aeration, temperature, humidity, solid material characteristics). The purification of the product is an additional element that has an impact on SSF production feasibility. Throughout the fermentation heat accumulation and the heterogeneous nature of the substrate (a complex gas-liquid-solid multiphase system) are two of the main SSF problems to overcome in scale-up. Beyond the use of SSF for biopesticides and biostimulant production—discussed in detail in the following sections—SSF demonstrate to be a valid process in several fields. SSF is commonly utilized in the production of metabolites (e.g., antibiotics, aromas, biosurfactants, enzymes, organic acids) biofuels, and environmental purposes (e.g., bioremediation) [62,67]. These productions are carried out employing different types of bioreactors. The following section describes the most common types of devices and the different types of configurations from laboratory to industrial scale.

3. Solid-State Fermentation Bioreactors

The laboratory-scale SSF devices consist of inert supports (e.g., Petri dishes, flasks, and bottles) that can be used to process a few grams of matrix to carry out rapid screenings (e.g., inoculum-to-matrix ratios, optimal temperature). Generally, at this scale, the temperature is the only parameter controlled and no forced aeration or agitation is applied [68]. Once reached the optimal lab scale conditions (e.g., inoculum rate, matrix quantities, optimal temperature) pilot and industrial scale processes are studied and optimized in bioreactors with sophisticated control systems. There are many types of bioreactors that mainly differ based on the presence or absence of agitation and forced aeration [69]. The simplest type is the tray bioreactor (Figure 1A) in which the solid material is laid on trays constructed with inert material (e.g., metal, wood, plastic). Trays are placed in a tray chamber with a suitable gap among them in which a circulating air controls temperature and humidity. In tray bioreactors, the air is not forced, and agitation might occur occasionally based on the process carried out [69].

In the presence of occasional agitation and forced aeration we can find packed-bed bioreactors (Figure 1B), glass or plastic column reactors in which the solid material is packed inside it. Aeration is guaranteed by fluxing air from the bottom and the temperature is maintained by external devices (e.g., heat exchangers or cooling/heating jackets). Packed-bed bioreactors can be used also in the presence of intermittent mixing and forced aeration, providing the agitation by a mechanical stirrer or airflow [70].

For SSF that need slow continuous agitation and no forced aeration, there are two types of stirred drum bioreactors (Figure 1C). In these types of bioreactors, the solid material is filled within the drum, and air is blown in it. The agitation is assured by a rotating drum in the rotating-drum bioreactor (above) and by paddles inside the drum unit in the stirred-drum bioreactor (below).

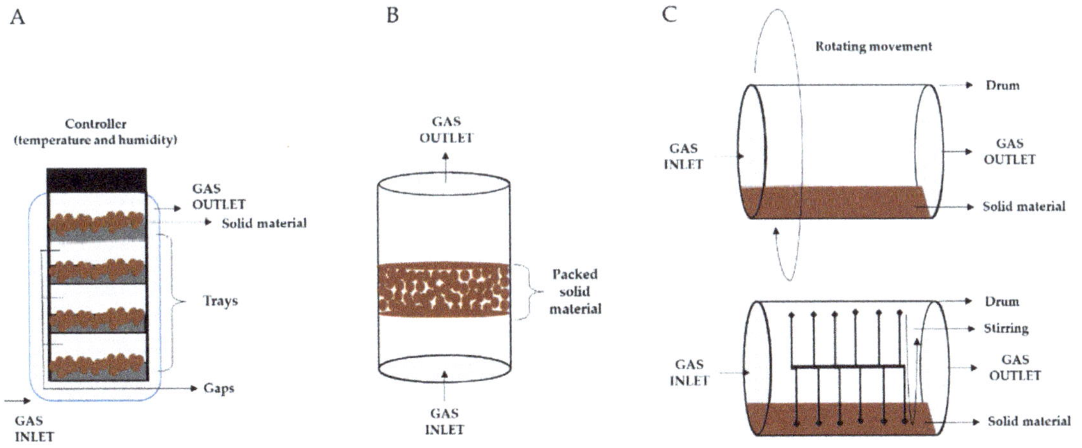

Figure 1. SSF bioreactors with occasional agitation and without forced aeration ((**A**) Tray bioreactor); with occasional agitation and forced aeration ((**B**) Packed-bed bioreactor); and with slow continuous agitation and without forced aeration ((**C**) Two models of stirred drum bioreactors) [69].

In SSF with slow continuous agitation and forced aeration, there are three types of bioreactors that can be used including stirred-aerated bioreactors (Figure 2A), gas-solid fluidized beds bioreactors (Figure 2B), and rocking drums bioreactors (Figure 2C). These reactors vigorously blow air through the bed while agitating it. Depending on the type of mixing, such a bioreactor can normally be operated in one of two modes: continuously mixed or intermittently mixed bioreactors. Thanks to the addition of water to the bed, the mixing system reduces the cooling demand. The sensitivity of the microorganisms to shear effects from mixing as well as the mechanical and sticky characteristics of the substrate particles will determine whether continuous or intermittent mixing should be used [68–70]. Although several studies used and set up a wide variety of alternatives to these fermenters, the tray- or drum-type bioreactor still serves as the starting model to design them [69].

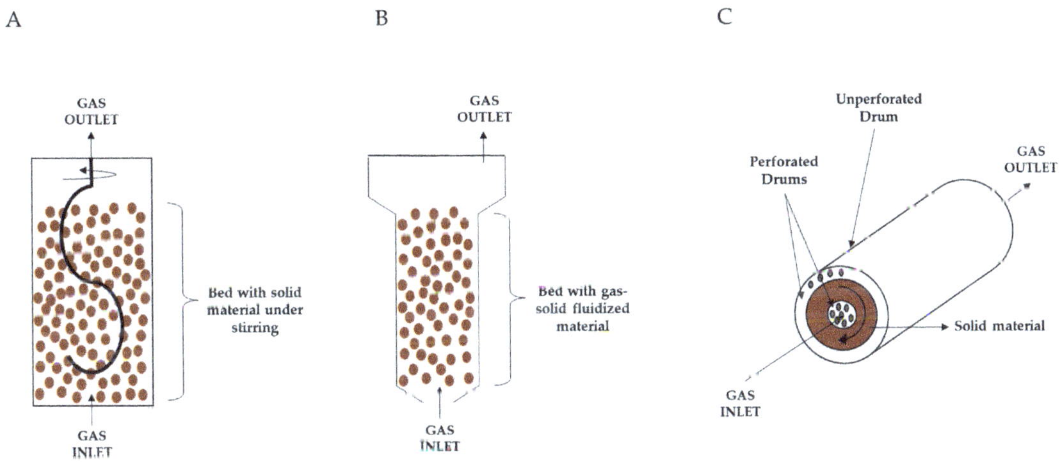

Figure 2. SSF bioreactors with continuous agitation and forced aeration. (**A**) Stirred aired bioreactor; (**B**) Gas-solid fluidized bed bioreactor; (**C**) Rocking drum bioreactor [69].

4. Solid-State Fermentation for Biostimulants Production

Many works focused the investigations on the SSF application to produce biostimulant agents. Table 1 summarises the existing literature on the subject, focusing on the last 10 years. Only a few studies evaluated the suitability of SSF to produce biostimulant agents and tested the SSF products on plants. Almost the entirety of the reports focused on the study of *Trichoderma* spp., considering a wide range of substrates, and evaluating the biostimulant effects of SSF products on different horticultural and officinal plants. Among these studies on *Trichoderma* spp. noteworthy is the recent study of Liu et al., in which it has been demonstrated the suitability of rice straw in combination with other natural-derived amino acids as SSF substrate to improve the population of *T. guizhouense* NJAU4742 and develop an innovative biostimulant capable to improve pepper growth and development [71]. Biostimulant production through SSF was also valid in other studies involving other fungal and bacterial strains. Romano et al., for example, demonstrated the suitability of SSF with vermiculite, exhausted yeasts, and vinasse to produce a *Kosakonia pseudosacchari*-based biostimulant able to induce plant growth and development on maize [72]. Most of the studies on these other microbes are single reports or described by only one group of scholars. The exception is *Bacillus* spp.; for the latter, several optimized SSF processes have been used to produce spores useful for biostimulating action on various crops. Other environmental useful applications were described for *Bacillus* spores produced by SSF processes. Rodriguez-Morgado et al., for example, efficiently used *Bacillus licheniformis* spores derived from SSF of sewage sludge to improve soil biochemical characteristics (e.g., ergosterol concentration and enzymatic activity).

Table 1. Literature on solid-state fermentation (SSF) use to produce biostimulant agents.

Species	Substrate	T (°C)	Days	Maximum Biomass Yield	Plants Promoted	Ref.
Trichoderma spp.	agricultural digestate	26 °C	6	689.80 ± 80.53 mg mycelium/g substrate	cress	[73]
Trichoderma spp.	apple, banana, and grapefruits wastes	26 °C	6	689.8 ± 80.5 mg/g substrate	cress and tomato	[74]
Purpureocillium lilacinum	hair waste	28 °C	8	-	tomato	[75]
Trichoderma atroviride strain MUCL45632	wheat bran	-	-	-	melon, pepper, tomato, and zucchini	[76]
Aspergillus flavipes	soybean (most suitable)	-	-	-	*Eucalyptus* clone IPB2	[77]
Trichoderma guizhouense NJAU4742	rice straw + amino acids	28 °C	7	4.62×10^{10} conidia	pepper	[71]
Fusarium redolens KY992586 (RF1), *Phialemoniopsis cornearis* MK408657 (SF1), and *Macrophomina pseudophaseolina* MF351729 (SF2)	wheat bran	28 °C	10	38×10^{12} (RF1), 14×10^{11} (SF1), and 21×10^{12} (SF2) CFU g^{-1}	*Coleus forskohlii*	[78]
Kosakonia pseudosacchari TL13	vermiculite, exausted yeasts and vinasse	15 °C	30	7–6.9 log CFU g^{-1} or mL^{-1}	maize	[72]
Trichoderma asperellum	silica-rich spent mushroom	28 °C	31	12.37×10^{13} cfu/g bioformulation	tomato	[61]
Bacillus circulans Xue-113168	food waste and feldspar	30 °C	7	8–10 CFU g^{-1}	rapeseed	[79]

5. Solid-State Fermentation for Biopesticides Production

Biopesticides are a wide category of compounds of biological origin that can exert an antagonistic effect against other organisms, with deterring, competitive, or biocidal effects. The definition is debated in the literature, as different authors and regulatory agencies might disagree on the inclusion of certain products depending on their method of action or source [80]. The Food and Agriculture Organization defines biopesticides as "A generic term generally applied to a substance derived from nature, such as a microorganism or botanical or semiochemical, that may be formulated and applied in a manner similar to a conventional chemical pesticide and that is normally used for short-term pest control" [81]. The European Union defines pesticides or biological control products as "substances used to suppress, eradicate and prevent organisms that are considered harmful", including both plant protection and biocidal products, and specifies that biopesticides are a subcategory "comprised of substances that are derived from living organisms and certain minerals" [82]. The Environmental Protection Agency of the United States (EPA) divides biopesticides into three types: biochemical, microbial, and plant-incorporated-protectants.

Microbial biopesticides are the most common, and the subjects of this review. They consist of microorganisms such as bacteria, fungi, and protists, which can directly suppress target organisms, or produce compounds that have properties that are useful to control or eliminate them [83]. Certain authors include viruses too [84,85]. The rationale behind the use of biopesticides is the idea that they are generally less persistent and more degradable than traditional pesticides, but also more target-selective, thus reducing their environmental impact. They are expected to be more economical than conventional pesticides, which is the main interest for poor countries, although certain limitations discussed in the following sections can increase costs. The intended goal is to allow to effectively fight pests without significant crop losses. There is increasing scientific literature showing interest in their potential applications, their production, their enhancements, their limitations to be overcome, the technical side of their employment, and the legal and economic aspects [86–91]. The EPA particularly suggest them as a component of integrated pest management that is capable of effectively reducing the use of conventional pesticides while keeping high yields [92].

The scientific literature contains many reports on the use of SSF to produce biopesticides. Supplementary Table S1 (Supplementary Material) and Table 2 summarise the existing literature on the subject focused on the last 10 years, with some notable cases from the previous period. Most of the scientific papers deal with the use of SSF to grow fungi and oomycetes (Supplementary Table S1, Refs. [93–170]) and produce biopesticides targeting insects, nematodes, molds, and weeds. A wide variety of growth substrates can be employed. The most common substrates are wheat bran and straw, rice, and barley. Coffee husks, sugarcane bagasse, and sorghum, also are often used. More exotic substrates such as palm kernel cakes or forage cactus pears gave interesting results, which can be important for tropical countries. Even if cereals are the most common (e.g., wheat, rice, wheat bran), several works employed food and agricultural wastes. In the recent work of Ghoreishi et al., for example, grass clippings and pruning waste are used as substrate in SSF to grow *Trichoderma harzianum* and produce biopesticides (i.e., conidial spores) useful against phytopathogenic molds [170]. In the same work, the optimization of the growth parameters (mainly moisture and fermentation time) and ratios of the substrates (i.e., tryptophan, grass, and pruning waste) allowed the enhancement of 3-indole acetic acid and spores recovery.

As summarized in Table 2, the production of biopesticides through SSF is also possible with the use of bacterial inocula. *Bacillus* spp. are the commonly used inocula to produce biopesticides targeting phytopathogenic bacteria, molds, mosquitos, and insects using diverse substrates (including different wastes). Specific targets are *Culex pipiens*, *Rhizoctonia solani*, *Agrobacterium tumefaciens*, *Colletotrichum lini*, *Fusarium oxysporum*, and *Phytophthora palmivora*. *Bacillus thuringiensis* is by far the most successful bacterial biopesticide, due to its efficacy against insects and readily available, with strains selected for specific targets [171–174]. Most of its success depends also on the fact that it is considered safe for mammals, humans included, although it has been reported that in rare cases some strains might affect human health through enterotoxins [175]. A few studies also reported the inoculation of actinomycetes for antimicrobial agents. The literature survey also showed promising applications of bacterial strains SSFs for biopesticides production. *Rhizopus oligosporum* SSF, for example, is commonly used to produce food (tempeh) [176]. However, the same technology has not been yet applied for the optimization of potential biopesticides released by this species (e.g., antifungal chitinases) [177]. Also, the recent work of Widyastuti et al. described the *Pseudonocardia antitumoralis* 18D36-A1 SSF using shrimp shell wastes to produce antifungal biopesticides against Malassezia globose (a mammalian pathogen) [178].

Table 2. Literature on solid-state fermentation (SSF) use to produce biopesticides employing bacterial inocula.

Species	Tg	Growth Substrates	T (°C)	D	Maximum Biomass Yield	Tested Against	Ref.
Bacillus amyloliquefaciens BS20	B	feed-grade soybean meal, corn flour, and wheat bran	37 °C	2	7.24 × 10 ^{10} CFU/g	-	[179]
Bacillus amyloliquefaciens B-1895	B	corncobs	37 and 32 °C	4	47 × 10 ^{10} spores g^{-1} biomass	-	[180]
Bacillus amyloliquefaciens HM618	M	food waste "mainly included rice, vegetables, and small amounts of soup"	-	-			[181]
Bacillus sphaericus NRC69	Q	wheat bran, rice hull, wheat straw, corn stover, corn cobs, cotton stakes, olive meal, date stone, pea peels, potato peels	30 °C	6	116 × 10 ^{9} CFU/g	*Culex pipiens*	[182]
Bacillus sphaericus 14N1 *Lysinibacillus sphaericus*	Q	wheat germ meal, linen meal (4.5% each), and 0.2% yeast extract with fine sand as the carrying material	30 °C	5	-	*Culex pipiens*	[183]
Bacillus subtilis RB14-CS	M	soybean curd residue (okara) + some other nutrients	25 °C	5–10	2.75 × 10 ^{9} CFU/g dry soil	*Rhizoctonia solani*	[184]
Bacillus subtilis RB14-CS	M	soybean curd residue (okara).	25 °C	4	10 ^{10} to 10 ^{11}	*Rhizoctonia solani*	[185]
Bacillus subtilis RB14-CS	M	soybean curd residue (okara).	25 °C	6	-	-	[186]
Bacillus subtilis SPB1	B	dried and grinded seeds of Aleppo pine	37 °C	2	27.59 ± 1.63 mg/g	*Agrobacterium tumefaciens* C58	[187]
Bacillus thuringiensis var *israelensis* CECT 5904	I	bulking agent was mixed with digestate and biowaste	30, 37, and 45 °C	3	4 × 10 ^{8} spores g^{-1} DM	-	[188]
Bacillus thuringiensis var *kurstaki* NRRL HD-73 (CECT 4497)	I	organic fraction of municipal solid waste	27 °C	4	10 ^{9} CFU g DM^{-1}	-	[189]
Bacillus thuringiensis var *kurstaki* HD-73 (ATCC-35866)	I	polyurethane foam	30 °C	21–36 h	8 × 10^{9} mL^{-1}	-	[190]
Bacillus thuringiensis var *israelensis* CECT 5904	I	OFMSW, three cosubstrates	30 °C	3	1.1 × 10^{9} spores g^{-1}	-	[191]
Bacillus thuringiensis var *israelensis* CECT 5904	I	digested sewage sludge and digested OFMSW cosubstrates on the solid residue after enzymatic hydrolysis of OFMSW	39/42	4	10^{8}–10^{11} spores g DM^{-1}	-	[192]

Table 2. *Cont.*

Species	Tg	Growth Substrates	T (°C)	D	Maximum Biomass Yield	Tested Against	Ref.
Pseudomonas aeruginosa LYT-4	M	tung tree (*Vernicia fordii*)	30 °C	6	-	*Colletotrichum lini*, *Rhizoctonia solani* and *Fusarium oxysporum*	[193]
Streptomyces sp. K61	M	silica, cornsteep solids, dolomite lime, lactose	22–28 °C	5 + 5	4.5×10^9 CFU g^{-1}	-	[169]
Streptomyces gilvosporeus Z28	M	blend of rapeseed cake, rice hull, wheat bran and crude glycerol	28 °C	10	-	-	[194]
Streptomyces griseorubens JSD-1	B	peat soil with 2% *w*/*w* rice husk	32 °C	7	1.69×10^9 CFU g^{-1}	-	[195]
Streptomyces hygroscopicus B04	M	various combinations of rapeseed meal, wheat bran, and vermicompost	28 °C	5–7	1.34×10^9	*Fusarium oxysporum*	[196]
Streptomyces similanensis 9X166	M	rice bran, cassava chips, and coconut husks	33 ± 2 °C	7	151×10^7 CFU/g^{-1}	*Phytophthora palmivora*	[197]

In the table, Tg, Targets:B, bacteria; M, molds; I, insects; Q, mosquitos; D, SSF days; DM, dry matter.

Even if these products have a promising future, there literature survey also unveiled some issues that must be discussed. Several formulations have limited effectiveness and consistency compared to traditional pesticides (they can be slow to kill and take time to reduce pest populations), shorter shelf life both in storage and in the field (ironically a side effect of their high biodegradability), the differing standard method of preparations and guidelines, more labor-intensive application management, more difficult storage and handling, difficulty in scaling-up for large production, consequential increase in costs in refined products despite the initial cheapness of starting ingredients [87,88]. Biopesticides are considered at risk of selecting resistance in target species, mainly because of a lack of management compared to other methods, but the concern is inferior to that for conventional pesticides, particularly when dealing with biopesticides based on infection means rather than toxins. To reduce this risk, it has been proposed to use a wide array of biopesticides in a heterogeneous landscape, valorizing diversity to reduce chances for selection [198]. There are some reports about possible underrecognized out-of-target toxicity of biopesticide formulations, that require further attention, management, and investigation [199–201]. Another recent source of concern is that certain microbial biopesticides could be reservoirs of antibiotic resistance, but this field is still mostly unknown and under investigation [202]. Lack of information and awareness can also limit the diffusion of biopesticides in third-world countries [203]. Recent reports unveiled that several Indian microbial biopesticide-based solutions have quality problems (e.g., impurities, the excessive moisture content in solid formulations, or fewer colony propagules than stated). More than 50% of these products do not fulfill Central Insecticides Board and Registration Committee (CIBRC) standards [204]. However, as the demand for more sustainable and environmentally friendly pest management solutions continues to grow, research into new and improved biopesticides is likely to increase, and several formulations are already currently commercialized, standards of production are being defined, and new regulations are being approved. The current studies investigating their applications and reviewing their benefits, limits, and possible developments, are already showing that research is increasing [205–211]. Their main application will be within integrated pest management strategies [210]. There are also proposals for genetically engineering biopesticides [211–213].

6. SSF for Sustainable Agriculture: Advantages and Limitations

As presented in the previous section, SSF has been used to cultivate a wide range of microorganisms because it offers several potential advantages. SSF mimics the natural condition of growth for many microorganisms. It is considered cheap and of low impact, as it can rely on easily available substrates such as food discards, agriculture wastes, and urban wastes. Cost productions are one of the major limiting factors for biopesticides. Furthermore, SSF can be a way of recycling such wastes, as they are a source of pollution that must be disposed of. Another factor that could reduce costs is the fact that it does not require to use energy to heat the process [66,214–216]. As with other techniques, SSF has limitations. The first significant SSF constraint is directly related to heating. Fermentation depends heavily on temperature because many microbes need specific temperatures to develop, while the process itself heats things up. Since solid substrates have a low heat conductivity, air convection is the most common method of heat dissipation; however, this results in increased moisture loss and substrate drying, both of which have an impact on fermentation. Alternately, water can be added to reduce heat or drying, but this requires a mixing device that is unsuitable for the cultivation of filamentous fungi and may release nutrients that contaminants could use. Some studies have been directed towards optimizing the parameters to avoid heat accumulation. The study by Figueroa-Montero et al., for example, used the combination of mathematical models with internal air circulation by forced convection to modify the transfer of heat and water and to allow dissipation of the heat generated in the bioprocess [217]. Several other studies unveiled how matrix porosity should be studied to optimize air penetration, heat transfer, and effective air diffusion and decrease the heterogeneous nature of the substrate's negative effects on the bioprocess [218].

Product recovery is SSF's second significant drawback. Sometimes the finished product can be utilized right away, but metabolites frequently diffuse through solid substrates and need to be extracted, typically using large amounts of organic solvents, which greatly raises expenses and negates the initial cost savings. The substance that has been used up becomes waste that must be disposed of. When extracting metabolites, purification is a problem as well because it can be difficult and expensive. The product isolation and purification procedure depend on the product type, (e.g., intracellular or extracellular cell metabolite or whole cell biomass). Extracellular products are extracted directly from the fermented solid substrate, while intracellular products are extracted after the cell wall rapture (e.g., high-pressure homogenizer) [219]. Nevertheless, SSF uses a low water content to produce a higher concentration of products than SmSF [62]. This allows SSF to be a more cost-effective and lower solvents usage process than SmSF. The SSF constraints, the numerous approaches being used to solve them, and the potential future directions for further developing this biotechnology to increase its competitiveness in the worldwide market were all covered in detail by Oiza et al. [216].

Some other limitations can be linked to the lack of reports describing the growth behavior of some microbes. As underlined in the previous sections, the main microbes that are cultivated through SSF are filamentous fungi. They are particularly suited for SSF as they naturally colonize and decompose organic residuals and wastes. They are also competing against most contaminants; thus, it is usually not necessary to use aseptic bioreactors. They also do not require high amounts of water, their spores are particularly resistant to hostile environmental conditions, and the fermentation process nets high productivity and the final concentration of stable products. These factors render SSF more advantageous compared to SmSF for fungi. The most common issue is scaling up, as many processes are effective only on a laboratory scale and not for mass production [94,220–222]. Conidia produced using solid substrate fermentation are often more stable and resistant to stresses caused by drying than those produced in liquid culture [221]. No reports on the SSF cultivation of microorganisms for biopesticides of the domain of Archaea can be found. Nevertheless, these microbes are useful to produce enzymes, degrade agricultural wastes, and production of compounds valuable in food preservation and medical fields [223–228].

7. Conclusions and Future Perspectives

Diverse microorganisms can grow on solid surfaces without or almost without free-flowing water thanks to SSF. The food, pharmaceutical, energy, and chemical sectors all use this useful process. However, this technology's use in creating formulations useful for agriculture is still in its infancy and several key issues need to be addressed. The literature survey revealed that there are several substrates that have been tested for use in SSF. Many of them can also be useful in tropical countries, where certain local agriculture products leave wastes that could be exploited. But the most common and promising substrates are the global-wide cereals, since they are among the most cultivated crops in the world, and their biochemical profile is composed of easily available nutrients for the growth of microorganisms. This can pose a supplying problem when we consider that, as a staple food, cereals are first and foremost destined for the human diet, in a similar issue to biofuel crops. Therefore, research should improve production sustainability, focusing on recycling food residuals or agro-industrial wastes. Recent literature also shows that SSF has a good potential for producing biostimulants and biopesticides that are beneficial to agriculture, but there is still much work to do to solve the current operational limitations: mainly decreasing the costs of extraction, increasing the shelf-life of the final products, increasing the final returns in cell or spore production. Improving yields can be done either by increasing productivity for a single area or by allocating more land to crop fields. Surface intensification can consume soil, but land use is a major problem for the reduction of ecosystems and biodiversity. Therefore, the challenge of the XXI century will be set in the balance between these different requirements.

New tools for plant stimulation and protection, to produce more while consuming less soil and resources, are a welcome path to pursue. Biostimulants are the subject of intense research to mitigate the excess of nitrogen in soil and water systems and reduce our dependence on fertilizers, which are needed to sustain crop yields, yet their abuse is environmentally polluting. Biopesticides currently are mainly directed against insects and molds, which are sources of major interest in agriculture, both to contrast deleterious pests and to avoid damaging useful organisms that could offer many ecosystem services like pollination, predation of phytophages, biomonitoring, etc. Bioinsecticides, as evidenced in our search, are important for sanitary reasons too, to contain mosquitoes which are vectors of dangerous diseases such as malaria, dengue, or zika. Agroecosystems are sensible to the spread of mosquitoes, as they reproduce in water sources that are easily and plenty available in farmlands, from irrigation systems to stagnations, and particularly where rice is cultivated. In third world countries, dichlorodiphenyltrichloroethane (DDT) is still allowed and used as the most effective compound against mosquitoes to counter their spread, but it is environmentally persistent. The development of alternative larvicidal formulations can be helpful in reducing the need for DDT where terrible diseases are endemic. Categories like bioherbicides instead are lacking, despite SSF being a suitable method to cultivate fungi with herbicidal properties. Thus, it is necessary more research on practical and cheap strains to develop competitive products to reduce the usage of the more effective and widespread conventional herbicides.

As far as we know there is also currently no available commercial product based on protists and Archaea for the biostimulant and biopesticide markets. These overlooked lineages can be further studied for the possibility of discovering species with potentially useful capabilities and developing biotechnological processes and products that exploit them. This applied field that we discussed will therefore rely also on basic research, whose importance is evident from the fact that the world of microorganisms is still largely unknown and in the process of being discovered in its taxonomical, genetic, enzymatic, metabolic, and symbiotic components. The discovery of new microbial species or strains that could be potentially useful will play a pivotal role in the development of new formulations. Given the good number of reports that demonstrated their suitability, filamentous fungi (mainly *Trichoderma* spp.) and *Bacillus* spp. are the most promising inoculants for biostimulant and biopesticide SSF production. Considering the growing interest in bio-based products, the need for more sustainable agricultural practices, and the SSF potential for agricultural applications, the subject is worth to be investigated. Still more research is needed.

Supplementary Materials: The following supporting information can be downloaded at: https://www.mdpi.com/article/10.3390/microorganisms11061408/s1, Table S1: Literature on solid-state fermentation (SSF) used to produce biopesticides employing inocula of fungi and oomycetes.

Author Contributions: Conceptualization, M.P.; methodology, C.E., P.C. and M.D.G.; investigation, A.M., B.F., E.S., D.M. and R.D.; resources, C.E. and P.C.; data curation, A.M., E.S. and M.P.; writing—original draft preparation, B.F., R.D., D.M. and E.S.; writing—review and editing, A.M. and M.P.; supervision, M.D.G. and M.P. All authors have read and agreed to the published version of the manuscript.

Funding: This research received no external funding.

Institutional Review Board Statement: Not applicable.

Informed Consent Statement: Not applicable.

Data Availability Statement: Not applicable.

Conflicts of Interest: The authors declare no conflict of interest.

References

1. Swilling, M. Introduction: Change in the Age of Sustainabilit. In *The Age of Sustainability*; Routledge: London, UK, 2019; pp. 1–34, ISBN 9780429057823.
2. Nicholls, C.I.; Altieri, M.A.; Vazquez, L. Agroecological Principles for the Conversion of Farming Systems. In *Agroecological Practices for Sustainable Agriculture: Principles, Applications, and Making the Transition*; World Scientific: Singapore, 2017; pp. 1–18.
3. Jägermeyr, J. Agriculture's Historic Twin-Challenge Toward Sustainable Water Use and Food Supply for All. *Front. Sustain. Food Syst.* **2020**, *4*, 36. [CrossRef]
4. Rouphael, Y.; Colla, G. Toward a Sustainable Agriculture Through Plant Biostimulants: From Experimental Data to Practical Applications. *Agronomy* **2020**, *10*, 1461. [CrossRef]
5. Hijri, M. Microbial-Based Plant Biostimulants. *Microorganisms* **2023**, *11*, 686. [CrossRef] [PubMed]
6. Sutton, M.A.; Oenema, O.; Erisman, J.W.; Leip, A.; van Grinsven, H.; Winiwarter, W. Too Much of a Good Thing. *Nature* **2011**, *472*, 159–161. [CrossRef] [PubMed]
7. War, A.R.; Paulraj, M.G.; Ahmad, T.; Buhroo, A.A.; Hussain, B.; Ignacimuthu, S.; Sharma, H.C. Mechanisms of Plant Defense against Insect Herbivores. *Plant. Signal. Behav.* **2012**, *7*, 1306–1320. [CrossRef] [PubMed]
8. Walling, L.L. The Myriad Plant Responses to Herbivores. *J. Plant. Growth Regul.* **2000**, *19*, 195–216. [CrossRef]
9. Lucas, P. Mechanical Defences to Herbivory. *Ann. Bot.* **2000**, *86*, 913–920. [CrossRef]
10. Hanley, M.E.; Lamont, B.B.; Fairbanks, M.M.; Rafferty, C.M. Plant Structural Traits and Their Role in Anti-Herbivore Defence. *Perspect. Plant. Ecol. Evol. Syst.* **2007**, *8*, 157–178. [CrossRef]
11. Yactayo-Chang, J.P.; Tang, H.V.; Mendoza, J.; Christensen, S.A.; Block, A.K. Plant Defense Chemicals against Insect Pests. *Agronomy* **2020**, *10*, 1156. [CrossRef]
12. Ballhorn, D.J.; Kautz, S.; Heil, M.; Hegeman, A.D. Analyzing Plant Defenses in Nature. *Plant. Signal. Behav.* **2009**, *4*, 743–745. [CrossRef]
13. Wittstock, U.; Gershenzon, J. Constitutive Plant Toxins and Their Role in Defense against Herbivores and Pathogens. *Curr. Opin. Plant Biol.* **2002**, *5*, 300–307. [CrossRef] [PubMed]
14. Peñaflor, M.F.G.V.; Bento, J.M.S. Herbivore-Induced Plant Volatiles to Enhance Biological Control in Agriculture. *Neotrop. Entomol.* **2013**, *42*, 331–343. [CrossRef] [PubMed]
15. Kaur, S.; Samota, M.K.; Choudhary, M.; Choudhary, M.; Pandey, A.K.; Sharma, A.; Thakur, J. How Do Plants Defend Themselves against Pathogens-Biochemical Mechanisms and Genetic Interventions. *Physiol. Mol. Biol. Plants* **2022**, *28*, 485–504. [CrossRef]
16. Bednarek, P.; Osbourn, A. Plant-Microbe Interactions: Chemical Diversity in Plant Defense. *Science* **2009**, *324*, 746–748. [CrossRef]
17. Conrath, U. Systemic Acquired Resistance. *Plant Signal. Behav.* **2006**, *1*, 179–184. [CrossRef] [PubMed]
18. Purugganan, M.D.; Fuller, D.Q. The Nature of Selection during Plant Domestication. *Nature* **2009**, *457*, 843–848. [CrossRef] [PubMed]
19. Denison, R.F.; Kiers, E.T.; West, S.A. Darwinian Agriculture: When Can Humans Find Solutions Beyond The Reach of Natural Selection? *Q. Rev. Biol.* **2003**, *78*, 145–168. [CrossRef]
20. Milla, R.; Osborne, C.P.; Turcotte, M.M.; Violle, C. Plant Domestication through an Ecological Lens. *Trends Ecol. Evol.* **2015**, *30*, 463–469. [CrossRef]
21. Chaudhary, B. Plant Domestication and Resistance to Herbivory. *Int. J. Plant. Genomics* **2013**, *2013*, 572784. [CrossRef]
22. Chen, Y.H.; Gols, R.; Benrey, B. Crop Domestication and Its Impact on Naturally Selected Trophic Interactions. *Annu. Rev. Entomol.* **2015**, *60*, 35–58. [CrossRef]
23. Turcotte, M.M.; Lochab, A.K.; Turley, N.E.; Johnson, M.T.J. Plant Domestication Slows Pest Evolution. *Ecol. Lett.* **2015**, *18*, 907–915. [CrossRef]
24. Meyer, R.S.; DuVal, A.E.; Jensen, H.R. Patterns and Processes in Crop Domestication: An Historical Review and Quantitative Analysis of 203 Global Food Crops. *New Phytol.* **2012**, *196*, 29–48. [CrossRef]
25. Bebber, D.P.; Gurr, S.J. Crop-Destroying Fungal and Oomycete Pathogens Challenge Food Security. *Fungal Genet. Biol.* **2015**, *74*, 62–64. [CrossRef] [PubMed]
26. Oerke, E.-C. Crop Losses to Pests. *J. Agric. Sci.* **2006**, *144*, 31–43. [CrossRef]
27. Soliman, T.; Mourits, M.C.M.; Oude Lansink, A.G.J.M.; van der Werf, W. Economic Impact Assessment in Pest Risk Analysis. *Crop. Prot.* **2010**, *29*, 517–524. [CrossRef]
28. El-Sayed, R.A.; Jebur, A.B.; Kang, W.; El-Demerdash, F.M. An Overview on the Major Mycotoxins in Food Products: Characteristics, Toxicity, and Analysis. *J. Future Foods* **2022**, *2*, 91–102. [CrossRef]
29. Deutsch, C.A.; Tewksbury, J.J.; Tigchelaar, M.; Battisti, D.S.; Merrill, S.C.; Huey, R.B.; Naylor, R.L. Increase in Crop Losses to Insect Pests in a Warming Climate. *Science* **2018**, *361*, 916–919. [CrossRef] [PubMed]
30. Elad, Y.; Pertot, I. Climate Change Impacts on Plant Pathogens and Plant Diseases. *J. Crop. Improv.* **2014**, *28*, 99–139. [CrossRef]
31. Perrone, G.; Ferrara, M.; Medina, A.; Pascale, M.; Magan, N. Toxigenic Fungi and Mycotoxins in a Climate Change Scenario: Ecology, Genomics, Distribution, Prediction and Prevention of the Risk. *Microorganisms* **2020**, *8*, 1496. [CrossRef]
32. Cooper, J.; Dobson, H. The Benefits of Pesticides to Mankind and the Environment. *Crop. Prot.* **2007**, *26*, 1337–1348. [CrossRef]
33. Tudi, M.; Daniel Ruan, H.; Wang, L.; Lyu, J.; Sadler, R.; Connell, D.; Chu, C.; Phung, D.T. Agriculture Development, Pesticide Application and Its Impact on the Environment. *Int. J. Environ. Res. Public Health* **2021**, *18*, 1112. [CrossRef] [PubMed]
34. van der Werf, H.M.G. Assessing the Impact of Pesticides on the Environment. *Agric. Ecosyst. Environ.* **1996**, *60*, 81–96. [CrossRef]

35. Pimentel, D. Green Revolution Agriculture and Chemical Hazards. *Sci. Total Environ.* **1996**, *188*, S86–S98. [CrossRef] [PubMed]
36. Kim, K.-H.; Kabir, E.; Jahan, S.A. Exposure to Pesticides and the Associated Human Health Effects. *Sci. Total Environ.* **2017**, *575*, 525–535. [CrossRef]
37. Serrão, J.E.; Plata-Rueda, A.; Martínez, L.C.; Zanuncio, J.C. Side-Effects of Pesticides on Non-Target Insects in Agriculture: A Mini-Review. *Sci. Nat.* **2022**, *109*, 17. [CrossRef]
38. Heap, I. Global Perspective of Herbicide-Resistant Weeds. *Pest. Manag. Sci.* **2014**, *70*, 1306–1315. [CrossRef]
39. Araújo, M.F.; Castanheira, E.M.S.; Sousa, S.F. The Buzz on Insecticides: A Review of Uses, Molecular Structures, Targets, Adverse Effects, and Alternatives. *Molecules* **2023**, *28*, 3641. [CrossRef]
40. Miller, S.A.; Ferreira, J.P.; LeJeune, J.T. Antimicrobial Use and Resistance in Plant Agriculture: A One Health Perspective. *Agriculture* **2022**, *12*, 289. [CrossRef]
41. Bourguet, D.; Guillemaud, T. The Hidden and External Costs of Pesticide Use. In *Sustainable Agriculture Reviews*; Lichtfouse, E., Ed.; Springer: Cham, Switzerland, 2016; pp. 35–120.
42. Wilson, C.; Tisdell, C. Why Farmers Continue to Use Pesticides despite Environmental, Health and Sustainability Costs. *Ecol. Econ.* **2001**, *39*, 449–462. [CrossRef]
43. Delcour, I.; Spanoghe, P.; Uyttendaele, M. Literature Review: Impact of Climate Change on Pesticide Use. *Food Res. Int.* **2015**, *68*, 7–15. [CrossRef]
44. Lenné, J. Pests and Poverty: The Continuing Need for Crop Protection Research. *Outlook Agric.* **2000**, *29*, 235–250. [CrossRef]
45. Forget, G. Pesticides and the Third World. *J. Toxicol. Environ. Health* **1991**, *32*, 11–31. [CrossRef] [PubMed]
46. Roberts, D.; Mattoo, A. Sustainable Agriculture—Enhancing Environmental Benefits, Food Nutritional Quality and Building Crop Resilience to Abiotic and Biotic Stresses. *Agriculture* **2018**, *8*, 8. [CrossRef]
47. Zilberman, D.; Schmitz, A.; Casterline, G.; Lichtenberg, E.; Siebert, J.B. The Economics of Pesticide Use and Regulation. *Science* **1991**, *253*, 518–522. [CrossRef] [PubMed]
48. Lykogianni, M., Bempelou, E.; Karamaouna, F.; Aliferis, K.A. Do Pesticides Promote or Hinder Sustainability in Agriculture? The Challenge of Sustainable Use of Pesticides in Modern Agriculture. *Sci. Total Environ.* **2021**, *795*, 148625. [CrossRef]
49. Jepson, P.C.; Murray, K.; Bach, O.; Bonilla, M.A.; Neumeister, L. Selection of Pesticides to Reduce Human and Environmental Health Risks: A Global Guideline and Minimum Pesticides List. *Lancet Planet Health* **2020**, *4*, e56–e63. [CrossRef]
50. Ulrich, E.M.; Morrison, C.N.; Goldsmith, M.R.; Foreman, W.T. Chiral Pesticides: Identification, Description, and Environmental Implications. In *Reviews of Environmental Contamination and Toxicology*; Whitacre, D., Ed.; Springer: Boston, MA, USA, 2012; pp. 1–74.
51. Jeschke, P. Propesticides and Their Use as Agrochemicals. *Pest. Manag. Sci.* **2016**, *72*, 210–225. [CrossRef]
52. Tooker, J.F.; O'Neal, M.E.; Rodriguez-Saona, C. Balancing Disturbance and Conservation in Agroecosystems to Improve Biological Control. *Annu. Rev. Entomol.* **2020**, *65*, 81–100. [CrossRef]
53. Meena, M.; Swapnil, P.; Divyanshu, K.; Kumar, S.; Harish; Tripathi, Y.N.; Zehra, A.; Marwal, A.; Upadhyay, R.S. PGPR-Mediated Induction of Systemic Resistance and Physiochemical Alterations in Plants against the Pathogens: Current Perspectives. *J. Basic. Microbiol.* **2020**, *60*, 828–861. [CrossRef]
54. Poore, J.; Nemecek, T. Reducing Food's Environmental Impacts through Producers and Consumers. *Science* **2018**, *360*, 987–992. [CrossRef]
55. Lopes, M.J.d.S.; Dias-Filho, M.B.; Gurgel, E.S.C. Successful Plant Growth-Promoting Microbes: Inoculation Methods and Abiotic Factors. *Front. Sustain. Food Syst.* **2021**, *5*, 606454. [CrossRef]
56. Bashan, Y.; De-Bashan, L.E.; Prabhu, S.R.; Hernandez, J.-P. Advances in Plant Growth-Promoting Bacterial Inoculant Technology: Formulations and Practical Perspectives (1998–2013). *Plant Soil* **2014**, *378*, 1–33. [CrossRef]
57. Kumawat, K.C.; Keshani; Nagpal, S.; Sharma, P. Present Scenario of Bio-Fertilizer Production and Marketing around the Globe. In *Biofertilizers*; Elsevier: Amsterdam, The Netherlands, 2021; pp. 389–413.
58. Parnell, J.J.; Berka, R.; Young, H.A.; Sturino, J.M.; Kang, Y.; Barnhart, D.M.; DiLeo, M.V. From the Lab to the Farm: An Industrial Perspective of Plant Beneficial Microorganisms. *Front. Plant Sci.* **2016**, *7*, 1110. [CrossRef] [PubMed]
59. Schmidt, F.R. Optimization and Scale up of Industrial Fermentation Processes. *Appl. Microbiol. Biotechnol.* **2005**, *68*, 425–435. [CrossRef] [PubMed]
60. Berestetskiy, A. Development of Mycoherbicides. In *Encyclopedia of Mycology*; Elsevier: Amsterdam, The Netherlands, 2021; pp. 629–640.
61. Singhania, R.R.; Sukumaran, R.K.; Patel, A.K.; Larroche, C.; Pandey, A. Advancement and Comparative Profiles in the Production Technologies Using Solid-State and Submerged Fermentation for Microbial Cellulases. *Enzyme Microb. Technol.* **2010**, *46*, 541–549. [CrossRef]
62. Lizardi-Jiménez, M.A.; Hernández-Martínez, R. Solid State Fermentation (SSF): Diversity of Applications to Valorize Waste and Biomass. *3 Biotech* **2017**, *7*, 44. [CrossRef]
63. Ojo, A.O.; de Smidt, O. Lactic Acid: A Comprehensive Review of Production to Purification. *Processes* **2023**, *11*, 688. [CrossRef]
64. Sadh, P.K.; Duhan, S.; Duhan, J.S. Agro-Industrial Wastes and Their Utilization Using Solid State Fermentation: A Review. *Bioresour. Bioprocess.* **2018**, *5*, 1. [CrossRef]
65. Rodríguez Couto, S. Exploitation of Biological Wastes for the Production of Value-added Products under Solid-state Fermentation Conditions. *Biotechnol. J.* **2008**, *3*, 859–870. [CrossRef]

66. Yafetto, L. Application of Solid-State Fermentation by Microbial Biotechnology for Bioprocessing of Agro-Industrial Wastes from 1970 to 2020: A Review and Bibliometric Analysis. *Heliyon* **2022**, *8*, e09173. [CrossRef]
67. Wang, E.-Q.; Li, S.-Z.; Tao, L.; Geng, X.; Li, T.-C. Modeling of Rotating Drum Bioreactor for Anaerobic Solid-State Fermentation. *Appl. Energy* **2010**, *87*, 2839–2845. [CrossRef]
68. Krishna, C. Solid-State Fermentation Systems—An Overview. *Crit. Rev. Biotechnol.* **2005**, *25*, 1–30. [CrossRef] [PubMed]
69. Mitchell, D.A.; Berovič, M.; Krieger, N. *Solid-State Fermentation Bioreactors*; Mitchell, D.A., Berovič, M., Krieger, N., Eds.; Springer: Berlin/Heidelberg, Germany, 2006; ISBN 978-3-540-31285-7.
70. Vandenberghe, L.P.S.; Pandey, A.; Carvalho, J.C.; Letti, L.A.J.; Woiciechowski, A.L.; Karp, S.G.; Thomaz-Soccol, V.; Martínez-Burgos, W.J.; Penha, R.O.; Herrmann, L.W.; et al. Solid-State Fermentation Technology and Innovation for the Production of Agricultural and Animal Feed Bioproducts. *Syst. Microbiol. Biomanufacturing* **2021**, *1*, 142–165. [CrossRef]
71. Liu, Q.; Meng, X.; Li, T.; Raza, W.; Liu, D.; Shen, Q. The Growth Promotion of Peppers (*Capsicum annuum* L.) by Trichoderma Guizhouense NJAU4742-Based Biological Organic Fertilizer: Possible Role of Increasing Nutrient Availabilities. *Microorganisms* **2020**, *8*, 1296. [CrossRef]
72. Romano, I.; Ventorino, V.; Ambrosino, P.; Testa, A.; Chouyia, F.E.; Pepe, O. Development and Application of Low-Cost and Eco-Sustainable Bio-Stimulant Containing a New Plant Growth-Promoting Strain *Kosakonia pseudosacchari* TL13. *Front. Microbiol.* **2020**, *11*, 2044. [CrossRef] [PubMed]
73. Alias, C.; Bulgari, D.; Gobbi, E. It Works! Organic-Waste-Assisted Trichoderma Spp. Solid-State Fermentation on Agricultural Digestate. *Microorganisms* **2022**, *10*, 164. [CrossRef]
74. Bulgari, D.; Alias, C.; Peron, G.; Ribaudo, G.; Gianoncelli, A.; Savino, S.; Boureghda, H.; Bouznad, Z.; Monti, E.; Gobbi, E. Solid-State Fermentation of *trichoderma* spp.: A New Way to Valorize the Agricultural Digestate and Produce Value-Added Bioproducts. *J. Agric. Food Chem.* **2023**, *71*, 3994–4004. [CrossRef]
75. Cavello, I.A.; Crespo, J.M.; García, S.S.; Zapiola, J.M.; Luna, M.F.; Cavalitto, S.F. Plant Growth Promotion Activity of Keratinolytic Fungi Growing on a Recalcitrant Waste Known as "Hair Waste" . *Biotechnol. Res. Int.* **2015**, *2015*, 952921. [CrossRef]
76. Colla, G.; Rouphael, Y.; Di Mattia, E.; El-Nakhel, C.; Cardarelli, M. Co-Inoculation of *Glomus intraradices* and *Trichoderma atroviride* Acts as a Biostimulant to Promote Growth, Yield and Nutrient Uptake of Vegetable Crops. *J. Sci. Food Agric.* **2015**, *95*, 1706–1715. [CrossRef]
77. Zanoni do Prado, D.; Oliveira, S.L.; Okino-Delgado, C.H.; Auer, S.; Ludwig-Müller, J.; Ribeiro da Silva, M.; Júnior da Costa Fernandes, C.; Carbonari, C.A.; Zambuzzi, W.F.; Fleuri, L.F. *Aspergillus flavipes* as a Novel Biostimulant for Rooting-Enhancement of Eucalyptus. *J. Clean. Prod.* **2019**, *234*, 681–689. [CrossRef]
78. Mastan, A.; Rane, D.; Dastager, S.G.; Vivek Babu, C.S. Development of Low-Cost Plant Probiotic Formulations of Functional Endophytes for Sustainable Cultivation of Coleus Forskohlii. *Microbiol. Res.* **2019**, *227*, 126310. [CrossRef] [PubMed]
79. Shengping, X.; Liantian, M.; Yujie, M.; Yan, D.; Hongbo, Y. Optimizing *Bacillus circulans* Xue-113168 for Biofertilizer Production and Its Effects on Crops. *Afr. J. Biotechnol.* **2016**, *15*, 2795–2803. [CrossRef]
80. Morán-Diez, M.E.; Glare, T.R. What Are Microbial-Based Biopesticides? In *Methods in Molecular Biology*; Glare, T., Moran-Diez, M., Eds.; Humana Press: Clifton, NJ, USA, 2016; pp. 1–10.
81. FAO. *WHO Guidelines for the Registration of Microbial, Botanical and Semiochemical Pest Control Agents for Plant Protection and Public Health Uses*; FAO: Rome, Italy, 2017; ISBN 978-92-5-130015-2.
82. Bourguignon, D. *EU Policy and Legislation on Pesticides: Plant Protection Products and Biocides*; European Union: Strasbourg, France, 2017.
83. Seenivasagan, R.; Babalola, O.O. Utilization of Microbial Consortia as Biofertilizers and Biopesticides for the Production of Feasible Agricultural Product. *Biology* **2021**, *10*, 1111. [CrossRef] [PubMed]
84. Sun, X. History and Current Status of Development and Use of Viral Insecticides in China. *Viruses* **2015**, *7*, 306–319. [CrossRef]
85. Szewczyk, B.; Hoyos-Carvajal, L.; Paluszek, M.; Skrzecz, I.; Lobo de Souza, M. Baculoviruses—Re-Emerging Biopesticides. *Biotechnol. Adv.* **2006**, *24*, 143–160. [CrossRef] [PubMed]
86. Saxena, S.; Pandey, A.K. Microbial Metabolites as Eco-Friendly Agrochemicals for the next Millennium. *Appl. Microbiol. Biotechnol.* **2001**, *55*, 395–403. [CrossRef]
87. Essiedu, J.A.; Adepoju, F.O.; Ivantsova, M.N. Benefits and Limitations in Using Biopesticides: A Review. In *Proceedings of the VII International Young Researchers' Conference—Physics, Technology, Innovations (PTI-2020) AIP Con-ference Proceedings 2313, Ekaterinburg, Russia, 18–22 May 2022*; Vladimir, A., Volkovich, I.V., Kashin, A.A., Smirnov, E.D.N., Eds.; AIP Publishing: Ekaterinburg, Russia, 2020; pp. 080002-1–080002-6.
88. Glare, T.R.; Gwynn, R.L.; Moran-Diez, M.E. Development of Biopesticides and Future Opportunities. *Methods Mol. Biol.* **2016**, *1477*, 211–221. [CrossRef]
89. Rodgers, P.B. Potential of Biopesticides in Agriculture. *Pestic. Sci.* **1993**, *39*, 117–129. [CrossRef]
90. Vero, S.; Garmendia, G.; Allori, E.; Sanz, J.M.; Gonda, M.; Alconada, T.; Cavello, I.; Dib, J.R.; Diaz, M.A.; Nally, C.; et al. Microbial Biopesticides: Diversity, Scope, and Mechanisms Involved in Plant Disease Control. *Diversity* **2023**, *15*, 457. [CrossRef]
91. Kumar, S. Biopesticides: A Need for Food and Environmental Safety. *J. Biofertil. Biopestic.* **2012**, *3*, 4. [CrossRef]
92. EPA What Are Biopesticides? Available online: https://www.epa.gov/ingredients-used-pesticide-products/what-are-biopesticides (accessed on 1 April 2023).

93. Méndez-González, F.; Figueroa-Montero, A.; Saucedo-Castañeda, G.; Loera, O.; Favela-Torres, E. Addition of Spherical-style Packing Improves the Production of Conidia by *Metarhizium robertsii* in Packed Column Bioreactors. *J. Chem. Technol. Biotechnol.* **2022**, *97*, 1517–1525. [CrossRef]
94. De Vrije, T.; Antoine, N.; Buitelaar, R.M.; Bruckner, S.; Dissevelt, M.; Durand, A.; Gerlagh, M.; Jones, E.E.; Lüth, P.; Oostra, J.; et al. The Fungal Biocontrol Agent *Coniothyrium minitans*: Production by Solid-State Fermentation, Application and Marketing. *Appl. Microbiol. Biotechnol.* **2001**, *56*, 58–68. [CrossRef]
95. Devi, V. Mass Production of *Bacillus Thuringiensis* and *Nomuraea Rileyi*. In *Handbook of Integrated Pest Management*; Chattopadhyay, C., Tanwar, R.K., Sehgal, M., Birah, A., Bhagat, S., Ahmad, N., Mehta, N., Eds., ICAR- DKMA Publication: New Delhi, India, 2018; pp. 353–357.
96. Devi, P.S.V.; Chowdary, A.; Prasad, Y.G. Cost-Effective Multiplication of the Entomopathogenic Fungus *Nomuraea rileyi* (F) Samson *. *Mycopathologia* **2000**, *151*, 35–39. [CrossRef] [PubMed]
97. Ferreira, J.M.; Pinto, S.M.N.; Soares, F.E.F. *Metarhizium robertsii* Protease and Conidia Production, Response to Heat Stress and Virulence against Aedes Aegypti Larvae. *AMB Express* **2021**, *11*, 166. [CrossRef]
98. Xie, L.; Han, J.H.; Kim, J.J.; Lee, S.Y. Effects of Culture Conditions on Conidial Production of the Sweet Potato Whitefly Pathogenic Fungus *Isaria javanica*. *Mycoscience* **2016**, *57*, 64–70. [CrossRef]
99. Ye, S.; Shang, L.; Xie, X.; Cao, Y.; Chen, C. Optimization of In Vitro Culture Conditions for Production of *Cordyceps bassiana* Spores (Ascomycetes) and the Effect of Spore Extracts on the Health of *Caenorhabditis elegans*. *Int. J. Med. Mushrooms* **2021**, *23*, 45–55. [CrossRef] [PubMed]
100. Velozo, S.G.M.; Velozo, M.R.; Domingues, M.M.; Becchi, L.K.; de Carvalho, V.R.; de Souza Passos, J.R.; Zanuncio, J.C.; Serrão, J.E.; Stephan, D.; Wilcken, C.F. From the Dual Cyclone Harvest Performance of Single Conidium Powder to the Effect of *Metarhizium anisopliae* on the Management of *Thaumastocoris peregrinus* (*Hemiptera: Thaumastocoridae*). *PLoS ONE* **2023**, *18*, e0283543. [CrossRef]
101. Thakre, M.; Thakur, M.; Malik, N.; Ganger, S. Mass Scale Cultivation of Entomopathogenic Fungus *Nomuraea rileyi* Using Agricultural Products and Agro Wastes. *J. Biopestic.* **2011**, *4*, 176–179.
102. Weber, F.J.; Tramper, J.; Rinzema, A. A Simplified Material and Energy Balance Approach for Process Development and Scale-Up of *Coniothyrium minitans* Conidia Production by Solid-State Cultivation in a Packed-Bed Reactor. *Biotechnol. Bioeng.* **1999**, *65*, 447–458. [CrossRef]
103. Vogelgsang, S.; Watson, A.K.; Ditommaso, A. Effect of Moisture, Inoculum Production, and Planting Substrate on Disease Reaction of Field Bindweed (*Convolvulus arvensis* L.) to the Fungal Pathogen, *Phomopsis convolvulus*. *Eur. J. Plant. Pathol.* **1998**, *104*, 253. [CrossRef]
104. Sala, A.; Barrena, R.; Meyling, N.V.; Artola, A. Conidia Production of the Entomopathogenic Fungus *Beauveria bassiana* Using Packed-Bed Bioreactor: Effect of Substrate Biodegradability on Conidia Virulence. *J. Environ. Manag.* **2023**, *341*, 118059. [CrossRef] [PubMed]
105. Ingle, Y.V. Effect of Different Growing Media on Mass Production of *Nomuraea rileyi*. *Int. J. Environ. Sci.* **2014**, *4*, 1006–1014.
106. Sala, A.; Barrena, R.; Sánchez, A.; Artola, A. Fungal Biopesticide Production: Process Scale-up and Sequential Batch Mode Operation with *Trichoderma harzianum* Using Agro-Industrial Solid Wastes of Different Biodegradability. *Chem. Eng. J.* **2021**, *425*, 131620. [CrossRef]
107. Gandarilla-Pacheco, F.L.; Morales-Ramos, L.H.; Pereyra-Alférez, B.; Elías-Santos, M.; Quintero-Zapata, I. Production of Infectious Units of *Isaria fumosorosea* (*Hypocreales*: *Cordycipitaceae*) from Different Indigenous Isolates of Northeastern Mexico Using 3 Propagation Strategies. *Rev. Argent. Microbiol.* **2018**, *50*, 81–89. [CrossRef] [PubMed]
108. Oostra, J.; Tramper, J.; Rinzema, A. Model-Based Bioreactor Selection for Large-Scale Solid-State Cultivation of *Coniothyrium minitans* Spores on Oats. *Enzyme Microb. Technol.* **2000**, *27*, 652–663. [CrossRef] [PubMed]
109. Naeimi, S.; Khosravi, V.; Varga, A.; Vágvölgyi, C.; Kredics, L. Screening of Organic Substrates for Solid-State Fermentation, Viability and Bioefficacy of *Trichoderma harzianum* As12-2, a Biocontrol Strain against Rice Sheath Blight Disease. *Agronomy* **2020**, *10*, 1258. [CrossRef]
110. Leão, R.; De Oliveira, S.; Da, A.C.; Santos, S.; Félix, A.; Costa, D.A.; Tiago, P.V. Substrates for the Production of Entomopathogenic Isolates of *Dusarium caatingaense* to Control *Dactylopius opuntiae*. *Anais Acad. Pernambucana Ciência Agronômica* **2023**, *19*, 1–11.
111. Rayhane, H.; Josiane, M.; Gregoria, M.; Yiannis, K.; Nathalie, D.; Ahmed, M.; Sevastianos, R. From Flasks to Single Used Bioreactor: Scale-up of Solid State Fermentation Process for Metabolites and Conidia Production by *Trichoderma asperellum*. *J. Environ. Manag.* **2019**, *252*, 109496. [CrossRef]
112. Singh, J.; Pandey, A.; Pandey, A.K. Solid Substrate Fermentation of Mycoherbicidal Agent *Alternaria alternata* Fgcc#25. *Recent Res. Sci. Technol.* **2010**, *2*, 2076–5061.
113. Zhang, Y.; Liu, J.; Zhou, Y.; Ge, Y. Spore Production of Clonostachys Rosea in a New Solid-State Fermentation Reactor. *Appl. Biochem. Biotechnol.* **2014**, *174*, 2951–2959. [CrossRef]
114. Viccini, G.; Mannich, M.; Capalbo, D.M.F.; Valdebenito-Sanhueza, R.; Mitchell, D.A. Spore Production in Solid-State Fermentation of Rice by Clonostachys Rosea, a Biopesticide for Gray Mold of Strawberries. *Process. Biochem.* **2007**, *42*, 275–278. [CrossRef]
115. Hamrouni, R.; Claeys-Bruno, M.; Molinet, J.; Masmoudi, A.; Roussos, S.; Dupuy, N. Challenges of Enzymes, Conidia and 6-Pentyl-Alpha-Pyrone Production from Solid-State-Fermentation of Agroindustrial Wastes Using Experimental Design and *T. asperellum* Strains. *Waste Biomass Valorization* **2020**, *11*, 5699–5710. [CrossRef]

116. Cando-Narvaez, A.; Méndez-Hernández, J.E. Rice Recycling: A Simple Strategy to Improve Conidia Production in Solid-State Cultures. *Lett. Appl. Microbiol.* **2021**, *74*, 385–394. [CrossRef]
117. Shi, Y.; Xu, X.; Zhu, Y. Optimization of *Verticillium lecanii* Spore Production in Solid-State Fermentation on Sugarcane Bagasse. *Appl. Microbiol. Biotechnol.* **2009**, *82*, 921–927. [CrossRef]
118. De la Cruz-Quiroz, R.; Roussos, S.; Hernandez-Castillo, D.; Rodríguez-Herrera, R.; López-López, L.I.; Castillo, F.; Aguilar, C.N. Solid-State Fermentation in a Bag Bioreactor: Effect of Corn Cob Mixed with Phytopathogen Biomass on Spore and Cellulase Production by *Trichoderma asperellum*. In *Fermentation Processes*; IntechOpen: London, UK, 2017.
119. de Lima, I.J.; Leão, M.P.C.; da Silva Santos, A.C.; da Costa, A.F.; Tiago, P.V. Production of Conidia by Entomopathogenic Isolates of *Fusarium caatingaense* on Different Vegetable Substrates. *Biocontrol Sci. Technol.* **2021**, *31*, 206–218. [CrossRef]
120. Awan, U.A.; Meng, L.; Xia, S.; Raza, M.F.; Zhang, Z.; Zhang, H. Isolation, Fermentation, and Formulation of Entomopathogenic Fungi Virulent against Adults of *Diaphorina citri*. *Pest. Manag. Sci.* **2021**, *77*, 4040–4053. [CrossRef]
121. Cuadrado-Osorio, P.D.; Castillo-Saldarriaga, C.R.; Cubides-Cárdenas, J.A.; Álvarez, M.I.G.; Bautista, E.J. Chlamydospores Production Enhancement of *Duddingtonia flagrans* in a Solid-State Fermentation System. *Dyna* **2021**, *88*, 152–159. [CrossRef]
122. Klaic, R.; Sallet, D.; Foletto, E.L.; Jacques, R.J.S.; Guedes, J.V.C.; Kuhn, R.C.; Mazutti, M.A. Optimization of Solid-State Fermentation for Bioherbicide Production by *Phoma* sp. *Braz. J. Chem. Eng.* **2017**, *34*, 377–384. [CrossRef]
123. Sargin, S.; Gezgin, Y.; Eltem, R.; Vardar, F. Micropropagule Production from *Trichoderma harzianum* EGE-K38 Using Solid-State Fermentation and a Comparative Study for Drying Methods. *Turk. J. Biol.* **2013**, *37*, 139–146. [CrossRef]
124. Masangkay, R.F.; Paulitz, T.C.; Hallett, S.G.; Watson, A.K. Solid Substrate Production of *Alternaria alternata* f. sp. *Sphenocleae conidia*. *Biocontrol Sci. Technol.* **2000**, *10*, 399–409. [CrossRef]
125. Aita, B.C.; Spannemberg, S.S.; Schmaltz, S.; Zabot, G.L.; Tres, M.V.; Kuhn, R.C.; Mazutti, M.A. Production of Cell-Wall Degrading Enzymes by Solid-State Fermentation Using Agroindustrial Residues as Substrates. *J. Environ. Chem. Eng.* **2019**, *7*, 103193. [CrossRef]
126. Rustiguel, C.B.; Jorge, J.A.; Guimarães, L.H.S. Optimization of the Chitinase Production by Different *Metarhizium anisopliae* Strains under Solid-State Fermentation with Silkworm Chrysalis as Substrate Using CCRD. *Adv. Microbiol.* **2012**, *2*, 268–276. [CrossRef]
127. Nalini, S.; Parthasarathi, R. Production and Characterization of Rhamnolipids Produced by *Serratia rubidaea* SNAU02 under Solid-State Fermentation and Its Application as Biocontrol Agent. *Bioresour. Technol.* **2014**, *173*, 231–238. [CrossRef] [PubMed]
128. Reddy, S.G.E.; Sahotra, S. Multiplication of Entomopathogenic Fungus (*Lecanicillium lecanii*) on Apple Pomace and Its Toxicity against Aphid (*Aphis craccivora*). *Toxin Rev.* **2020**, *39*, 252–257. [CrossRef]
129. Türkölmez, Ş.; Özer, G.; Derviş, S. Clonostachys Rosea Strain ST1140: An Endophytic Plant-Growth-Promoting Fungus, and Its Potential Use in Seedbeds with Wheat-Grain Substrate. *Curr. Microbiol.* **2023**, *80*, 36. [CrossRef]
130. Adetunji, C.O.; Oloke, J.K. Effect of Wild and Mutant Strain of *Lasiodiplodia pseudotheobromae* Mass Produced on Rice Bran as a Potential Bioherbicide Agents for Weeds under Solid State Fermentation. *J. Appl. Biol. Biotechnol.* **2013**, *1*, 18–23.
131. Loera-Corral, O.; Porcayo-Loza, J.; Montesinos-Matias, R.; Favela-Torres, E. Production of Conidia by the Fungus *Metarhizium anisopliae* Using Solid-State Fermentation. *Methods Mol. Biol.* **2016**, *1477*, 61–69.
132. Camargo, A.F.; Dalastra, C.; Ulrich, A.; Scapini, T.; Bonatto, C.; Klanovicz, N.; Michelon, W.; Lerin, L.; Júnior, S.L.A.; Mossi, A.J.; et al. The Bioherbicidal Potential of Isolated Fungi Cultivated in Microalgal Biomass. *Bioprocess. Biosyst. Eng.* **2023**, *46*, 665–679. [CrossRef]
133. Adetunji, C.O.; Adejumo, I.O.; Oloke, J.K.; Akpor, O.B. Production of Phytotoxic Metabolites with Bioherbicidal Activities from *Lasiodiplodia pseudotheobromae* Produced on Different Agricultural Wastes Using Solid-State Fermentation. *Iran. J. Sci. Technol. Trans. A Sci.* **2018**, *42*, 1163–1175. [CrossRef]
134. Mascarin, G.M.; Pereira-Junior, R.A.; Fernandes, É.K.K.; Quintela, E.D.; Dunlap, C.A.; Arthurs, S.P. Phenotype Responses to Abiotic Stresses, Asexual Reproduction and Virulence among Isolates of the Entomopathogenic Fungus *Cordyceps javanica* (*Hypocreales: Cordycipitaceae*). *Microbiol. Res.* **2018**, *216*, 12–22. [CrossRef]
135. Mascarin, G.M.; Kobori, N.N.; Quintela, E.D.; Delalibera, I. The Virulence of Entomopathogenic Fungi against *Bemisia tabaci* Biotype B (*Hemiptera*: *Aleyrodidae*) and Their Conidial Production Using Solid Substrate Fermentation. *Biol. Control.* **2013**, *66*, 209–218. [CrossRef]
136. Baldoni, D.B.; Antoniolli, Z.I.; Mazutti, M.A.; Jacques, R.J.S.; Dotto, A.C.; de Oliveira Silveira, A.; Ferraz, R.C.; Soares, V.B.; de Souza, A.R.C. Chitinase Production by *Trichoderma koningiopsis* UFSMQ40 Using Solid State Fermentation. *Braz. J. Microbiol.* **2020**, *51*, 1897–1908. [CrossRef] [PubMed]
137. Guo, Q.Y.; Cheng, L.; Zhu, H.X.; Li, W.; Wei, Y.H.; Chen, H.Y.; Guo, L.Z.; Weng, H.; Wang, J. Herbicidal Activity of *Aureobasidium pullulans* PA-2 on Weeds and Optimization of Its Solid-State Fermentation Conditions. *J. Integr. Agric.* **2020**, *19*, 173–182. [CrossRef]
138. Nagy, V.; Toke, E.R.; Keong, L.C.; Szatzker, G.; Ibrahim, D.; Omar, I.C.; Szakács, G.; Poppe, L. Kinetic Resolutions with Novel, Highly Enantioselective Fungal Lipases Produced by Solid State Fermentation. *J. Mol. Catal. B Enzym.* **2006**, *39*, 141–148. [CrossRef]
139. Ghaffari, S.; Karimi, J.; Kamali, S.; Moghadam, E.M. Biocontrol of *Planococcus citri* (*Hemiptera*: *Pseudococcidae*) by *Lecanicillium longisporum* and *Lecanicillium lecanii* under Laboratory and Greenhouse Conditions. *J. Asia Pac. Entomol.* **2017**, *20*, 605–612. [CrossRef]
140. Brand, D.; Roussos, S.; Prochmann, F.A.; Pohl, J.; Soccol, C.R. Production of a Biocompost by Solid State Fermentation Against the Coffee Nematodes. In *New. Horizons in Biotechnology*; Springer: Dordrecht, The Netherlands, 2003; pp. 403–412.

141. Corrêa, B.; da Silveira Duarte, V.; Silva, D.M.; Mascarin, G.M.; Júnior, I.D. Comparative Analysis of Blastospore Production and Virulence of *Beauveria bassiana* and *Cordyceps fumosorosea* against Soybean Pests. *BioControl* **2020**, *65*, 323–337. [CrossRef]
142. Adetunji, C.O.; Oloke, J.K.; Osemwegie, O.O.; Ehis-Eriakha, C.B. Use of Agro-Wastes for *Lasiodiplodia pseudotheobromae* (C1136) Production with Sustainable Bioefficacy. *Environ. Dev. Sustain.* **2022**, *24*, 7794–7809. [CrossRef]
143. Dos Reis, C.B.L.; Sobucki, L.; Mazutti, M.A.; Guedes, J.V.C.; Jacques, R.J.S. Production of Chitinase from *Metarhizium anisopliae* by Solid-State Fermentation Using Sugarcane Bagasse as Substrate. *Ind. Biotechnol.* **2018**, *14*, 230–234. [CrossRef]
144. Ganassi, S.; Di Domenico, C.; Altomare, C.; Samuels, G.J.; Grazioso, P.; Di Cillo, P.; Pietrantonio, L.; Cristofaro, A. De Potential of Fungi of the Genus Trichoderma for Biocontrol of *Philaenus spumarius*, the Insect Vector for the Quarantine Bacterium Xylella Fastidosa. *Pest. Manag. Sci.* **2023**, *79*, 719–728. [CrossRef]
145. Teshler, M.P.; Ash, G.J.; Zolotarov, Y.; Watson, A.K. Increased Shelf Life of a Bioherbicide through Combining Modified Atmosphere Packaging and Low Temperatures. *Biocontrol Sci. Technol.* **2007**, *17*, 387–400. [CrossRef]
146. Nguyen, H.C.; Tran, T.V.A.; Nguyen, Q.L.; Nguyen, N.N.; Nguyen, M.K.; Nguyen, N.T.T.; Su, C.H.; Lin, K.H. Newly Isolated *Paecilomyces lilacinus* and *Paecilomyces javanicus* as Novel Biocontrol Agents for *Plutella xylostella* and *Spodoptera litura*. *Not. Bot. Horti Agrobot. Cluj Napoca* **2017**, *45*, 280–286. [CrossRef]
147. Medina, J.D.; Granada, D.; Bedoya, J.C.; Cardona, N. Production of *Akanthomyces lecanii* by Submerged Fermentation to Control *Tetranychus urticae* (Acari: Tetranychidae) in *Gerbera jamesonii* Crops. *Biocontrol Sci. Technol.* **2021**, *31*, 1377–1387. [CrossRef]
148. Das, M.M.; Haridas, M.; Sabu, A. Process Development for the Enhanced Production of Bio-Nematicide *Purpureocillium lilacinum* KU8 under Solid-State Fermentation. *Bioresour. Technol.* **2020**, *308*, 123328. [CrossRef]
149. Mcquilken, M.P.; Whipps, J.M. Production, Survival and Evaluation of Solid-Substrate Inocula of *Coniothyrium minitans* against Sclerotinia Sclerotiorum. *Eur. J. Plant. Pathol.* **1995**, *101*, 101–110. [CrossRef]
150. Méndez-González, F.; Loera, O.; Saucedo-Castañeda, G.; Favela-Torres, E. Forced Aeration Promotes High Production and Productivity of Infective Conidia from *Metarhizium robertsii* in Solid-State Fermentation. *Biochem. Eng. J.* **2020**, *156*, 107492. [CrossRef]
151. do Nascimento Silva, J.; Mascarin, G.M.; dos Santos Gomes, I.C.; Tinoco, R.S.; Quintela, E.D.; dos Reis Castilho, L.; Freire, D.M.G. New Cost-Effective Bioconversion Process of Palm Kernel Cake into Bioinsecticides Based on *Beauveria bassiana* and *Isaria javanica*. *Appl. Microbiol. Biotechnol.* **2018**, *102*, 2595–2606. [CrossRef]
152. Babu, R.; Sajeena, A.; Seetharaman, K. Solid Substrate for Production of Alternaria Alternata Conidia: A Potential Mycoherbicide for the Control of *Eichhornia Crassipes* (Water Hyacinth). *Weed Res.* **2004**, *44*, 298–304. [CrossRef]
153. Zhang, Y.; Gao, X.; Liu, J.; Ge, Y. Pilot Production of Clonostachys Rosea Conidia in a Solid-State Fermentor Optimized Using Response Surface Methodology. *Eng. Life Sci.* **2015**, *15*, 772–778. [CrossRef]
154. Qiu, L.; Li, J.J.; Li, Z.; Wang, J.J. Production and Characterization of Biocontrol Fertilizer from Brewer's Spent Grain via Solid-State Fermentation. *Sci. Rep.* **2019**, *9*, 480. [CrossRef]
155. Suminto, S.; Takatsuji, E.; Iguchi, A.; Kanzaki, H.; Okuda, T.; Nitoda, T. A New *Asteltoxin analog* with Insecticidal Activity from *Pochonia suchlasporia* TAMA 87. *J. Pestic. Sci.* **2020**, *45*, 81–85. [CrossRef]
156. El-Beltagi, H.S.; El-Mahdy, O.M.; Mohamed, H.I.; El-Ansary, A.E. Antioxidants, Antimicrobial, and Anticancer Activities of Purified Chitinase of *Talaromyces funiculosus* Strain CBS 129594 Biosynthesized Using Crustacean Bio-Wastes. *Agronomy* **2022**, *12*, 2818. [CrossRef]
157. Song, X.Y.; Xie, S.T.; Chen, X.L.; Sun, C.Y.; Shi, M.; zhong Zhang, Y. Solid-State Fermentation for Trichokonins Production from *Trichoderma koningii* SMF2 and Preparative Purification of Trichokonin VI by a Simple Protocol. *J. Biotechnol.* **2007**, *131*, 209–215. [CrossRef] [PubMed]
158. Bhosale, N.; Ingle, D.Y.V.; Goramnagar, H.B. Influence of Different Growing Media on Mass Production of *Aschersonia aleyrodis*, a Fungal Pathogen of Citrus Black Fly. *Int. J. Curr. Microbiol. Appl. Sci.* **2020**, *9*, 1583–1591. [CrossRef]
159. Mar, T.T.; Lumyong, S. Conidial Production of Entomopathogenic Fungi in Solid State Fermentation. *KKU Res. J.* **2012**, *17*, 762–768.
160. Das, M.M.; Aguilar, C.N.; Haridas, M.; Sabu, A. Production of Bio-Fungicide, *Trichoderma harzianum* CH1 under Solid-State Fermentation Using Coffee Husk. *Bioresour. Technol. Rep.* **2021**, *15*, 100708. [CrossRef]
161. Sharma, A.; Sharma, S.; Sabir, N.; El-Sheikh, M.A.; Alyemeni, M. Impact Assessment of Karanja Deoiled Cake and Sundried Biogas Slurry as a Mixed Substrate on the Nematicidal Potential of *Purpureocillium lilacinum*. *J. King Saud. Univ. Sci.* **2021**, *33*, 101399. [CrossRef]
162. May, B.A.; Vandergheynst, J.S. A Predictor Variable for Efficacy of *Lagenidium giganteum* Produced in Solid-State Cultivation. *J. Ind. Microbiol. Biotechnol.* **2001**, *27*, 203–207. [CrossRef]
163. Prakash, G.V.S.B.; Padmaja, V.; Kiran, R.R.S. Statistical Optimization of Process Variables for the Large-Scale Production of *Metarhizium anisopliae* Conidiospores in Solid State Fermentation. *Bioresour. Technol.* **2008**, *99*, 1530–1537. [CrossRef]
164. Barra, P.; Barros, G.; Etcheverry, M.; Nesci, A. Mass Production Studies in Solid Substrates with the Entomopathogenic Fungus, *Purpureocillium lilacinum*. *Int. J. Adv. Agric. Res.* **2018**, *6*, 78–84.
165. Brand, D.; Roussos, S.; Pandey, A.; Zilioli, P.C.; Pohl, J.; Soccol, C.R. Development of a Bionematicide With *Paecilomyces lilacinus* to Control Meloidogyne Incognita. *Appl. Biochem. Biotechnol.* **2004**, *118*, 81–88. [CrossRef]
166. Mathulwe, L.L.; Malan, A.P.; Stokwe, N.F. Mass Production of Entomopathogenic Fungi, *Metarhizium robertsii* and *Metarhizium pinghaense*, for Commercial Application against Insect Pests. *J. Vis. Exp.* **2022**, *2022*, e63246. [CrossRef]

167. Tenkegna, T.A.; Belay, Y.C. Use of Solid-State Fermentation on Selected Agricultural Wastes for Mass Production of *Beauveria bassiana*. *Pest. Manag. J. Ethiop.* **2017**, *20*, 25–39.
168. Hamrouni, R.; Molinet, J.; Dupuy, N.; Taieb, N.; Carboue, Q.; Masmoudi, A.; Roussos, S. The Effect of Aeration for 6-Pentyl-Alpha-Pyrone, Conidia and Lytic Enzymes Production by *Trichoderma asperellum* Strains Grown in Solid-State Fermentation. *Waste Biomass Valorization* **2020**, *11*, 5711–5720. [CrossRef]
169. Virtanen, V.; Nyyssölä, A.; Leisola, M.; Seiskari, P. An Aseptically Operatable Static Solid State Bioreactor Consisting of Two Units. *Biochem. Eng. J.* **2008**, *39*, 594–597. [CrossRef]
170. Ghoreishi, G.; Barrena, R.; Font, X. Using Green Waste as Substrate to Produce Biostimulant and Biopesticide Products through Solid-State Fermentation. *Waste Manag.* **2023**, *159*, 84–92. [CrossRef]
171. Melo, A.L.d.A.; Soccol, V.T.; Soccol, C.R. *Bacillus thuringiensis*: Mechanism of Action, Resistance, and New Applications: A Review. *Crit. Rev. Biotechnol.* **2016**, *36*, 317–326. [CrossRef]
172. Bravo, A.; Likitvivatanavong, S.; Gill, S.S.; Soberón, M. *Bacillus thuringiensis*: A Story of a Successful Bioinsecticide. *Insect Biochem. Mol. Biol.* **2011**, *41*, 423–431. [CrossRef]
173. Sanahuja, G.; Banakar, R.; Twyman, R.M.; Capell, T.; Christou, P. *Bacillus thuringiensis*: A Century of Research, Development and Commercial Applications. *Plant. Biotechnol. J.* **2011**, *9*, 283–300. [CrossRef]
174. Sanchis, V.; Bourguet, D. *Bacillus thuringiensis*: Applications in Agriculture and Insect Resistance Management. A Review. *Agron. Sustain. Dev.* **2008**, *28*, 11–20. [CrossRef]
175. Biggel, M.; Jessberger, N.; Kovac, J.; Johler, S. Recent Paradigm Shifts in the Perception of the Role of *Bacillus thuringiensis* in Foodborne Disease. *Food Microbiol.* **2022**, *105*, 104025. [CrossRef]
176. Wronkowska, M.; Christa, K.; Ciska, E.; Soral-Śmietana, M. Chemical Characteristics and Sensory Evaluation of Raw and Roasted Buckwheat Groats Fermented by R *Hizopus oligosporus*. *J. Food Qual.* **2015**, *38*, 130–138. [CrossRef]
177. Terakawa, T.; Takaya, N.; Horiuchi, H.; Koike, M.; Takagi, M. A Fungal Chitinase Gene From *Rhizopus oligosporus* Confers Antifungal Activity to Transgenic Tobacco. *Plant Cell. Rep.* **1997**, *16*, 439–443. [CrossRef] [PubMed]
178. Widyastuti, W.; Setiawan, F.; Al Afandy, C.; Irawan, A.; Laila, A.; Juliasih, N.L.G.R.; Setiawan, W.A.; Arai, M.; Hendri, J.; Setiawan, A. Antifungal Agent Chitooligosaccharides Derived from Solid-State Fermentation of Shrimp Shell Waste by *Pseudonocardia antitumoralis* 18D36-A1. *Fermentation* **2022**, *8*, 353. [CrossRef]
179. Su, Y.; Liu, C.; Long, Z.; Ren, H.; Guo, X. Improved Production of Spores and Bioactive Metabolites from *Bacillus amyloliquefaciens* in Solid-State Fermentation by a Rapid Optimization Process. *Probiotics Antimicrob. Proteins* **2019**, *11*, 921–930. [CrossRef] [PubMed]
180. Berikashvili, V.; Sokhadze, K.; Kachlishvili, E.; Elisashvili, V.; Chikindas, M.L. *Bacillus amyloliquefaciens* Spore Production Under Solid-State Fermentation of Lignocellulosic Residues. *Probiotics Antimicrob. Proteins* **2018**, *10*, 755–761. [CrossRef]
181. Miao, C.-H.; Wang, X.-F.; Qiao, B.; Xu, Q.-M.; Cao, C.-Y.; Cheng, J.-S. Artificial Consortia of *Bacillus amyloliquefaciens* HM618 and *Bacillus subtilis* for Utilizing Food Waste to Synthetize Iturin A. *Environ. Sci. Pollut. Res.* **2022**, *29*, 72628–72638. [CrossRef] [PubMed]
182. El-Bendary, M.A. Production of *Mosquitocidal bacillus sphaericus* by Solid State Fermentation Using Agricultural Wastes. *World J. Microbiol. Biotechnol.* **2010**, *26*, 153–159. [CrossRef]
183. El-Bendary, M.A.; Moharam, M.E. Formulation of Spore Toxin Complex of *Bacillus thuringiensis* and *Lysinibacillus sphaericus* Grown under Solid State Fermentation. *Biol. Control* **2019**, *131*, 54–61. [CrossRef]
184. Mizumoto, S.; Shoda, M. Medium Optimization of *Antifungal lipopeptide*, Iturin A, Production by *Bacillus subtilis* in Solid-State Fermentation by Response Surface Methodology. *Appl. Microbiol. Biotechnol.* **2007**, *76*, 101–108. [CrossRef]
185. Mizumoto, S.; Hirai, M.; Shoda, M. Production of *Lipopeptide antibiotic* Iturin A Using Soybean Curd Residue Cultivated with *Bacillus subtilis* in Solid-State Fermentation. *Appl. Microbiol. Biotechnol.* **2006**, *72*, 869–875. [CrossRef]
186. Khan, A.W.; Zohora, U.S.; Rahman, M.S.; Okanami, M.; Ano, T. Production of Iturin A through Glass Column Reactor (GCR) from Soybean Curd Residue (Okara) by *Bacillus subtilis* RB14-CS under Solid State Fermentation (SSF). *Adv. Biosci. Biotechnol.* **2012**, *3*, 143–148. [CrossRef]
187. Bouassida, M.; Mnif, I.; Hammami, I.; Triki, M.-A.; Ghribi, D. *Bacillus subtilis* SPB1 Lipopeptide Biosurfactant: Antibacterial Efficiency against the Phytopathogenic Bacteria *Agrobacterium tumefaciens* and Compared Production in Submerged and Solid State Fermentation Systems. *Food Sci. Biotechnol.* **2023**. [CrossRef]
188. Mejias, L.; Estrada, M.; Barrena, R.; Gea, T. A Novel Two-Stage Aeration Strategy for *Bacillus thuringiensis* Biopesticide Production from Biowaste Digestate through Solid-State Fermentation. *Biochem. Eng. J.* **2020**, *161*, 107644. [CrossRef]
189. Rodríguez, P.; Cerda, A.; Font, X.; Sánchez, A.; Artola, A. Valorisation of Biowaste Digestate through Solid State Fermentation to Produce Biopesticides from *Bacillus thuringiensis*. *Waste Manag.* **2019**, *93*, 63–71. [CrossRef] [PubMed]
190. Lima-Pérez, J.; López-Pérez, M.; Viniegra-González, G.; Loera, O. Solid-State Fermentation of *Bacillus thuringiensis* Var Kurstaki HD-73 Maintains Higher Biomass and Spore Yields as Compared to Submerged Fermentation Using the Same Media. *Bioprocess. Biosyst. Eng.* **2019**, *42*, 1527–1535. [CrossRef]
191. Molina-Peñate, E.; Sánchez, A.; Artola, A. Enzymatic Hydrolysis of the Organic Fraction of Municipal Solid Waste: Optimization and Valorization of the Solid Fraction for *Bacillus thuringiensis* Biopesticide Production through Solid-State Fermentation. *Waste Manag.* **2022**, *137*, 304–311. [CrossRef]

192. Molina-Peñate, E.; del Carmen Vargas-García, M.; Artola, A.; Sánchez, A. Filling in the Gaps in Biowaste Biorefineries: The Use of the Solid Residue after *Enzymatic hydrolysis* for the Production of Biopesticides through Solid-State Fermentation. *Waste Manag.* **2023**, *161*, 92–103. [CrossRef] [PubMed]
193. Ma, J.; Jiang, H.; Li, P.; Li, C.; Liu, R.; Li, J.; Xiao, Z.; Pi, B.; Zhao, M.; Hu, W.; et al. Production of Free Amino Acid Fertilizer from Tung Meal by the Newly Isolated *Pseudomonas aeruginosa* LYT-4 Strain with Simultaneous Potential Biocontrol Capacity. *Renew. Energy* **2020**, *166*, 245–252. [CrossRef]
194. Zeng, X.; Miao, W.; Zeng, H.; Zhao, K.; Zhou, Y.; Zhang, J.; Zhao, Q.; Tursun, D.; Xu, D.; Li, F. Production of Natamycin by *Streptomyces gilvosporeus* Z28 through Solid-State Fermentation Using Agro-Industrial Residues. *Bioresour. Technol.* **2019**, *273*, 377–385. [CrossRef]
195. Liu, H.; Zhang, D.; Zhang, X.; Zhou, C.; Zhou, P.; Zhi, Y. Medium Optimization for Spore Production of a Straw-Cellulose Degrading Actinomyces Strain under Solid-State Fermentation Using Response Surface Method. *Sustainability* **2020**, *12*, 8893. [CrossRef]
196. Shen, T.; Wang, C.; Yang, H.; Deng, Z.; Wang, S.; Shen, B.; Shen, Q. Identification, Solid-State Fermentation and Biocontrol Effects of *Streptomyces hygroscopicus* B04 on Strawberry Root Rot. *Appl. Soil. Ecology* **2016**, *103*, 36–43. [CrossRef]
197. Sakdapetsiri, C.; Fukuta, Y.; Aramsirirujiwet, Y.; Shirasaka, N.; Tokuyama, S.; Kitpreechavanich, V. Solid State Fermentation, Storage and Viability of *Streptomyces similanensis* 9X166 Using Agro-Industrial Substrates against Phytophthora Palmivora -Induced Black Rot Disease in Orchids. *Biocontrol Sci. Technol.* **2019**, *29*, 276–292. [CrossRef]
198. Mangan, R.; Bussière, L.F.; Polanczyk, R.A.; Tinsley, M.C. Increasing Ecological Heterogeneity Can Constrain Biopesticide Resistance Evolution. *Trends Ecol. Evol.* **2023**. [CrossRef]
199. de Souza Machado, A.A.; Zarfl, C.; Rehse, S.; Kloas, W. Low-Dose Effects: Nonmonotonic Responses for the Toxicity of a *Bacillus thuringiensis* Biocide to *Daphnia magna*. *Environ. Sci. Technol.* **2017**, *51*, 1679–1686. [CrossRef]
200. De Bock, T.; Zhao, X.; Jacxsens, L.; Devlieghere, F.; Rajkovic, A.; Spanoghe, P.; Höfte, M.; Uyttendaele, M. Evaluation of B. Thuringiensis-Based Biopesticides in the Primary Production of Fresh Produce as a Food Safety Hazard and Risk. *Food Control* **2021**, *130*, 108390. [CrossRef]
201. Rousidou, C.; Papadopoulou, E.S.; Kortsinidou, M.; Giannakou, I.O.; Singh, B.K.; Menkissoglu-Spiroudi, U.; Karpouzas, D.G. Bio-Pesticides: Harmful or Harmless to Ammonia Oxidizing Microorganisms? The Case of a *Paecilomyces lilacinus*-Based Nematicide. *Soil. Biol. Biochem.* **2013**, *67*, 98–105. [CrossRef]
202. Kaze, M.; Brooks, L.; Sistrom, M. Antimicrobial Resistance in Bacillus-Based Biopesticide Products. *Microbiology* **2021**, *167*. [CrossRef]
203. Coulibaly, O.; Cherry, A.J.; Nouhoheflin, T.; Aitchedji, C.C.; Al-Hassan, R. Vegetable Producer Perceptions and Willingness to Pay for Biopesticides. *J. Veg. Sci.* **2007**, *12*, 27–42. [CrossRef]
204. Kumar, K.K.; Sridhar, J.; Murali-Baskaran, R.K.; Senthil-Nathan, S.; Kaushal, P.; Dara, S.K.; Arthurs, S. Microbial Biopesticides for Insect Pest Management in India: Current Status and Future Prospects. *J. Invertebr. Pathol.* **2019**, *165*, 74–81. [CrossRef]
205. Fenibo, E.O.; Ijoma, G.N.; Matambo, T. Biopesticides in Sustainable Agriculture: Current Status and Future Prospects. In *New and Future Development in Biopesticide Research: Biotechnological Exploration*; Springer Nature: Singapore, 2022; pp. 1–53.
206. Samada, L.H.; Tambunan, U.S.F. Biopesticides as Promising Alternatives to Chemical Pesticides: A Review of Their Current and Future Status. *Online J. Biol. Sci.* **2020**, *20*, 66–76. [CrossRef]
207. Fenibo, E.O.; Ijoma, G.N.; Matambo, T. Biopesticides in Sustainable Agriculture: A Critical Sustainable Development Driver Governed by Green Chemistry Principles. *Front. Sustain. Food Syst.* **2021**, *5*, 619058. [CrossRef]
208. Fravel, D.R. Commercialization and Implementation of Biocontrol. *Annu. Rev. Phytopathol.* **2005**, *43*, 337–359. [CrossRef] [PubMed]
209. Verma, D.K.; Guzmán, K.N.R.; Mohapatra, B.; Talukdar, D.; Chávez-González, M.L.; Kumar, V.; Srivastava, S.; Singh, V.; Yulianto, R.; Malar, S.E.; et al. Recent Trends in Plant- and Microbe-Based Biopesticide for Sustainable Crop Production and Environmental Security. In *Recent. Developments in Microbial Technologies. Environmental and Microbial Biotechnology*; Prasad, R., Kumar, V., Singh, J., Upadhyaya, C.P., Eds.; Springer: Singapore, 2021; pp. 1–37.
210. Chandler, D.; Bailey, A.S.; Tatchell, G.M.; Davidson, G.; Greaves, J.; Grant, W.P. The Development, Regulation and Use of Biopesticides for Integrated Pest Management. *Philos. Trans. R. Soc. B Biol. Sci.* **2011**, *366*, 1987–1998. [CrossRef]
211. Glare, T.; Caradus, J.; Gelernter, W.; Jackson, T.; Keyhani, N.; Köhl, J.; Marrone, P.; Morin, L.; Stewart, A. Have Biopesticides Come of Age? *Trends Biotechnol.* **2012**, *30*, 250–258. [CrossRef] [PubMed]
212. Bharti, V.; Ibrahim, S. Biopesticides: Production, Formulation and Application Systems. *Int. J. Curr. Microbiol. Appl. Sci.* **2020**, *9*, 3931–3946. [CrossRef]
213. Lovett, B.; St. Leger, R.J. Genetically Engineering Better Fungal Biopesticides. *Pest. Manag. Sci.* **2018**, *74*, 781–789. [CrossRef] [PubMed]
214. Kumar, V.; Ahluwalia, V.; Saran, S.; Kumar, J.; Patel, A.K.; Singhania, R.R. Recent Developments on Solid-State Fermentation for Production of Microbial Secondary Metabolites: Challenges and Solutions. *Bioresour. Technol.* **2021**, *323*, 124566. [CrossRef]
215. Chilakamarry, C.R.; Mimi Sakinah, A.M.; Zularisam, A.W.; Sirohi, R.; Khilji, I.A.; Ahmad, N.; Pandey, A. Advances in Solid-State Fermentation for Bioconversion of Agricultural Wastes to Value-Added Products: Opportunities and Challenges. *Bioresour. Technol.* **2022**, *343*, 126065. [CrossRef]

216. Oiza, N.; Moral-Vico, J.; Sánchez, A.; Oviedo, E.R.; Gea, T. Solid-State Fermentation from Organic Wastes: A New Generation of Bioproducts. *Processes* **2022**, *10*, 2675. [CrossRef]
217. Figueroa-Montero, A.; Esparza-Isunza, T.; Saucedo-Castañeda, G.; Huerta-Ochoa, S.; Gutiérrez-Rojas, M.; Favela-Torres, E. Improvement of Heat Removal in Solid-State Fermentation Tray Bioreactors by Forced Air Convection. *J. Chem. Technol. Biotechnol.* **2011**, *86*, 1321–1331. [CrossRef]
218. Wang, L.; Liu, Y.; Chen, H.-Z. Advances in Porous Characteristics of the Solid Matrix in Solid-State Fermentation. In *Current Developments in Biotechnology and Bioengineering*; Elsevier: Amsterdam, The Netherlands, 2018; pp. 19–29.
219. Martins, S.; Mussatto, S.I.; Martínez-Avila, G.; Montañez-Saenz, J.; Aguilar, C.N.; Teixeira, J.A. Bioactive Phenolic Compounds: Production and Extraction by Solid-State Fermentation. A Review. *Biotechnol. Adv.* **2011**, *29*, 365–373. [CrossRef] [PubMed]
220. Sala, A.; Barrena, R.; Artola, A.; Sánchez, A. Current Developments in the Production of Fungal Biological Control Agents by Solid-State Fermentation Using Organic Solid Waste. *Crit. Rev. Environ. Sci. Technol.* **2019**, *49*, 655–694. [CrossRef]
221. Hölker, U.; Höfer, M.; Lenz, J. Biotechnological Advantages of Laboratory-Scale Solid-State Fermentation with Fungi. *Appl. Microbiol. Biotechnol.* **2004**, *64*, 175–186. [CrossRef] [PubMed]
222. De la Cruz Quiroz, R.; Roussos, S.; Hernández, D.; Rodríguez, R.; Castillo, F.; Aguilar, C.N. Challenges and Opportunities of the Bio-Pesticides Production by Solid-State Fermentation: Filamentous Fungi as a Model. *Crit. Rev. Biotechnol.* **2015**, *35*, 326–333. [CrossRef] [PubMed]
223. Atanasova, N.S.; Pietilä, M.K.; Oksanen, H.M. Diverse Antimicrobial Interactions of Halophilic Archaea and Bacteria Extend over Geographical Distances and Cross the Domain Barrier. *Microbiol. Open* **2013**, *2*, 811–825. [CrossRef]
224. Martin del Campo, M.; Camacho, R.M.; Mateos-Díaz, J.C.; Müller-Santos, M.; Córdova, J.; Rodríguez, J.A. Solid-State Fermentation as a Potential Technique for Esterase/Lipase Production by Halophilic Archaea. *Extremophiles* **2015**, *19*, 1121–1132. [CrossRef]
225. Hensley, S.A.; Moreira, E.; Holden, J.F. Hydrogen Production and Enzyme Activities in the Hyperthermophile Thermococcus Paralvinellae Grown on Maltose, Tryptone, and Agricultural Waste. *Front. Microbiol.* **2016**, *7*, 174. [CrossRef]
226. Kumar, V.; Singh, B.; van Belkum, M.J.; Diep, D.B.; Chikindas, M.L.; Ermakov, A.M.; Tiwari, S.K. Halocins, Natural Antimicrobials of Archaea: Exotic or Special or Both? *Biotechnol. Adv.* **2021**, *53*, 107834. [CrossRef]
227. O'Connor, E.; Shand, R. Halocins and Sulfolobicins: The Emerging Story of Archaeal Protein and Peptide Antibiotics. *J. Ind. Microbiol. Biotechnol.* **2002**, *28*, 23–31. [CrossRef]
228. Ellen, A.F.; Rohulya, O.V.; Fusetti, F.; Wagner, M.; Albers, S.-V.; Driessen, A.J.M. The Sulfolobicin Genes of *Sulfolobus scidocaldarius* Encode Novel Antimicrobial Proteins. *J. Bacteriol.* **2011**, *193*, 4380–4387. [CrossRef]

microorganisms

MDPI

Article

Effects of Below-Ground Microbial Biostimulant *Trichoderma harzianum* on Diseases, Insect Community, and Plant Performance in *Cucurbita pepo* L. under Open Field Conditions

Pierluigi Forlano [1,†], Stefania Mirela Mang [1,*,†], Vittoria Caccavo [1], Paolo Fanti [2], Ippolito Camele [1], Donatella Battaglia [1] and Vincenzo Trotta [1,*]

[1] School of Agricultural, Forestry, Food and Environmental Sciences (SAFE), University of Basilicata, Viale dell'Ateneo Lucano 10, 85100 Potenza, Italy
[2] Department of Science, University of Basilicata, Viale dell'Ateneo Lucano 10, 85100 Potenza, Italy
* Correspondence: stefania.mang@unibas.it (S.M.M.); vincenzo.trotta@unibas.it (V.T.)
† These authors contributed equally to this work.

Abstract: Agrochemicals are generally used in agriculture to maximize yields and product quality, but their overuse can cause environmental pollution and human health problems. To reduce the off-farm input of chemicals, numerous biostimulant products based on beneficial symbiont plant fungi are receiving a great deal of attention. The evolution of plant diseases and the performance of insects are influenced by plant chemical defences, both of which are, in turn, influenced by below-ground symbionts. Direct and indirect plant defences mediated by belowground symbionts against plant diseases and insect herbivores were demonstrated in greenhouses experiments. However, little attention has been paid to the use of *Trichoderma* under open field conditions, and no data are available for zucchini (*Cucurbita pepo* L.) plants in the field. To determine the effects of a commercial *Trichoderma harzianum* strain T22 on plant viruses, powdery mildew, the arthropod community, and on the agronomic performance associated with zucchini plants, an experiment was conducted in 2022 under open field conditions in South Italy. Our results indicate that *T. harzianum* T22 makes zucchini plants more attractive to aphids and to Hymenoptera parasitoid but failed to control zucchini pathogens. The complex plant–disease–arthropod–microorganism interactions that occurred in the field during the entire plant cycle are discussed to enrich our current information on the possibilities of using these microorganisms as a green alternative in agriculture.

Keywords: zucchini squash; plant diseases; viruses; powdery mildew; aphids; parasitoids; integrated pest management

Citation: Forlano, P.; Mang, S.M.; Caccavo, V.; Fanti, P.; Camele, I.; Battaglia, D.; Trotta, V. Effects of Below-Ground Microbial Biostimulant *Trichoderma harzianum* on Diseases, Insect Community, and Plant Performance in *Cucurbita pepo* L. under Open Field Conditions. *Microorganisms* **2022**, *10*, 2242. https://doi.org/10.3390/microorganisms10112242

Academic Editors: Vishnu D. Rajput, Anuj Ranjan and Abhishek Chauhan

Received: 14 October 2022
Accepted: 10 November 2022
Published: 12 November 2022

1. Introduction

Agrochemicals are at the base of intensive agricultural systems, and the increasing demand of food for humans has enhanced their use worldwide [1]. Agrochemicals include pesticides and other product categories that promote plant growth and preserve plant health are used to maximize crop yield. However, the use of agrochemicals, particularly synthetic agrochemicals and inorganic fertilizers, causes toxicity to humans and ecosystems [2,3]. Another problem related to the use of synthetic agrochemicals is the increasing emergence of resistant strains of pests and pathogens [4–6].

Beneficial soil microbes, enhancing crop yield and promoting plant defences, such as non pathogenic bacteria [7], mycorrhizal fungi [8] and plant growth promoting fungi [9], are a possible alternative to the use of agrochemicals. The application of beneficial microbial inoculants in agriculture has increased over the past two decades [10]. Fungi of the genus *Trichoderma* are among the most effective plant growth promoters in cultivated plant species [11]. The induction of plant resistance against pests and pathogens by fungi of the genus *Trichoderma* has been much studied in the tomato [12–19]. Some *Trichoderma* strains were found to activate the plant systemic acquired resistance (SAR) and/or induce systemic resistance (ISR) against biotic and abiotic stress agents [11,20–23]. For example, tomato defence responses against the green stink bug *Nezara viridula* L. were enhanced by *T. harzianum* strain T22 through an early increase in transcript levels of jasmonic acid (JA) marker genes [14].

Zucchini (*Cucurbita pepo* L.) is the most important economically and globally widespread species among the cultivated Cucurbitaceae [24]. Zucchini includes a wide assortment of varieties and cultivars [25] and is one of the most important and consumed vegetables worldwide. However, there are few research studies investigating the interaction between *Trichoderma* spp. and zucchini pathogens [26,27].

In the field, aphids, mainly *Aphis gossypii* Glover (Homoptera: Aphididae), and some pathogens, such as phytoviruses and the powdery mildew fungal agents, namely, *Golovinomyces cichoracearum* (DC.) V.P. Heluta, *Podosphaera fusca* (Fr.) U. Braun & Shishkoff and *Leveillula taurica* (Lév.) G. Arnaud, are the most harmful organisms that cause plant damage and production losses in zucchini crops [28–30]. *Aphis gossypii* is considered the major pest of cucurbits. It is a polyphagous and destructive pest of more than twenty crop plant species. In hot regions, during the prolonged dry seasons, it produces large colonies on Cucurbitaceae, and it may survive on an ample variety of plant species, including cultivated and spontaneous Graminaceae [31]. In colder temperate regions, it is restricted to glasshouses, where it is a major pest. For some plant species, its direct feeding can cause serious damage to plant tissues, such as curled leaves and stunted shoots [32]. In zucchini plants, *A. gossypii* infestation causes a transcriptional up-regulation of genes underlying the biosynthesis of salicylic acid (SA) and of genes that modulate the SA-mediated defence response. As a consequence, aphids actively disperse on the plant, rather than starting their feeding activity where they were originally deposited, as observed in controls [33].

Although *A. gossypii* can cause direct damage to zucchini, the main damage is related to phytovirus transmission [34]. *A. gossypii* can transmit more than 50 phytoviruses, including non-persistent viruses of cucurbits, such as the Cucumber mosaic virus (CMV), the Zucchini yellow mosaic virus (ZYMV), the Papaya ringspot virus (PRSV), and the Watermelon mosaic virus (WMV) [31]. These viruses can infect zucchini plants and, as in the case of ZYMV, can cause 40 to 50% of yield losses [35]. Transmission of these viruses occurs during intracellular stylet punctures of aphids in epidermal or mesophyll cells, concomitant with saliva ejection [32].

In recent decades, the most used strategies to control aphid infestations and pathogens in zucchinis have been primarily focused on the selection of resistant genotypes [28,36,37] and on the use of pesticides [38,39]. Although pesticides may effectively reduce aphid populations in field, their use may improve the dispersion of viruses transmitted by aphids. This is due to the dispersive effect of some pesticides on aphids that survive the pesticide treatment [40,41]. Moreover, in literature, the development of resistance to insecticides in *A. gossypii* in several world regions is well reported [38,39,42–44].

The below-ground interactions between plants and microorganisms are very complex and it remains to be understood whether microbial biostimulants such as *Trichoderma* can be used to control harmful organisms. The aim of this study was to investigate the possibility of controlling zucchini pests and the most relevant virus diseases and powdery mildew in the field using the *Trichoderma harzianum* strain T22. The effects of inoculation of the commercial *T. harzianum* strain T22 on the arthropod community, on the above-mentioned plant diseases, and on the agronomic performance of zucchini squash was studied in detail for the first time in an experimental field.

2. Materials and Methods

2.1. Crop Cultivation

The present study was performed in an experimental field located in Pignola (40°34′06.2″ N, 15°45′35.4″ E; 780 m above sea level), Potenza, Italy, during the period June to September 2022.

The soil was left fallow the year before the experiment and then ploughed to a depth of 25 cm, rotavated, and levelled before planting the crop. The soil characteristics are listed below: particles smaller than 2 mm in size, 935 g/kg; particles larger than 2 mm, 65 g/kg; apparent density, 1.294 kg/dm^3; texture composition of sand, 481 g/kg; clay, 149 g/kg; silt, 370 g/kg at a depth of 0–30 cm. The content of total carbonate and total organic matter was of 16 g/kg and 32.8 g/kg, respectively. The composition of the soil was as follows: total N, 2 g/kg; P, 29 mg/kg; Ca 11.1 meq/100 g; Mg, 4.6 meq/100 g; Na, 1.8 meq/100 g; soil pH (H_2O), 6.2. According to the world reference base for soil resources, the soil was a dystric cambisol (Bd68-2bc).

Zucchini seedlings (*Cucurbita pepo* L.) of the San Pasquale cultivar (Pagano Domenico & Figli, Scafati, Salerno, Italy) were used in this experiment. Zucchini plants placed in alveolate containers were purchased from a nursery and transplanted to the field on 6 June 2022. No fertilizers were used during the present experiment and the zucchini plants were not treated with any type of agrochemical during the entire field trial. Water irrigation was applied through the drip irrigation system.

2.2. Meteorological Data

The temperature and rainfall data recorded during the experiment are shown in Figure S1. During the period of interest, the average temperature was about 20 °C. The temperatures reached a maximum of 34 °C in August. The precipitations recorded in June, July, and in the first 15 days of August were very low. The Agrometeorological Service of the "Agenzia Lucana per lo Sviluppo e l'Innovazione in Agricoltura (ALSIA)" of the Basilicata Region provided the meteorological data for the area in which the experimental farm is located.

2.3. Experimental Design

The effects of *T. harzianum* T22 on diseases, insect community, and plant performance in zucchini plants were investigated. The first treatment consisted of non-inoculated zucchini plants (control), while the microbial biostimulant used in the present experiment to inoculate the plants was *Trichoderma harzianum* Rifai strain KRL-AG2 (T-22) (KOPPERT B.V., Berkel en Rodenrijs, the Netherlands), a purified strain which disperses in water.

The experiment was carried out on a strip of soil of about 40 m long and 14 m wide, divided into 6 plots of 22.5 m^2 (9 m × 2.5 m) separated from each other by a strip 3 m wide left without plants. Thus, 3 plots treated with *T. harzianum* T22 and 3 control plots were obtained, alternating along the length of the field. The plants in each plot were manually transplanted on 6 June 2022 in 2 rows 1.8 m apart from each other. Each row was 8 m long with a plant spacing of 66 cm, with a total of 12 plants/row (24 plants/plot).

2.4. Fungal Inoculation

Before the experiments, the viability of the commercial formulation of *T. harzianum* T22 was evaluated in the laboratory by serial dilution. The dilutions were placed on Petri plates (9 cm in diameter) containing Potato Dextrose Agar (PDA) medium (Oxoid Ltd., Hants, UK) amended with the antibiotic streptomycin sulphate 40 mg/L (MerckKGaA, Darmstadt, Germany) until growth could be detected. As suggested by a previous study [45], the number of colony forming units (CFU) was counted after 24 h of incubation at 25 °C in the dark.

Once the viability of the commercial product has been confirmed, zucchini seedlings were inoculated with *T. harzianum* T22 following the manufacturer's instructions 5 days before transplantation. The alveolate containers with 24 seedlings were then watered with 3 g of commercial *T. harzianum* T22 (containing 1 × 10^9 CFU/g of *T. harzianum* T22) dissolved in 3 litres of water. Each plant was watered with about 41 mL of the fungal suspension. The treatment was repeated after 4 days and then the seedlings were transplanted. A total of 72 inoculated zucchini plants and 72 non inoculated control plants were transplanted into the field. No fungal inoculation was performed after the transplantation.

In addition, 9 control and 9 treated plants were transplanted into pots and placed in a greenhouse to determine the presence of *T. harzianum* T22 in the roots. Twenty-four days after the last treatment, root samples were accurately collected and gently washed to remove soil residues. The colonization of *T. harzianum* T22 was confirmed by microscopy observations on squashed fine root hairs. In this study, the coloration of the zucchini hairy roots was performed following a rapid protocol set up using Tryptan blue 0.4% (Sigma Aldrich, Darmstadt, Germany) and commercial Pelikan blue (Hannover, Germany) dyes, glycerol 50%, and heating. To ensure the presence of *T. harzianum* T22 throughout the experiment, fungal colonization was also confirmed by strain isolations on PDA medium. Zucchini roots were sampled in the field 25 days (1 July), 54 days (30 July), and 85 days (30 August) after transplantation. For each sampling date, roots were collected from three control and three inoculated plants and transported to the laboratory. The roots were washed under running tap water to remove soil, then superficially sterilised using a 70% hydroalcoholic solution followed by a sodium hypochlorite solution at 1%. The roots were finally washed with sterile distilled water and dried on sterile paper. Each sample (a piece of root of about 2 cm^2 in size) was placed on Petri plates with PDA medium amended with the streptomycin sulfate (0.05%). Plates were incubated at 25 $\pm$ 1 °C for 7 days and the presence of the fungus was then determined.

2.5. Arthropod Sampling

During the first month after transplantation, the plants were very small, with few leaves and very few specimens of insects were found. From the second month after transplantation, the arthropod community on zucchini plants increased and was studied. An arthropod community survey was carried out by adopting two different sampling techniques: sampling of zucchini leaves, which mainly provided information on arthropod community colonising the plant, and capturing insects with colored pan trap sets, which mainly provided information on the winged arthropod community visiting the zucchini plant.

2.5.1. Arthropod Sampling on Zucchini Leaves

To investigate the arthropod community on zucchini plants, within each plot, six plants were randomly sampled at 9:00 a.m., for a total of 36 plants/date. Sampled leaves, fully unfolded and of about 20 centimetres, were taken from the middle part of the zucchini plant. Each leaf was gently inserted in a transparent zip lock plastic bag (40 cm $\times$ 30 cm) and then cut at the insertion with it stem. This procedure allows an accurate sampling of small arthropods, also collecting the ones that drop and/or jump when the plant is touched, thus obtaining quantitative data on their abundance. The plastic bags were kept in darkness at 5 °C and transported to the laboratory for the identification of the arthropods. The collected arthropods were transferred to 50 mL Falcon tubes filled with ~30 mL of 70% hydroalcoholic solution and refrigerated at 4 °C until identification. The samples were then observed under a stereomicroscope. Arthropods from each sample were counted and classified at order, family, and, when possible, at the species level. Furthermore, the presence of damage caused by leaf miners on the leaves were noted and analysed. Five different leaf samples were carried out on 14 and 26 July, 9 and 26 August, and 8 September (that is, 38, 50, 64, 81, and 94 days after transplantation).

2.5.2. Arthropod Sampling with Coloured Pan Traps

Arthropods were also sampled using pan trap sets consisting of one blue, one yellow, and one white bowl. Pan trap is a passive sampling method that provides an ample return of data for relatively short periods of time and is particularly appropriate for faunal surveys [46], without a collector effect [47]. The traps were made by painting plastic bowls (17 cm in diameter, 4.5 cm deep), with blue (RAL standard colour codes: 5015) or yellow (RAL standard colour codes: 1023) acrylic paint sprays or left white. The pan traps were placed on the ground, in the middle of each experimental plot, as close as possible to the plants of each experimental plot. Each trap was filled with 400 mL of water and 4 mL of dishwashing detergent with no fragrance added to break surface tension. Traps were set out early at 8:00 a.m. and collected three days later, at the same time. In case of rain, pan traps were removed and the sampled specimens were not considered; the traps were replaced 24 h after rain stopped. Traps were collected in the order they were placed to ensure that all traps were available to insects for a similar time. Pan trap survey was carried out in four different data, with an interval of time of about two weeks among them: on 14 July, 1 and 16 August, and 8 September (that is, 38, 56, 71, and 94 days after transplantation). The arthropods were removed from the soap–water solution using a fine mesh colander and gently transferred with a soft paintbrush in 50 mL Falcon tubes, filled with 70% ethanol. Falcon tubes were stocked at 4 °C until the identification of the arthropods. The samples were then observed under a stereomicroscope. Subsequently, the arthropods of each sample were counted and classified according to their order, family, and, when possible, at the species level.

2.6. *Evaluation of Diseases in Zucchini Plants*

The present study focused on the presence of zucchini viruses and powdery mildew since these are the most important zucchini diseases in the considered area. All plants were visually inspected for the presence/absence of diseases on 6 and 24 July, 4, 8, 19, and 26 August, and 7 September. The presence of virus (chlorosis, severe mosaic, deformation, blistering, and reduced leaf size) and/or powdery mildew (white powdery growth and subsequently spots or patches preferably on the leaf or on plant stems) symptoms was recorded.

The zucchini viruses in plants were identified by enzyme-linked immunosorbent assay (ELISA), while the identification of the causal agent of the powdery mildew was done by microscopic observations.

2.6.1. Evaluation of Zucchini Viruses in the Field

The presence and development of four of the most common zucchini plant viruses, Zucchini yellow mosaic virus (ZYMV), Papaya ringspot virus (PRSV), Cucumber mosaic virus (CMV) and Watermelon zucchini virus (WMV), were evaluated throughout the entire cultivation period. The symptoms of the viruses were visually observed in the field on all parts of the plant. To assess the degree of viral attack, each plant was examined individually and the degree of attack per plot was estimated using the following formula:

$$Degree\ of\ attack\ (DA\%) = \frac{Number\ of\ symptomatic\ plants\ /\ plot}{Total\ number\ of\ plants/\ plot} \times 100$$

The *DA*% for viruses was determined for all six plots during the experimental trial.

2.6.2. ELISA Assay

The presence of the four viruses investigated (CMV, ZYMV, PRSV, and WMV) was assessed twenty days before the last harvest. For each virus, the ELISA tests were performed using specific antibodies and kits (Loewe® Biochemica GmbH, Sauerlach, Germany) following the manufacturer's instructions. Briefly, a DAS ELISA [48] was performed using a polyclonal antiserum rabbit for each virus. Leaves and fruits from the experimental plot were placed in plastic bags, transported to the laboratory, and stored at 4 °C. For the ELISA test, the sap was extracted by homogenizing 1 g of sample in 10 mL of Conjugate/Sample buffer (ELISA kit) in plastic BIOREBA extraction bags (BIOREBA AG, Reinach, Switzerland) using a commercial homogenizer. Sap samples were collected in Eppendorf tubes and stored at −20 °C. The DAS ELISA assays consisted in coating the Nunc™ MicroWell 96-Well Microplates (ThermoFisher Scientific Inc., Waltham, MA, USA) plates (200 µL/well) with antigen-specific antibodies (IgG) 1:200 diluted in coating buffer, incubation of the plates at 37 °C for 4 h, followed by four manual washings with washing buffer at room temperature (RT), followed by samples application and overnight incubation at 4 °C. Subsequently, antibody–AP–conjugate application (200 µL/well) 1:200 diluted in conjugated buffer, incubation of the plates at 37 °C for 4 h, four washings, and enzymatic assays using substrate buffer added with 1 mg/mL of PNPP tablets were performed. The results were evaluated by comparing the visual reaction, determined as a yellow colour development, in the plate between the control (positive/negative) and samples. After 1 and 2 h of substrate incubation, plates were read photometrically at 405 nm wavelength using an ELISA Reader model A3 (DAS, Rome, Italy). All samples were run in duplicate.

2.6.3. Powdery Mildew Evaluation Assay

The percentage of powdery mildew disease attack was assessed by field observations. All leaves and fruits from control and *T. harzianum* T22 treated zucchini plants were individually observed. To assess the percentage of powdery mildew attack in the field, the following scale was used: 0 (not infected) = 0% attack; 1 (low) = 1–25% infected tissue; 2 (medium) = 25.1–50% infected tissue; 3 (high) = 50.1–75% infected tissue; 4 (very high) = >75.1% infected tissue. Furthermore, to identify the possible pathogen causal agent responsible for the observed symptoms on zucchini plants, 25 symptomatic leaves and fruits were randomly collected from plants in the field and, on the same day, used to identify the causal agent of the powdery mildew in the laboratory. For species identification, conidia were directly obtained from the infected zucchini leaves and fruits collected in the field. The conidia were then observed under a light microscope (Axioscope, Zeiss, Germany) and also other morphological characteristics reported in literature for zucchini powdery mildew causal fungus were considered [49,50].

2.7. Evaluation of Plant Growth and Productivity

Plant growth was estimated by measuring the stem length of zucchini plants, excluding the leaf. This survey was carried out on four different data: on 22 June, 6 and 22 July, and 4 August. The stem length was measured in four plants per plot.

The zucchini fruits were first harvested on 9 July (33 days from transplanting) and were successively collected every two days until 22 August. From 22 August to 30 August, zucchini fruits were collected every four days. For each harvest, marketable fruits were counted and weighed for each experimental plot. The mean values of the weight of the zucchini fruits and the cumulative number and weight of the zucchini fruits per plant harvested from the plots inoculated with *T. harzianum* T22 and from the controls from 9 July to 30 August were then calculated.

2.8. Statistical Analysis

The number of arthropods sampled over time on the leaves, in the pan traps, and the data relating to the disease symptoms of the powdery mildew were analysed with a Poisson generalized linear mixed models (GLMMs) with a log-link function fitted with ML (maximum likelihood) and Laplace approximation. The discrete Poisson distribution best approximates the process that generated the observed data. The *p*-values for the differences between the treatments, sampling dates, and their interactions were obtained through analyses of deviance (Type II Wald chi-square tests). The following general model was applied:

$$Y = \mu + Treatment + Date + Treatment \times Date + Plot\ \{Treatment\ \{Date\}\} + \varepsilon$$

where *Y* is the studied variable with a Poisson distribution, *Treatment* and *Date* are the fixed factors, and *Plot* is the random effect consisting of the three experimental plots nested in *Treatment* and *Date*. This model accounts for the non-independence of the data (pseudoreplication of measures) due to the different experimental plots (the random effect) that are part of the present design.

Data on plant length, fruit weight, cumulative number, and cumulative weight of zucchini fruits per plant were analysed using linear mixed-effects models (LMMs) fitted with REML (restricted maximum likelihood). The homoscedasticity and normality assumptions for these ANOVAs were checked and met on these data. The *p*-values for the differences between the treatments, sampling dates, and their interactions were obtained through ANOVAs (type II Wald chi-square tests). To better appreciate the (possible) differences in fruit wight over time, in this analysis the sampling dates were grouped into 4 periods: 9–19 July, 20–31 July, 1–12 August and 13–25 August. The general model applied for these analyses was the same as applied for the analysis of the insect community.

The percentage of virus infected plants per plot was analysed using a linear model (LM) after an arcsine transformation of the data. The following model was applied:

$$Y = \mu + Treatment + Date + Treatment \times Date + \varepsilon$$

where Y is the percentage (transformed) of virus-infected plants, and *Treatment* and *Date* (7 levels) are fixed effects. The *p*-values for were obtained by a factorial model ANOVA (Type II sum-square tests).

To test for the influence of virosis on powdery mildew symptoms, the following model was applied:

$$Y = \mu + Virus\ class + Treatment + Virus\ classes \times Treatment + \varepsilon$$

where Y is the quantification of the disease symptoms of the powdery mildew on a plant, *Virus class* (four levels of degree of virus attack, 0: 0 *DA*%, 1: 1–25 *DA*%, 2: 26–50 *DA*%, 3: 51–75 *DA*%, and 4: 76–100 *DA*%) and *Treatment* are fixed effects. The *p*-values were obtained by a factorial model ANOVA (Type II sum-square tests).

For all the analyses described so far, the model distributions were also chosen as the best fitting, based on AIC criteria [51] and the full models were presented. All statistical analyses were performed in R version 4.1.2 "Bird Hippie" [52], with lme4 [53], lmerTest [54] packages.

3. Results

3.1. *Trichoderma harzianum* T22 Inoculation

The zucchini plants transplanted into pots were inspected 24 days after inoculation to verify the success of the colonization of *T. harzianum* T22 in the roots. The 9 control and 9 inoculated plants were gently removed from the pots and photographed (Figure 1).

Figure 1. Zucchini plants 24 days after the inoculation. (**A**) *Trichoderma harzianum* T22 inoculated plant; (**B**) not inoculated control plant.

Compared to controls, the 9 zucchini plants inoculated with *T. harzianum* T22 showed an increased root development. In addition, light microscopy analyses showed that the colonization of the roots took place in all the 9 inoculated plants (100%). The presence of

T. harzianum T22 was detected by the observation of the coloured intracellular structures of the fungus in the zucchini roots. The intracellular structures (the vesicles produced by the fungus) were not present in control samples but only in *T. harzianum* T22-treated samples (Figure 2).

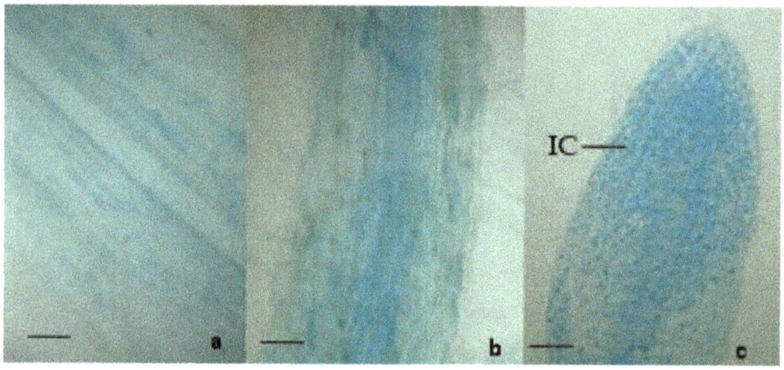

Figure 2. Light micrographs of zucchini roots stained with Tryptan blue and Pelikan blue dyes. (**a**,**b**): roots sections of control (not inoculated); (**c**): apex of primary root inoculated with *T. harzianum* T22, IC: intracellular structures; Bars = 100 μm.

In addition, *T. harzianum* T22 was isolated on PDA from all zucchini roots sampled in the treated plots 25 days, 54 days, and 85 days after transplantation. The presence of the fungus was not observed in the control Petri plates.

3.2. *Arthropods Sampling*

3.2.1. Arthropod Sampling on Zucchini Leaves

Leaf samples were collected for observation of the arthropods in the laboratory. During this sampling period, 256 arthropod specimens were collected on zucchini leaves, of which 107 and 149 were obtained from plants with *T. harzianum* T22 and control, respectively. The arthropods on zucchini leaves belonged to the families Aphididae (one species identified: apterous morph of *Aphis gossypii*), Cicadellidae, Thripidae, Chrysomelidae, Gryllidae, Coccinellidae (adults), Syrphidae (identified as eggs or adults), Braconidae (adults), and Miridae. We also collected eggs of Lepidoptera, 9 individuals belonging to the order of Araneae, 5 individuals of *Tetranychus urticae*, and 5 leaf mines of *Liriomyza trifolii* Burges (Diptera, Agromyzidae). The abundances of Cicadellidae, Thripidae, Gryllidae, Chrysomelidae, Coccinellidae, Miridae, Syrphidae, Braconidae, Araneae, *T. urticae*, and leaf miners were very low during the whole sampling period (Figure S2) and consequently were excluded from the analysis.

The abundance of *A. gossypii* and of eggs of Lepidoptera is shown in Figure 3.

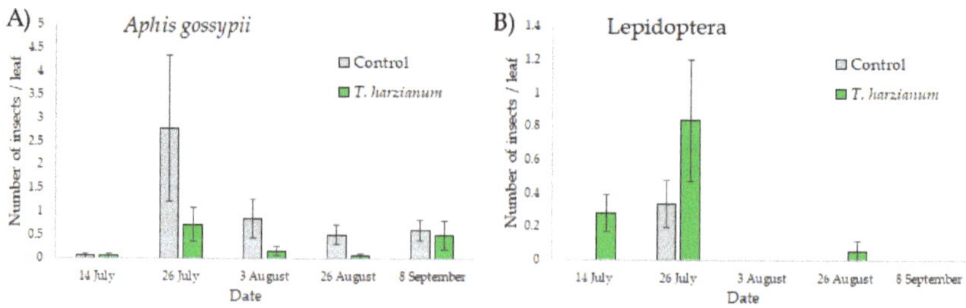

Figure 3. Mean values (±standard errors) of the number of apterous *Aphis gossypii* (**A**) and eggs of Lepidoptera (**B**) on zucchini leaves sampled from plants inoculated with *T. harzianum* T22 and control ones during the five sampling dates.

The GLMMs showed that the sampling dates influenced the abundance of apterous *Aphis gossypii* (χ^2 = 20.1, df = 4, $p < 0.001$) and of eggs of Lepidoptera (χ^2 = 10.2, df = 4, $p < 0.05$). The abundance of these insects was higher in July and then decreased in the following months. The abundance of *A. gossypii* was also affected by the treatment (χ^2 = 6.2, df = 1, $p < 0.05$), with more apterous individuals collected on control plants. No significant differences between control and plants inoculated with *T. harzianum* T22 in the number of Lepidoptera eggs were observed (χ^2 = 3.6, df = 1, p = 0.058). The interactions "treatment X date" were never found significant.

3.2.2. Arthropod Sampling with Coloured Pan Traps

During the sampling period, 3925 arthropod specimens were collected with the pan traps, of which 2307 and 1618 were obtained from plants inoculated with *T. harzianum* T22 and control, respectively. The arthropods collected in the traps belonged to the families Aphididae (winged morphs of *A. gossypii*), Cicadellidae, Thripidae, Chrysomelidae, Gryllidae, Coccinellidae, Miridae, and to the order Lepidoptera, Hymenoptera (Ichneumonoidea and Chalcidoidea), and Araneae. The abundances of Lepidoptera, Coccinellidae, Staphylinidae, Gryllidae, and Miridae were very low, and they have not been considered for the analysis (Figure S3).

The abundances of *A. gossypii*, Chrysomelidae, Thripidae, Cicadellidae, Hymenoptera parasitoids, and Araneae are shown in Figure 4.

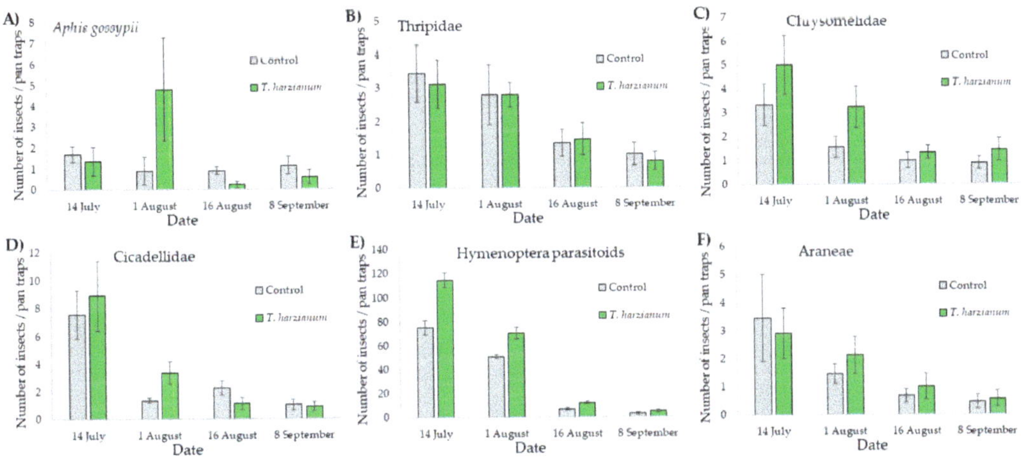

Figure 4. Mean values (± standard errors) of the number of winged *Aphis gossypii* (**A**), Thripidae (**B**), Chrysomelidae (**C**), Cicadellidae (**D**), Hymenoptera parasitoids (**E**), and Araneae (**F**) collected with pan trap sets placed near plants inoculated with *T. harzianum* T22 and control at the four sampling dates.

The GLMMs showed that the abundances of all the arthropods collected with the pan trap sets in the experimental field were affected by the sampling dates ($p < 0.001$ in all cases). The abundance of arthropods was higher in July and then decreased in the following months. Significant differences between treatments were found for the abundance of winged *A. gossypii* (χ^2 = 33.8, df = 1, $p < 0.001$), of Chrysomelidae (χ^2 = 5.1, df = 1, $p < 0.05$), and of Hymenoptera parasitoids (χ^2 = 61.9, df = 1, $p < 0.001$), with a higher number of insects collected in plots with zucchini inoculated with *T. harzianum* T22. The interaction "treatment X date" was only found significant for *A. gossypii* (χ^2 = 439,322, df = 3, $p < 0.001$) and for the family of Cicadellidae (and χ^2 = 10.1, df = 3, $p < 0.05$). Compared with the control, the abundance of aphids and Cicadellidae was higher on the 1 August on the *T. harzianum* T22 plots.

3.3. *Plant Diseases*

3.3.1. Field Evaluation of Zucchini Viral Diseases

The ANOVA performed on the data relating the viral infection gave significant differences among sampling dates ($F_{6,28} = 127.6$, $p < 0.001$) but not between treatment ($F_{1,28} = 2.4$, $p = 0.13$) or for the "treatment X date" interaction ($F_{6,28} = 0.76$, $p = 0.61$) as shown in Figure 5. The viral infections, in all plots, started on 24 July on both treated and untreated plants and continuously increased over time until the end of the cultivation period, reaching the 100% of infection at the beginning of September. On 7 September there was no difference in the viral symptoms observed in the field between the untreated (control) and treated (*T. harzianum* T22) plants per plot (Figure 5).

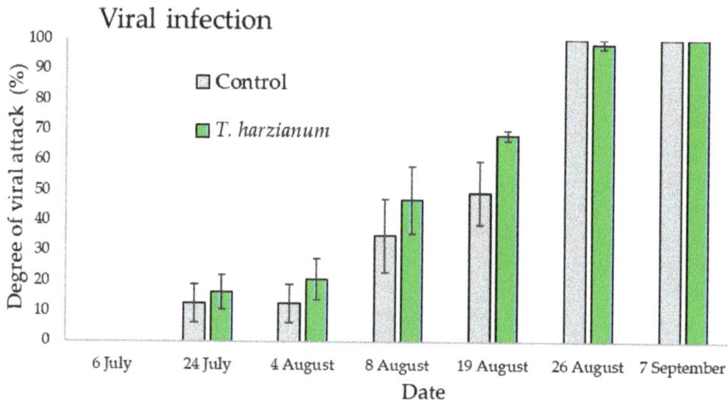

Figure 5. Mean degree of viral attack (±standard errors) in plots inoculated with *T. harzianum* T22 and control ones over time.

Regarding the influence of the virosis on powdery mildew symptoms of the zucchini plants, it was observed that in plants with the same symptoms of virosis, the powdery mildew infection was more evident for the control plants compared with the *T. harzianum* T22 inoculated ones (Figure 6). Furthermore, the ANOVA performed on these data gave significant differences related to the virosis classes ($F_{4,514} = 398.7$, $p < 0.001$) and between treatments ($F_{1,514} = 21.9$, $p < 0.001$) but not for the "virosis classes X treatment" interaction ($F_{4,514} = 0.6$, $p = 0.66$), as shown in Figure 6.

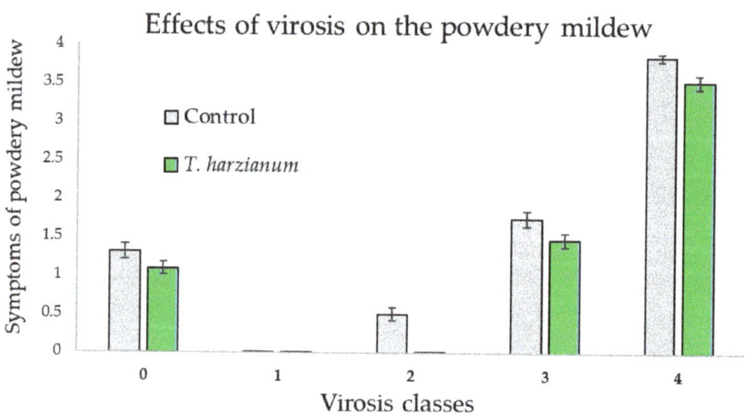

Figure 6. Mean values (±standard errors) of the symptoms the powdery mildew in relation to the virus classes in plants inoculated with *T. harzianum* T22 and control ones.

3.3.2. ELISA Test for Viruses in Zucchini Plants

The results of the ELISA serological assay demonstrated that of the four most common zucchini viruses (CMV, ZYMV, PRSNV, and WMV), only one virus (CMV) was not present in the experimental field, while all others were detected. In particular, ZYMV and PRSNV were detected at 100%, WMV had a 45% of incidence in control plants and a 44% incidence in the *T. harzianum* T22-treated ones (Figure 7). In summary, our results showed that zucchini plants were infected by the three of four most common viruses and the viral incidence was not much different between the control and *T. harzianum* T22-inoculated plants.

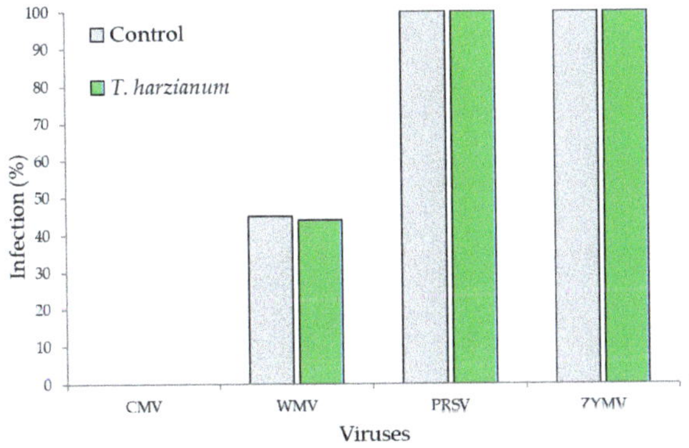

Figure 7. Percentage of infection for each virus determined by ELISA in zucchini plants.

3.3.3. Powdery Mildew

Microscopic analysis showed that the causal agent of powdery mildew attack on zucchini plants in the field was closely similar to the *P. fusca*. These results are based on the morphological features reported in literature [49,50] for zucchini powdery mildew and the fibrosin bodies' presence in the conidia.

The results regarding the symptoms of powdery mildew are shown in Figure 8. The symptoms were observed in the field after 19 August and the disease progressed in both controls and *T. harzianum* T22 plots, reaching 100% of infection on 7 September. Even if the disease symptom development was similar, a small delay was observed for the *T. harzianum* T22-treated plants compared to the control, at least in the initial and also during the disease development stages. However, the GLMMs showed that the symptoms of the powdery mildew were influenced by the sampling dates ($\chi2 = 200$, df = 3, $p < 0.001$), but not by the treatment ($\chi2 = 0.79$, df = 1, $p = 0.37$), nor by the interactions "treatment X date" ($\chi2 = 1.1$, df = 3, $p = 0.78$) as shown in Figure 8.

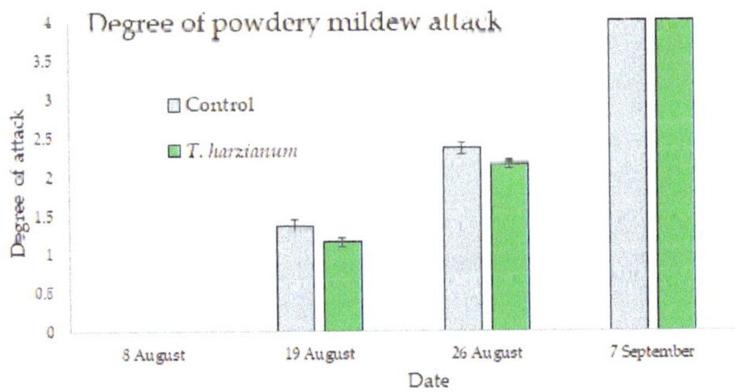

Figure 8. Mean values (±standard errors) of the powdery mildew degree of attack in plants inoculated with *T. harzianum* T22 and control over time.

3.4. Crop Sampling

3.4.1. Plant Length

Figure 9 shows the mean values of the length of zucchini plants inoculated with *T. harzianum* T22 and control on the four sampling dates.

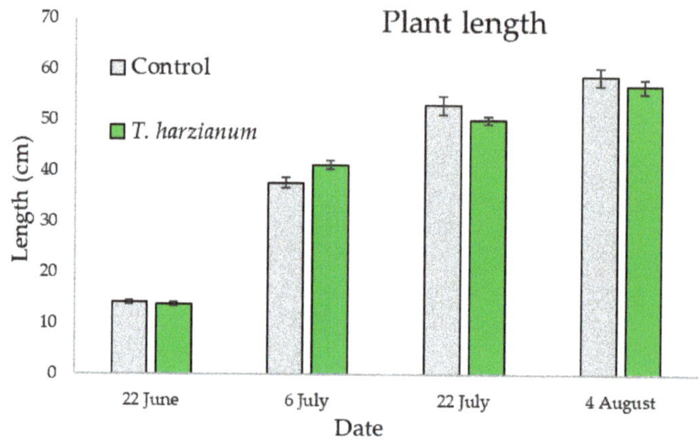

Figure 9. Mean values (±standard errors) of the length of zucchini plants inoculated with *T. harzianum* T22 and control on the four sampling dates.

For plant length, statistically significant differences were only found between sampling dates ($\chi^2 = 204$, df = 3, $p < 0.001$), indicating that plants inoculated with *T. harzianum* T22 or not inoculated have the same growth rate over time.

3.4.2. Plant Productivity

Figure 10 shows the mean values of the weight of the zucchini fruit harvested from plants inoculated with *T. harzianum* T22 and from the control during the four sampling periods.

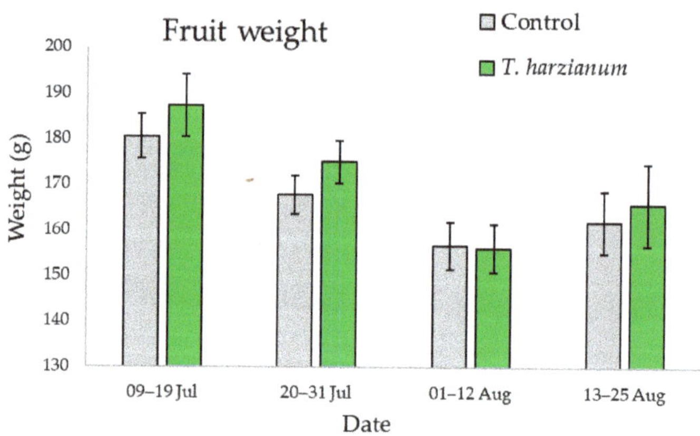

Figure 10. Mean values (±standard errors) of fruit weight from zucchini plants inoculated with *T. harzianum* T22 and control during the four sampling periods.

For fruit weight, statistically significant differences were only found among sampling periods ($\chi^2 = 18.1$, df = 3, $p < 0.001$), with heavier fruit produced during the first month. Even if the differences were not significant, during the first month plants inoculated with *T. harzianum* T22 showed a production of heavier fruit than controls.

Figure 11 shows the number and weight of fruits recorded after each harvest accumulatively from 9 July to 30 August from plants inoculated with *T. harzianum* T22 and from control ones.

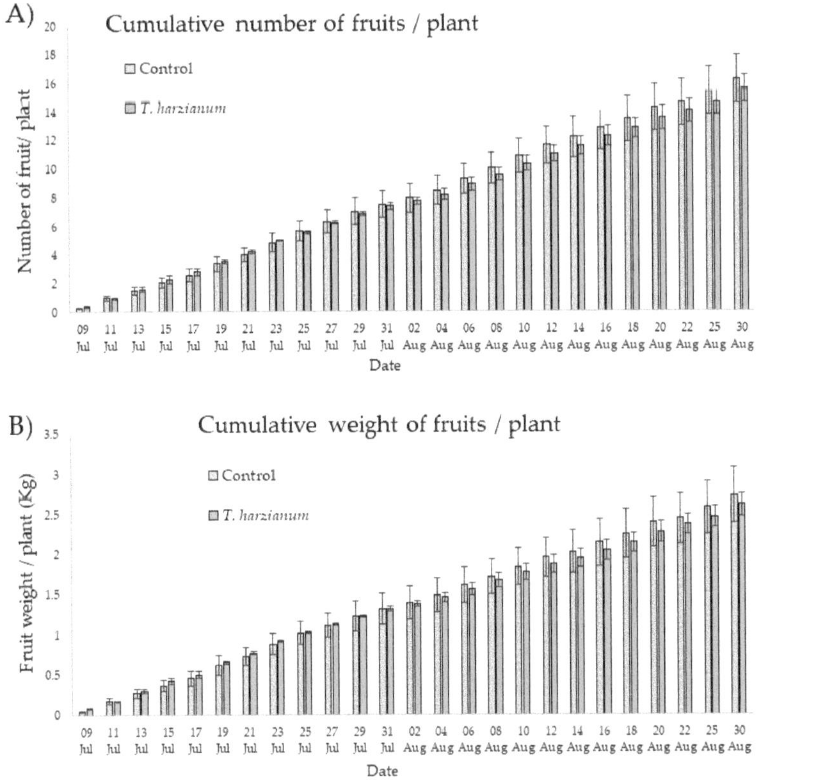

Figure 11. Mean values (±standard errors) of the cumulative number (**A**) and weight of zucchini fruits (**B**) recorded after each harvest from 9 July to 30 August from plants inoculated with *T. harzianum* T22 and from control.

The ANOVAs performed on these data show that the cumulative number of zucchini fruits/plant and the cumulative yield/plant were not affected by inoculation with *T. harzianum* T22 ($\chi^2 = 2.55$, df = 1, $p = 0.11$ and $\chi^2 = 0.74$, df = 1, $p = 0.39$, respectively). No significant interaction "date X treatment" was also found ($\chi^2 = 4.6$, df = 24, $p = 0.99$ and $\chi^2 = 3.2$, df = 24, $p = 0.99$, respectively).

4. Discussion

The use of beneficial microbial species in agriculture as biocontrol agents and plant growth promoters has increased in recent decades [10]. Among them, fungi of the genus *Trichoderma* are the most widespread and effective [11,55,56]. *Trichoderma* fungi can antagonize plant pathogens through competition, antibiosis, and mycoparasitism mechanisms. *Trichoderma* is known to induce metabolic and physiologic changes in the colonized plants [57] by activating the plant SAR and/or by inducing systemic resistance against biotic and abiotic stress agents [11,20–23]. It is well accepted that plants have sophisticated defence strategies and when attacked by pathogens or pests activate signalling defence mechanisms modulated by jasmonic JA, SA, or ethylene phytohormones (ET) [11,21,58–60]. However, these pathways can crosstalk and their synergistic interactions can play a fundamental role in the ISR activation [61–64].

Recently, the possibility of using fungi of the genus *Trichoderma* as pest biocontrol agents has been emphasized [65]. Most studies on below ground–above ground interactions involving *Trichoderma* and plant pests have used the tomato as a model plant in the laboratory [12,14,16–19,66]. Few studies have verified, under far more complex field conditions, the role of *Trichoderma* fungi as a pest biocontrol agent [15,56,67].

In the context of horticultural crops, Cucurbitaceae is the second family in terms of economic relevance after Solanaceae [68], and zucchini is the most important economically and globally widespread species among the cultivated Cucurbitaceae [24]. The main phytosanitary problems of zucchini are aphids, phytoviruses, and powdery mildews [29,30]. The management of these pest and pathogen infections on zucchini crops has so far been based mainly on use of resistant varieties [28,36,37] and pesticides [38,39]. The effectiveness of *Trichoderma* spp. as biocontrol agents against zucchini fungal pathogens is confirmed by several studies, especially on *Fusarium* spp. [69,70]. Inoculations with *Trichoderma* spp. inhibited *F. oxysporum* infection stimulating plant metabolism and increasing the activities of stress-resistance enzymes [70].

On the contrary, laboratory and field studies that report the effectiveness of *T. harzianum* as a biocontrol agent against aphids and phytoviruses in zucchini are not available. In this study, we measured the natural evolution of pests and diseases in a zucchini field by comparing plots inoculated or not inoculated with *T. harzianum* T22.

Throughout the cultivation period, *A. gossypii* was the only pest species worth mentioning. The highest occurrence on plants was observed on 26 July. Subsequently, the infestation decreased, probably due to the increase in temperature in mid-summer. Various studies on the ecology of aphid populations report a rapid population decline during the mid-summer, with host plants without aphids or with a lower abundance compared to the population abundance in early-summer and spring [71–73]. A major part of the aphids sampled on zucchini leaves were apterae, and in many cases, colonies were formed by just an adult aphid and a few nymphs. Interestingly, the abundance of aphids on leaves was significantly higher on control plants. *Trichoderma* is known to be involved in priming, the activation of plant defence prior to invasion, and up-regulated several Serine/threonine- and Leucine-rich repeat protein kinases that activate defence against pests [66]. *Trichoderma* colonization can generate a pre-alerted state of "priming" to face incoming pest attacks more efficiently [66,74], inhibiting the development and reproduction of aphids on the leaves of inoculated plants. In contrast, the winged aphids caught in pan traps were significantly more numerous in the plots inoculated with *T. harzianum* T22. These data seem to indicate that *T. harzianum* T22 makes zucchini plants more attractive to aphids, but this is followed by limited colony production. Winged aphids, while not producing colonies, could contribute to the spread of viruses.

Another significant result is the high number of Hymenoptera parasitoid captured in pan traps placed in plots inoculated with *T. harzianum* T22. The increased attractiveness to parasitoids and the reduced infestation of aphids, as a result of colonization by *Trichoderma*, confirm the results obtained in the laboratory, although with a different plant/aphid/parasitoid system [75]. *Trichoderma* influenced the quantity and quality of the volatile organic compound (VOC) blends released by plants [12]. The attractiveness to parasitoids is associated with an enhanced release of VOCs such as methyl-salicylate and β-caryophyllene, known to be among the most active compounds in promoting parasitoids flight orientation [12,75]. *Trichoderma* spp. promotes plant nitrogen uptake [20] giving the plant a higher nutritional value which can orient insects at the time of oviposition.

Trichoderma harzianum T22 failed to control zucchini pathogens investigated in the experimental field. Both viral diseases and powdery mildew equally attacked the control and *T. harzianum* T22-inoculated plants, starting from the end of July for viral diseases to middle of August in the case of powdery mildew. Furthermore, the severity of both the diseases worsened over time and the symptoms observed on zucchini plants changed from very mild to very strong, reaching the maximum peak at the beginning of September (expressed as 100% of infection). It may be useful to point out that laboratory experiments, testing the induction of resistance pathways by *Trichoderma* fungi, usually use young plants. In our experimental trial, in the field, it was observed that both viral infections and powdery mildew attack spread when plants had already begun to produce fruits. The ontogeny of resistance in plants has been approached with reference to insects [76–78], but still less is known about phytopathogens. For example, Vitti et al. [79], reported favourable effects of *T. harzianum* T22 in tomato seedling artificially inoculated with cucumber mosaic virus (CMV) in laboratory experiments. The authors showed that *T. harzianum* T22 was able to promote the induction of tomato defence responses against CMV and also demonstrated that this involves reactive oxygen species (ROS).

Another study by Shen et al. [28], investigating the dynamic distribution of *A. gossypii* on the incidence of viral disease in six zucchini cultivars, concluded that the ability of zucchini plants to resist aphids attack was not consistent with their capacity to resist viral diseases. Slow transformation rate varieties with a mild disease phenotype in the late growth stage showed strong resistance to the disease.

We cannot exclude that the resistance induced by *Trichoderma* fungi observed in previous studies could be influenced by the phenological stage of the plant. The present study showed that there is no resistance effect in the field. Probably, the lack of resistance observed can be due to the higher costs of resistance for a plant that is already at the stage of fruit production. This aspect deserves future investigations. In fact, Shen et al. [28] showed that the disease resistance ability of zucchini plants always differed among the different growth stages.

Overall, no differences in terms of fruit yields were found between zucchini inoculated with *T. harzianum* T22 and control plants. This is in contrast to Hazef et al. [80], Elsisi [81], and El-Sharkawy et al. [82] who found an increase in the yield of zucchini plants under greenhouse and filed conditions due to the inoculation of *T. harzianum* T22.

5. Conclusions

In this study, the effects of *Trichoderma harzianum* T22 as a biological control agent against zucchini the pests of were investigated for the first time under field conditions. The interaction among *T. harzianum* T22/zucchini plant/pests appeared to be complex. However, our results confirmed the ability of *T. harzianum* T22 to alter the arthropod community by increasing the attractiveness of zucchini to winged aphids and hymenopteran parasitoids. Unlike the outcomes of other studies conducted in the laboratory, a reduction in pathogen infestation was not observed in zucchini inoculated with *T. harzianum* T22. The discrepancies between our findings and the laboratory studies should be better investigated to understand how the abiotic factors affected the *Trichoderma*/plant interaction under open field conditions. It would also be interesting to investigate the ontogeny of the resistance mechanisms in function of the zucchini phenological stages. In our study, the presence of diseases, which were widely spread in the experimental field, and the lack of use of fertilizers and agrochemicals may have hidden the positive effect of *T. harzianum* T22 on plant production.

Supplementary Materials. The following supporting information can be downloaded at: https://www.mdpi.com/article/10.3390/microorganisms10112242/s1, Figure S1: Meteorological data. Figure S2: Leaf arthropods. Mean values (±standard errors) of insect abundance of Cicadellidae, Thripidae, Gryllidae, Coccinellidae, Miridae, Syrphidae, Araneae, leaf miners, and *Tetranychus urticae* on zucchini leaves sampled from plants inoculated with *T. harzianum* T22 and control ones during the five sampling dates. Figure S3: Pan traps samples. Mean values (± standard errors) of insect abundance of Lepidoptera, Coccinellidae, Staphylinidae, Cicadellidae, and Miridae collected with pan traps placed near zucchini plants inoculated with *T. harzianum* T22 and control ones at the four sampling dates.

Author Contributions: Conceptualization, V.T. and D.B.; methodology, P.F. (Pierluigi Forlano), S.M.M., V.C., P.F. (Paolo Fanti), I.C., D.B. and V.T.; software, V.T.; validation, P.F. (Pierluigi Forlano), S.M.M., V.C., P.F. (Paolo Fanti), I.C., D.B. and V.T.; formal analysis, V.T.; data curation, P.F. (Pierluigi Forlano), S.M.M. and V.T.; writing—original draft preparation, P.F. (Pierluigi Forlano), S.M.M. and V.T.; revising the work critically for important intellectual content and editing, P.F. (Pierluigi Forlano), S.M.M., V.C., P.F. (Paolo Fanti), I.C., D.B. and V.T.; visualization: P.F. (Pierluigi Forlano), S.M.M., V.C., P.F. (Paolo Fanti), I.C., D.B. and V.T.; supervision, I.C., D.B., S.M.M. and V.T.; resources: D.B. and I.C.; funding acquisition, D.B. and I.C. All authors have read and agreed to the published version of the manuscript.

Funding: This research was founded by the PhD project entitled "Sustainable control of insect vectors of phytopathogenic viruses in the context of climate change: the role of root symbionts"; by PON "Research and Innovation" 2014–2020, Action IV.6, the project entitled "New strategies for the eco-sustainable control of harmful insects".

Data Availability Statement: The data presented in this study are available on request from the corresponding author.

Acknowledgments: The authors want to thank the "Azienda Agricola Rosa Marisa", Pignola (PZ), Italy (40°34′ N, 15°45′ E) for the support in the realization of the experimental field and the "Agenzia Lucana per lo Sviluppo e l'Innovazione in Agricoltura (ALSIA)" of the Basilicata Region, Italy for providing the metereological data.

Conflicts of Interest: The authors declare no conflict of interest.

References

1. Foley, J.A.; Ramankutty, N.; Brauman, K.A.; Cassidy, E.S.; Gerber, J.S.; Johnston, M.; Mueller, N.D.; O'Connell, C.; Ray, D.K.; West, P.C.; et al. Solutions for a Cultivated Planet. *Nature* **2011**, *478*, 337–342. [CrossRef] [PubMed]
2. Marschner, P.; Kandeler, E.; Marschner, B. Structure and Function of the Soil Microbial Community in a Long-Term Fertilizer Experiment. *Soil Biol. Biochem.* **2003**, *35*, 453–461. [CrossRef]
3. Khanna, R.; Gupta, S. Agrochemicals as a Potential Cause of Ground Water Pollution: A Review. *Int. J. Chem. Stud.* **2018**, *6*, 985–990.
4. Bass, C.; Jones, C.M. Editorial Overview: Pests and Resistance: Resistance to Pesticides in Arthropod Crop Pests and Disease Vectors: Mechanisms, Models and Tools. *Curr. Opin. Insect Sci.* **2018**, *27*, 4–7. [CrossRef]
5. Hawkins, N.J.; Bass, C.; Dixon, A.; Neve, P. The Evolutionary Origins of Pesticide Resistance. *Biol. Rev.* **2019**, *94*, 135–155. [CrossRef]
6. Carletto, J.; Martin, T.; Vanlerberghe-Masutti, F.; Brévault, T. Insecticide Resistance Traits Differ among and within Host Races in *Aphis gossypii*. *Pest Manag. Sci.* **2010**, *66*, 301–307. [CrossRef] [PubMed]
7. Liu, H.; Brettell, L.E.; Qiu, Z.; Singh, B.K. Microbiome-Mediated Stress Resistance in Plants. *Trends Plant Sci.* **2020**, *25*, 733–743. [CrossRef]
8. Pozo de la Hoz, J.; Rivero, J.; Azcón-Aguilar, C.; Urrestarazu, M.; Pozo, M.J. Mycorrhiza-Induced Resistance against Foliar Pathogens Is Uncoupled of Nutritional Effects under Different Light Intensities. *J. Fungi* **2021**, *7*, 402. [CrossRef]
9. Alfiky, A.; Weisskopf, L. Deciphering *Trichoderma*–Plant–Pathogen Interactions for Better Development of Biocontrol Applications. *J. Fungi* **2021**, *7*, 61. [CrossRef]
10. Batista, B.D.; Singh, B.K. Realities and Hopes in the Application of Microbial Tools in Agriculture. *Microb. Biotechnol.* **2021**, *14*, 1258–1268. [CrossRef]
11. Macías-Rodríguez, L.; Contreras-Cornejo, H.A.; Adame-Garnica, S.G.; Del-Val, E.; Larsen, J. The Interactions of *Trichoderma* at Multiple Trophic Levels: Inter-Kingdom Communication. *Microbiol. Res.* **2020**, *240*, 126552. [CrossRef] [PubMed]
12. Battaglia, D.; Bossi, S.; Cascone, P.; Digilio, M.C.; Prieto, J.D.; Fanti, P.; Guerrieri, E.; Iodice, L.; Lingua, G.; Lorito, M.; et al. Tomato below Ground-above Ground Interactions: *Trichoderma longibrachiatum* Affects the Performance of *Macrosiphum euphorbiae* and Its Natural Antagonists. *Mol. Plant. Microbe. Interact.* **2013**, *26*, 1249–1256. [CrossRef] [PubMed]
13. Coppola, M.; Cascone, P.; Di Lelio, I.; Woo, S.L.; Lorito, M.; Rao, R.; Pennacchio, F.; Guerrieri, E.; Digilio, M.C. *Trichoderma atroviride* P1 Colonization of Tomato Plants Enhances Both Direct and Indirect Defense Barriers against Insects. *Front. Physiol.* **2019**, *10*, 813. [CrossRef] [PubMed]
14. Alınç, T.; Cusumano, A.; Peri, E.; Torta, L.; Colazza, S. *Trichoderma harzianum* Strain T22 Modulates Direct Defense of Tomato Plants in Response to *Nezara viridula* Feeding Activity. *J. Chem. Ecol.* **2021**, *47*, 455–462. [CrossRef] [PubMed]
15. Caccavo, V.; Forlano, P.; Mang, S.M.; Fanti, P.; Nuzzaci, M.; Battaglia, D.; Trotta, V. Effects of *Trichoderma harzianum* Strain T22 on the Arthropod Community Associated with Tomato Plants and on the Crop Performance in an Experimental Field. *Insects* **2022**, *13*, 418. [CrossRef]

16. Martínez-Medina, A.; Fernández, I.; Sánchez-Guzmán, M.J.; Jung, S.C.; Pascual, J.A.; Pozo, M.J. Deciphering the Hormonal Signaling Network behind the Systemic Resistance Induced by *Trichoderma harzianum* in Tomato. *Front. Plant Sci.* **2013**, *4*, 206. [CrossRef]
17. Jogaiah, S.; Abdelrahman, M.; Tran, L.S.P.; Ito, S.I. Different Mechanisms of *Trichoderma virens*-Mediated Resistance in Tomato against *Fusarium wilt* Involve the Jasmonic and Salicylic Acid Pathways. *Mol. Plant Pathol.* **2018**, *19*, 870–882. [CrossRef]
18. Herrera-Téllez, V.I.; Cruz-Olmedo, A.K.; Plasencia, J.; Gavilanes-Ruíz, M.; Arce-Cervantes, O.; Hernández-León, S.; Saucedo-García, M. The Protective Effect of *Trichoderma asperellum* on Tomato Plants against *Fusarium oxysporum* and *Botrytis cinerea* Diseases Involves Inhibition of Reactive Oxygen Species Production. *Int. J. Mol. Sci.* **2019**, *20*, 2007. [CrossRef]
19. Heflish, A.A.; Abdelkhalek, A.; Al-Askar, A.A.; Behiry, S.I. Protective and Curative Effects of *Trichoderma asperelloides* Ta41 on Tomato Root Rot Caused by *Rhizoctonia solani* Rs33. *Agronomy* **2021**, *11*, 1162. [CrossRef]
20. Harman, G.E.; Howell, C.R.; Viterbo, A.; Chet, I.; Lorito, M. *Trichoderma* Species–Opportunistic, Avirulent Plant Symbionts. *Nat. Rev. Microbiol.* **2004**, *2*, 43–56.
21. Shoresh, M.; Harman, G.E.; Mastouri, F. Induced Systemic Resistance and Plant Responses to Fungal Biocontrol Agents. *Annu. Rev. Phytopathol.* **2010**, *48*, 21–43. [CrossRef] [PubMed]
22. Studholme, D.J.; Harris, B.; Le Cocq, K.; Winsbury, R.; Perera, V.; Ryder, L.; Ward, J.L.; Beale, M.H.; Thornton, C.R.; Grant, M. Investigating the Beneficial Traits of *Trichoderma hamatum* GD12 for Sustainable Agriculture-Insights from Genomics. *Front. Plant Sci.* **2013**, *4*, 1–13. [CrossRef] [PubMed]
23. Contreras-Cornejo, H.A.; Macías-Rodríguez, L.; Del-Val, E.; Larsen, J. Ecological Functions of *Trichoderma* Spp. and Their Secondary Metabolites in the Rhizosphere: Interactions with Plants. *FEMS Microbiol. Ecol.* **2016**, *92*, fiw036. [CrossRef] [PubMed]
24. Andolfo, G.; Di Donato, A.; Darrudi, R.; Errico, A.; Cigliano, R.A.; Ercolano, M.R. Draft of Zucchini (*Cucurbita pepo* L.) Proteome: A Resource for Genetic and Genomic Studies. *Front. Genet.* **2017**, *8*, 1–5. [CrossRef]
25. Paris, H.S. A Proposed Subspecific Classification for Cucurbita Pepo. *Phytologia* **1986**, *61*, 133–138.
26. Gilardi, G.; Pugliese, M.; Gullino, M.L.; Garibaldi, A. Effect of Biocontrol Agents and Potassium Phosphite against *Phytophthora crown* Rot, Caused by *Phytophthora capsici*, on Zucchini in a Closed Soilless System. *Sci. Hortic.* **2020**, *265*, 109207. [CrossRef]
27. Formisano, L.; Miras-Moreno, B.; Ciriello, M.; El-Nakhel, C.; Corrado, G.; Lucini, L.; Colla, G.; Rouphael, Y. *Trichoderma* and Phosphite Elicited Distinctive Secondary Metabolite Signatures in Zucchini Squash Plants. *Agronomy* **2021**, *11*, 1205. [CrossRef]
28. Shen, J.; Li, X.; Wang, J.; Zhang, C.; Li, L.; Zhang, L. Dynamic Distribution of *Aphis gossypii* Glover (Homoptera: Aphididae) and Incidence of Viral Disease in Different Zucchini (*Cucurbita Pepo* L.) Cultivars. *Biochem. Syst. Ecol.* **2018**, *77*, 31–36. [CrossRef]
29. Hinds, J.; Hooks, C.R.R. Population Dynamics of Arthropods in a Sunn-Hemp Zucchini Interplanting System. *Crop Prot.* **2013**, *53*, 6–12. [CrossRef]
30. Koné, K.; Tuo, Y.; Yapo, M.L.; Soro, F.; Traoré, D.; Koua, K.H. Main Insect Pests of Zucchini (*Cucurbita pepo* L), in the Dry Season and Impact on Production in Northern Côte d ' Ivoire. *J. Entomol. Zool. Stud.* **2019**, *7*, 523–527.
31. Blackman, R.L.; Eastop, V.F. *Aphids on the World's Herbaceous Plants and Shrubs. Volume 1: Host Lists and Keys. Volume 2: The Aphids*; John Wiley & Sons, Ltd.: Chichester, UK, 2006; ISBN 9780471489733.
32. Ng, J.C.K.; Perry, K.L. Transmission of Plant Viruses by Aphid Vectors. *Mol. Plant Pathol.* **2004**, *5*, 505–511. [CrossRef] [PubMed]
33. Coppola, M.; Manco, E.; Vitiello, A.; Di Lelio, I.; Giorgini, M.; Rao, R.; Pennacchio, F.; Digilio, M.C. Plant Response to Feeding Aphids Promotes Aphid Dispersal. *Entomol. Exp. Appl.* **2018**, *166*, 386–394. [CrossRef]
34. Ebert, T.A.; Cartwright, B.O. Biology and Ecology of *Aphis gossypii* Glover (Homoptera: Aphididae). *Southwest. Entomol.* **1997**, *22*, 116–153.
35. Al-Shahwan, I.M.; Abdalla, O.A.; Al-Saleh, M.A. Response of Greenhouse-Grown Cucumber Cultivars to an Isolate of Zucchini Yellow Mosaic Virus (ZYMV). *Plant Dis.* **1995**, *79*, 898–901. [CrossRef]
36. Nováková, S.; Flores-Ramírez, G.; Glasa, M.; Danchenko, M.; Fiala, R.; Skultety, L. Partially Resistant *Cucurbita pepo* Showed Late Onset of the Zucchini Yellow Mosaic Virus Infection Due to Rapid Activation of Defense Mechanisms as Compared to Susceptible Cultivar. *Front. Plant Sci.* **2015**, *6*, 263. [CrossRef]
37. Zhang, S.; Liu, J.; Xu, B.; Zhou, J. Differential Responses of *Cucurbita pepo* to *Podosphaera xanthii* Reveal the Mechanism of Powdery Mildew Disease Resistance in Pumpkin. *Front. Plant Sci.* **2021**, *12*, 1–15. [CrossRef]
38. Cao, C.W.; Zhang, J.; Gao, X.W.; Liang, P.; Guo, H.L. Overexpression of Carboxylesterase Gene Associated with Organophosphorous Insecticide Resistance in Cotton Aphids, *Aphis gossypii* (Glover). *Pestic. Biochem. Physiol.* **2008**, *90*, 175–180. [CrossRef]
39. Herron, G.A.; Wilson, L.J. Can Resistance Management Strategies Recover Insecticide Susceptibility in Pests?: A Case Study with Cotton Aphid *Aphis gossypii* (Aphididae: Hemiptera) in Australian Cotton. *Austral Entomol.* **2017**, *56*, 1–13. [CrossRef]
40. Desbiez, C.; Lecoq, H. Zucchini Yellow Mosaic Virus. *Plant Pathol.* **1997**, *46*, 809–829. [CrossRef]
41. Yuan, C.; Ullman, D.E. Comparison of Efficiency and Propensity as Measures of Vector Importance in Zucchini Yellow Mosaic Potyvirus Transmission by *Aphis gossypii* and *A. craccivora*. *Phytopathology* **1996**, *86*, 698–703. [CrossRef]
42. Herron, G.; Powis, K.; Rophail, J. Baseline Studies and Preliminary Resistance Survey of Australian Populations of Cotton Aphid *Aphis gossypii* Glover (Hemiptera: Aphididae). *Aust. J. Entomol.* **2000**, *39*, 33–38. [CrossRef]
43. Herron, G.A.; Powis, K.; Rophail, J. Insecticide Resistance in *Aphis gossypii* Glover (Hemiptera: Aphididae), a Serious Threat to Australian Cotton. *Aust. J. Entomol.* **2001**, *40*, 85–91. [CrossRef]
44. Ahmad, M.; Iqbal Arif, M. Susceptibility of Pakistani Populations of Cotton Aphid *Aphis gossypii* (Homoptera: Aphididae) to Endosulfan, Organophosphorus and Carbamate Insecticides. *Crop Prot.* **2008**, *27*, 523–531. [CrossRef]

45. Pocurull, M.; Fullana, A.M.; Ferro, M.; Valero, P.; Escudero, N.; Saus, E.; Gabaldón, T.; Sorribas, F.J. Commercial Formulates of *Trichoderma* Induce Systemic Plant Resistance to *Meloidogyne incognita* in Tomato and the Effect Is Additive to That of the Mi-1.2 Resistance Gene. *Front. Microbiol.* **2020**, *10*, 1–10. [CrossRef] [PubMed]
46. Southwood, T.R.E.; Henderson, P.A. *Ecological Methods*, 3rd ed.; Blackwell Science: Oxford, UK, 2000; Volume 278.
47. Leong, J.M.; Thorp, R.W. Colour-Coded Sampling: The Pan Trap Colour Preferences of Oligolectic and Nonoligolectic Bees Associated with a Vernal Pool Plant. *Ecol. Entomol.* **1999**, *24*, 329–335. [CrossRef]
48. Clark, M.F.; Adams, A.N. Characteristics of the Microplate Method of Enzyme-Linked Immunosorbent Assay for the Detection of Plant Viruses. *J. Gen. Virol.* **1977**, *34*, 475–483. [CrossRef]
49. Miazzi, M.; Laguardia, C.; Faretra, F. Variation in *Podosphaera xanthii* on Cucurbits in Southern Italy. *J. Phytopathol.* **2011**, *159*, 538–545. [CrossRef]
50. Braun, U.; Cook, R.T.A. *Taxonomic Manual of the Erysiphales (Powdery Mildews)*; CBS Biodiv.; CBS-KNAW Fungal Biodiversity Centre: Utrecht, The Netherlands, 2012; ISBN 978-90-70351-89-2.
51. Burnham, K.P.; Anderson, D.R. Multimodel Inference: Understanding AIC and BIC in Model Selection. *Sociol. Methods Res.* **2004**, *33*, 261–304. [CrossRef]
52. *R Core Team R: A Language and Environment for Statistical Computing*; R Foundation for Statistical Computing: Vienna, Austria, 2021; Available online: Https://Www.R-Project.Org/ (accessed on 15 December 2021).
53. Bates, D.; Mächler, M.; Bolker, B.M.; Walker, S.C. Fitting Linear Mixed-Effects Models Using Lme4. *J. Stat. Softw.* **2015**, *67*, 48. [CrossRef]
54. Kuznetsova, A.; Brockhoff, P.B.; Christensen, R.H.B. LmerTest Package: Tests in Linear Mixed Effects Models. *J. Stat. Softw.* **2017**, *82*, 1–26. [CrossRef]
55. Tyśkiewicz, R.; Nowak, A.; Ozimek, E.; Jaroszuk-ściseł, J. *Trichoderma*: The Current Status of Its Application in Agriculture for the Biocontrol of Fungal Phytopathogens and Stimulation of Plant Growth. *Int. J. Mol. Sci.* **2022**, *23*, 2329. [CrossRef]
56. Woo, S.L.; Ruocco, M.; Vinale, F.; Nigro, M.; Marra, R.; Lombardi, N.; Pascale, A.; Lanzuise, S.; Manganiello, G.; Lorito, M. *Trichoderma*-Based Products and Their Widespread Use in Agriculture. *Open Mycol. J.* **2014**, *8*, 71–126. [CrossRef]
57. Shoresh, M.; Harman, G.E. The Molecular Basis of Shoot Responses of Maize Seedlings to *Trichoderma harzianum* T22 Inoculation of the Root: A Proteomic Approach. *Plant Physiol.* **2008**, *147*, 2147–2163. [CrossRef] [PubMed]
58. Tucci, M.; Ruocco, M.; De Masi, L.; De Palma, M.; Lorito, M. The Beneficial Effect of *Trichoderma* Spp. on Tomato Is Modulated by the Plant Genotype. *Mol. Plant Pathol.* **2011**, *12*, 341–354. [CrossRef]
59. Ponzio, C.; Gols, R.; Pieterse, C.M.J.; Dicke, M. Ecological and Phytohormonal Aspects of Plant Volatile Emission in Response to Single and Dual Infestations with Herbivores and Phytopathogens. *Funct. Ecol.* **2013**, *27*, 587–598. [CrossRef]
60. Walling, L.L. The Myriad Plant Responses to Herbivores. *J. Plant Growth Regul.* **2000**, *19*, 195–216. [CrossRef]
61. Salas-Marina, M.A.; Silva-Flores, M.A.; Uresti-Rivera, E.E.; Castro-Longoria, E.; Herrera-Estrella, A.; Casas-Flores, S. Colonization of *Arabidopsis* Roots by *Trichoderma atroviride* Promotes Growth and Enhances Systemic Disease Resistance through Jasmonic Acid/Ethylene and Salicylic Acid Pathways. *Eur. J. Plant Pathol.* **2011**, *131*, 15–26. [CrossRef]
62. Contreras-Cornejo, H.A.; Macías-Rodríguez, L.; Beltrán-Peña, E.; Herrera-Estrella, A.; López-Bucio, J. *Trichoderma*-Induced Plant Immunity Likely Involves Both Hormonal- and Camalexin-Dependent Mechanisms in *Arabidopsis thaliana* and Confers Resistance against Necrotrophic Fungus *Botrytis cinerea*. *Plant Signal. Behav.* **2011**, *6*, 1554–1563. [CrossRef]
63. Bari, R.; Jones, J.D.G. Role of Plant Hormones in Plant Defence Responses. *Plant Mol. Biol.* **2009**, *69*, 473–488. [CrossRef]
64. Perazzolli, M.; Dagostin, S.; Ferrari, A.; Elad, Y.; Pertot, I. Induction of Systemic Resistance against *Plasmopara viticola* in Grapevine by *Trichoderma harzianum* T39 and Benzothiadiazole. *Biol. Control* **2008**, *47*, 228–234. [CrossRef]
65. Poveda, J. *Trichoderma* as Biocontrol Agent against Pests: New Uses for a Mycoparasite. *Biol. Control* **2021**, *159*, 104634. [CrossRef]
66. Coppola, M.; Diretto, G.; Digilio, M.C.; Woo, S.L.; Giuliano, G.; Molisso, D.; Pennacchio, F.; Lorito, M.; Rao, R. Transcriptome and Metabolome Reprogramming in Tomato Plants by *Trichoderma harzianum* Strain T22 Primes and Enhances Defense Responses against Aphids. *Front. Physiol.* **2019**, *10*, 745. [CrossRef] [PubMed]
67. Contreras-Cornejo, H.A.; Viveros-Bremauntz, F.; Del-Val, E.; Macías-Rodríguez, L.; López-Carmona, D.A.; Alarcón, A.; González-Esquivel, C.E.; Larsen, J. Alterations of Foliar Arthropod Communities in a Maize Agroecosystem Induced by the Root-Associated Fungus *Trichoderma harzianum*. *J. Pest Sci.* **2020**, *94*, 363–374. [CrossRef]
68. Afechtal, M.; Aarabe, A.; Chebli, B.; Mounir, M. The Occurrence of Major Viruses Infecting Zucchini Squash (*Cucurbita pepo* L.) in Morocco. *Eur. Sci. J.* **2019**, *15*, 188–196. [CrossRef]
69. El-Sharkawy, E.E.S.; Abdelrazik, E. Biocontrol of *Fusarium* Root Rot in Squash Using Mycorrhizal Fungi and Antagonistic Microorganisms. *Egypt. J. Biol. Pest Control* **2022**, *32*, 13. [CrossRef]
70. Li, M.; Ma, G.; Lian, H.; Su, X.; Tian, Y.; Huang, W.; Mei, J.; Jiang, X. The Effects of Trichoderma on Preventing Cucumber *Fusarium wilt* and Regulating Cucumber Physiology. *J. Integr. Agric.* **2019**, *18*, 607–617. [CrossRef]
71. Müller, C.B.; Adriaanse, I.C.T.; Belshaw, R.; Godfray, H.C.J. The Structure of an Aphid-Parasitoid Community. *J. Anim. Ecol.* **1999**, *68*, 346–370. [CrossRef]
72. Weisser, W.W. Metapopulation Dynamics in an Aphid-Parasitoid System. *Entomol. Exp. Appl.* **2000**, *97*, 83–92. [CrossRef]
73. Karley, A.J.; Parker, W.E.; Pitchford, J.W.; Douglas, A.E. The Mid-Season Crash in Aphid Populations: Why and How Does It Occur? *Ecol. Entomol.* **2004**, *29*, 383–388. [CrossRef]
74. Reimer-Michalski, E.-M.; Conrath, U. Innate Immune Memory in Plants. *Semin. Immunol.* **2016**, *28*, 319–327. [CrossRef]

75. Coppola, M.; Cascone, P.; Chiusano, M.L.; Colantuono, C.; Lorito, M.; Pennacchio, F.; Rao, R.; Woo, S.L.; Guerrieri, E.; Digilio, M.C. *Trichoderma harzianum* Enhances Tomato Indirect Defense against Aphids. *Insect Sci.* **2017**, *24*, 1025–1033. [CrossRef] [PubMed]
76. Boege, K.; Dirzo, R.; Siemens, D.; Brown, P. Ontogenetic Switches from Plant Resistance to Tolerance: Minimizing Costs with Age? *Ecol. Lett.* **2007**, *10*, 177–187. [CrossRef] [PubMed]
77. Boege, K.; Marquis, R.J. Facing Herbivory as You Grow up: The Ontogeny of Resistance in Plants. *Trends Ecol. Evol.* **2005**, *20*, 441–448. [CrossRef]
78. Trotta, V.; Toma, I.; Forlano, P.; Fanti, P.; Prieto, J.D.; Battaglia, D. The Age of Tomato Plants Affects the Development of *Macrosiphum euphorbiae* (Thomas, 1878) (Hemiptera) Colonies. *Agron. Colomb.* **2021**, *39*, 3–7. [CrossRef]
79. Vitti, A.; Pellegrini, E.; Nali, C.; Lovelli, S.; Sofo, A.; Valerio, M.; Scopa, A.; Nuzzaci, M. *Trichoderma harzianum* T-22 Induces Systemic Resistance in Tomato Infected by Cucumber Mosaic Virus. *Front. Plant Sci.* **2016**, *7*, 1–11. [CrossRef] [PubMed]
80. Hafez, Y.M.; El-Nagar, A.S.; Elzaawely, A.A.; Kamel, S.; Maswada, H.F. Biological Control of *Podosphaera xanthii* the Causal Agent of Squash Powdery Mildew Disease by Upregulation of Defense-Related Enzymes. *Egypt. J. Biol. Pest Control* **2018**, *28*. [CrossRef]
81. Elsisi, A.A. Evaluation of Biological Control Agents for Managing Squash Powdery Mildew under Greenhouse Conditions. *Egypt. J. Biol. Pest Control* **2019**, *29*, 89. [CrossRef]
82. El-Sharkawy, H.H.A.; Abbas, M.S.; Soliman, A.S.; Ibrahim, S.A.; El-Nady, I.A.I. Synergistic Effect of Growth-Promoting Microorganisms on Bio-Control of *Fusarium oxysporum* f. Sp. Pisi, Growth, Yield, Physiological and Anatomical Characteristics of Pea Plants. *Pestic. Biochem. Physiol.* **2021**, *178*, 104939. [CrossRef]

microorganisms

Article

Assessing the Sensitivity of *Plasmopara halstedii* Isolates to Mefenoxam through Host Responses

Nisha Nisha [1], Sergey Vinogradov [2], Katalin Körösi [1], Arbnora Berisha [1] and Rita Bán [1,*]

1 Department of Integrated Plant Protection, Institute of Plant Protection, Hungarian University of Agriculture and Life Sciences, H-2100 Godollo, Hungary; nisha27evs@gmail.com (N.N.); korosi.katalin.orsolya@uni-mate.hu (K.K.); a.berisha5@hotmail.com (A.B.)

2 Department of Agricultural Data Processing and Data Analysis, Institute of Agricultural and Food Economics, Hungarian University of Agriculture and Life Sciences, H-2100 Godollo, Hungary; vinogradov.szergej@uni-mate.hu

* Correspondence: ban.rita@uni-mate.hu

Abstract: Downy mildew caused by *Plasmopara halstedii* is responsible for significant economic losses in cultivated sunflowers. Field isolates of sunflower downy mildew resistant to mefenoxam, a previously effective active ingredient against the pathogen, have been found across Europe. The main goal of this study was to assess the sensitivity of *P. halstedii* isolates to mefenoxam through host responses to infection, such as symptoms measured by disease severity and growth reduction, and host tissue reactions, such as hypersensitive reaction and necrosis of invaded cells. Sunflower seeds were treated with Apron XL 350 FS at the European registered rate (3 mg/kg seeds). Seedlings were inoculated using the soil drench method with eight Hungarian *P. halstedii* isolates. Disease rates and plant heights were measured twice. Histological examinations of cross-sections of sunflower hypocotyls were performed using a fluorescence microscope. In our study, cluster analyses of sunflowers based on macroscopic and microscopic variables showed differentiation of groups of mefenoxam-treated sunflowers inoculated with different *P. halstedii* isolates. We first revealed a clear difference in host responses of mefenoxam-treated susceptible sunflowers. In addition, examining tissue reactions (e.g., hypersensitive reaction, necrosis) seems more accurate to estimate the sensitivity of *P. halstedii* isolates to mefenoxam than macroscopic symptoms.

Keywords: sunflower downy mildew; fungicide resistance; fluorescence microscopy; host reactions; hypersensitive reaction; mefenoxam

Citation: Nisha, N.; Vinogradov, S.; Körösi, K.; Berisha, A.; Bán, R. Assessing the Sensitivity of *Plasmopara halstedii* Isolates to Mefenoxam through Host Responses. *Microorganisms* **2023**, *11*, 821. https://doi.org/10.3390/microorganisms11040821

Academic Editors: Anuj Ranjan, Vishnu D. Rajput and Abhishek Chauhan

Received: 2 March 2023
Revised: 21 March 2023
Accepted: 21 March 2023
Published: 23 March 2023

1. Introduction

Sunflower (*Helianthus annuus* L.) is one of the essential crops in the world and the second most widely farmed oil seed in the European Union. Sunflower oil production worldwide was 19.2 million tons in 2020 [1]. Diseases can significantly compromise crop security by reducing yield and affecting oil content. For example, crop yield loss and quality degradation caused by plant pathogens can be up to 100% in sunflowers [2] (pp. 201–226). Sunflower downy mildew caused by the pathogen *Plasmopara halstedii* (Farl.) Berl. et de Toni is one of the most severe global diseases impacting production [3]. *Plasmopara halstedii* is an obligate biotrophic oomycete that needs a living host to complete its life cycle [4]. This pathogen is diploid and homothallic, able to reproduce asexually and sexually. Sunflower downy mildew spreads via wind and infected seeds; however, *P. halstedii* is mainly soil-borne [5]. The pathogen infects seedlings via their roots through zoospores, leading to systemic infection, and may cause local foliar lesions via airborne sporangia. Moreover, a local infection can turn systemic, resulting in the deformation of upper plant parts [6].

The symptoms of downy mildew in sunflower vary depending on the age of the tissue, the cultivars/genotypes utilized, and the environmental conditions at the time of infection [7]. Infected plants are underdeveloped and dwarfed, with chlorotic leaves coated with white sporangiophores and sporangia [8]. A high percentage of infection in the field, a short latent period, a high sporulation density, and a significant reduction in the hypocotyl length indicate the high aggressiveness of the pathogen [9]. Although severely diseased plants may die before or soon after emergence or during the seedling stage, most symptomatic plants survive but do not produce viable seeds. The potential yield loss after primary infection is often as high as 50% [2].

Aside from crop rotation, resistance breeding and chemical seed treatment are fundamental ways of controlling sunflower downy mildew [8]. Dominant *Pl* genes (downy mildew resistance genes) incorporated into sunflower hybrids confer resistance to the disease. However, several new virulent *P. halstedii* pathotypes have developed, overcoming the effect of those genes [10]. There are currently 50 pathotypes (virulence phenotypes) of the pathogen worldwide [3]. Regarding seed treatment, metalaxyl, a systemic phenyl amide, has been widely used to control different oomycetes because of its excellent preventive, curative, and eradicative effects [5]. Metalaxyl was later substituted with its stereoisomer, mefenoxam (metalaxyl-M), which is effective even at lower rates. Metalaxyl and mefenoxam are active ingredients of single-site fungicides that affect the specific metabolism of the target pathogen [11]. They block the rRNA biosynthesis (polymerase complex I) of pathogens, inhibiting mycelial growth and sporulation.

Shortly after the first field application of metalaxyl and mefenoxam, tolerant strains could be identified for several oomycetes (see Gisi and Sierotzki [12] (pp. 145–174) for review). First, Oros and Viranyi [13] showed resistance of *P. halstedii* to metalaxyl in greenhouse experiments in Hungary. Later, Delen et al. [14] also detected decreased pathogen sensitivity to this active ingredient in Turkey. Soon after, Lafon et al. [15] and Albourie et al. [16] in France, Gulya [17] (pp. 79–84) in the USA, Molinero-Ruiz et al. [18] in Spain, Körösi et al. [19] in Hungary, and Iwebor et al. [20] in Russia reported that some *P. halstedii* isolates were not controlled when the registered rate of mefenoxam was applied. The expression of fungicide resistance employed by the FRAC (Fungicide Resistance Action Committee) refers to an acquired, heritable reduction in the sensitivity of a pathogen to a particular fungicide [21]. The mechanism of resistance to mefenoxam has yet to be discovered. However, it has shown to be quantitative, i.e., the reduction in disease control and the loss of sensitivity of pathogen populations is gradual and partial. Furthermore, mefenoxam is an active ingredient with a high risk of resistance, according to the FRAC code list [22].

Since the causal agent of sunflower downy mildew is biotrophic, i.e., a living plant is necessary for its development, the sensitivity of *P. halstedii* isolates to mefenoxam can be measured through the plant's response. Earlier studies have established the sensitivity of *P. halstedii* to mefenoxam mainly based on symptoms (dwarfing, leaf chlorosis) and signs (sporulation) on the infected plants. In addition, the pathogen's development in the mefenoxam-treated susceptible seedlings was also studied with fluorescent microscopy by Mouzeyar et al. [23]. Similar host responses in the mefenoxam-treated plants to the pathogen, such as hypersensitive reactions, necrosis, and cell division, were found to be that of the genetically resistant, non-treated sunflowers. However, only a *P. halstedii* isolate sensitive to mefenoxam was included in that study. Furthermore, the studies that have tested the sensitivity of several *P. halstedii* isolates to mefenoxam are primarily in vivo tests examining macroscopic symptoms and signs of the treated and inoculated plants [16,17,19]. While understanding plant tissue responses to different pathogen variants is essential for safe crop production, the goals of this work were the following:

(1) to study the histopathology of hypocotyl infection in a susceptible sunflower cultivar inoculated with *P. halstedii* isolates with varying degrees of sensitivity to mefenoxam;
(2) to assess the sensitivity of *P. halstedii* isolates to mefenoxam through host responses to infection, such as symptoms measured by disease severity and growth reduction, and host tissue reactions, such as hypersensitive reaction and necrosis of invaded cells.

2. Materials and Methods

2.1. Plant Material and Treatment of Seeds with Mefenoxam

The sunflower cultivar Iregi szurke csikos was used in this experiment. It is susceptible to all pathotypes of *P. halstedii* because of the absence of *Pl* resistance genes. Seeds were disinfected via immersion in 1% sodium hypochlorite solution (42 g L^{-1} NaOCl) for 5 min and then rinsed with running tap water. Seeds were treated with Apron XL 350 FS (350 g/L mefenoxam, Syngenta AG, Switzerland) as per the European registered rate, i.e., 3 mg *a.i.* kg^{-1} seeds (*a.i.* = active ingredient) were evenly coated with the fungicide by mixing in a beaker. Treated seeds were kept for drying at room temperature for three days, then planted in pots (d = 8 cm, depth of sowing: 1.5 cm) containing clean, moistened perlite and, except for the sporulation induction period (for 24 h at 19 °C), kept in a growth chamber for 21 days (22 °C, relative humidity: 70%, 12 h photoperiod, light irradiance of 100 $\mu E \cdot m^{-2} \cdot s^{-1}$).

2.2. Experimental Design

Mefenoxam-treated inoculated and non-treated inoculated seeds were placed in pots (5 seeds per pot) and arranged in trays (10 pots per tray) for each *P. halstedii* isolate. Mefenoxam-treated non-inoculated and non-treated non-inoculated controls were also treated with the same arrangement as the inoculated ones to check normal plant growth. The experiment was repeated twice.

2.3. *Plasmopara halstedii* Isolates, Preparation of Inoculum, and Inoculation

Eight *P. halstedii* isolates collected in different years and locations were selected for this study from the collection of the Department of Integrated Plant Protection (Institute of Plant Protection, MATE, Godollo, Hungary) (Table 1). Isolates were stored at −70 °C on infected leaves in plastic Petri dishes. Pathotype identification of these isolates was made previously as a part of a survey between 2012 and 2019 in Hungary [3].

Table 1. List of *Plasmopara halstedii* isolates collected in Hungary (2012–2017) used in the experiment.

Isolate Code	Locality (County)	Year of Collection	Pathotype (Virulence Phenotype)
1	Mezőkovácsháza (Békés)	2017	724
4	Kömlő (Heves)	2014	704
5	Doboz (Békés)	2014	704
6	Körösladány (Békés)	2014	714
7	Szeghalom (Békés)	2017	724
8	Pély (Heves)	2017	704
9	Bonyhád (Tolna)	2017	724
11	Rákóczifalva (Jász-Nagykun-Szolnok)	2012	704

During inoculum preparation, frozen leaves with sporangia of *P. halstedii* were washed off in bidistilled water. The suspension concentration was adjusted to 50,000 sporangia per ml using a Bürker counting chamber. Three days after sowing, seedlings were inoculated using the soil drench method [24], i.e., the sporangial suspension (2 mL per seedling) was pipetted directly onto the perlite surface of each pot containing the seedlings. Non-inoculated control plants were included to ensure the damping-off symptom was attributable to the disease. For non-inoculated plants, bidistilled water was drenched over seedlings. Then, plants were kept at 16 °C in the dark for 24 h to ensure infection.

2.4. Disease Assessment

Nine days after inoculation, plants were sprayed with bidistilled water and covered with a dark polyethylene bag. Then, pots were placed in the dark for 24 h at 19 °C (relative humidity: 90–100%) to induce sporulation. The first evaluation was based on white coating (sporangiophores and sporangia) on cotyledons and pre-emergence damping-off, referring to Disease 1. Twenty-one days after inoculation, a second evaluation was made according to chlorosis along the veins of the true leaves, if they had them, or post-emergence damping-off, referring to Disease 2. Disease rates (%) for Disease 1 and 2 values were calculated as the percentage of diseased plants for all isolates. Plant heights were measured during each disease assessment (Height 1 and 2).

2.5. Microscopic Observations

Histological examinations of cross-sections of sunflower hypocotyls were performed using a fluorescence microscope (Olympus, Japan; filter block BX 50, transmission > 515 nm). Twenty-one days after inoculation, five sunflower hypocotyls were selected from each treatment (treated and non-treated with mefenoxam and inoculated with different *P. halstedii* isolates, respectively) and fixed in FAA solution (formalin: acetic acid: ethanol, 10:5:50 by vol.). Then, thin cross-sections (15–20 pieces) were cut with a razor blade from both the upper and lower parts of the hypocotyl and examined for pathogen structures (hyphae, haustoria) and tissue responses (hypersensitive reaction, cell necrosis). The hypersensitive reaction was defined by the autofluorescence of the cells and necrosis by the presence of brown, dead cells. In accordance with Bán et al. [25] (pp. 265–273), a 0–4 scale was used to observe pathogen structures and host responses. Briefly, the sections were divided theoretically into four quarters (both the cortical and pith parenchyma), and the presence of the pathogen and the plant responses were examined in each.

2.6. Statistical Analysis

Differences in disease rates, host characteristics (plant height), and host responses (HR and cell necrosis) were assessed by the one-way analysis of variance (ANOVA) followed by the Tukey HSD (honestly significant difference) multiple comparison post hoc test. In addition, the Kolmogorov–Smirnov test was used to test the normality of the distribution of the data within groups. Levene's test was used to determine whether variances were equal.

Two-way ANOVA was used to examine the interaction between treatment (non-treated, treated) and isolates. Using Ward's method, hierarchical cluster analysis was performed to group *P. halstedii* isolates based on their sensitivity to mefenoxam. To examine the correlation between variables, Pearson's correlation coefficient was performed for scale variables (disease rates, heights), and Spearman's correlation coefficient was used for ordinal variables (microscopic variables). IBM SPSS Statistics 27 software was used to conduct the statistical analysis.

3. Results

3.1. Disease Rates and Heights

Disease rates (%) and heights of mefenoxam-treated and non-treated sunflower plants inoculated with different *P. halstedii* isolates are shown in Figure 1. According to the sporulation of the pathogen on the cotyledons and pre-emergence damped off plants (Disease 1, Figure 1a), mefenoxam-treated sunflowers inoculated with isolates 1, 4, 5, 6, 7, and 9 showed significantly lower infection rates compared to non-treated ones. However, there were no significant differences in disease rates between treated and non-treated plants inoculated with isolates 8 and 11. The situation was similar with Disease 2 (ratio of chlorotic post-emergence damped-off plants and healthy sunflowers, Figure 1b), but there was no difference in the disease rate of treated and non-treated plants inoculated with isolates 7 in addition to isolates 8 and 11.

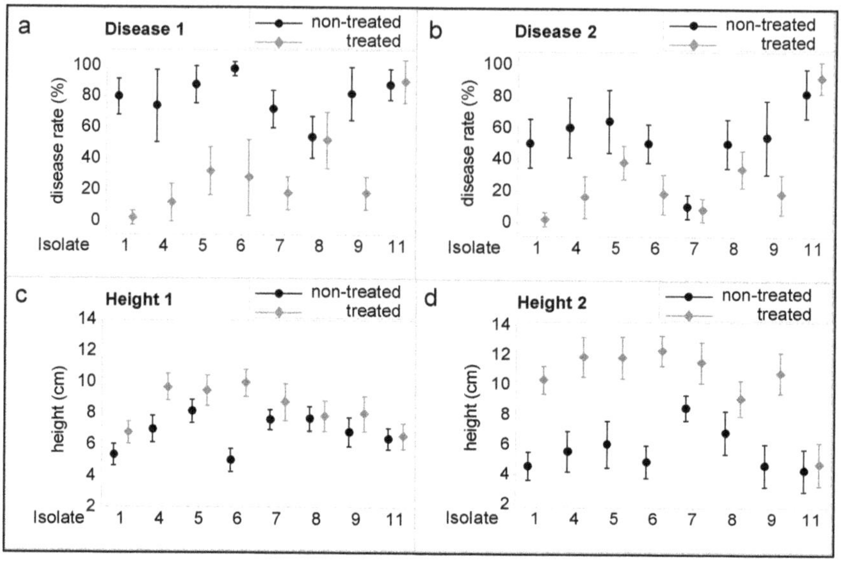

Figure 1. Disease rates (**a**,**b**) and heights (**c**,**d**) of mefenoxam-treated and non-treated sunflower plants inoculated with different *Plasmopara halstedii* isolates. Disease 1: ratio of sporulating pre-emergence damped-off plants and healthy sunflowers nine days after inoculation. Disease 2: ratio of chlorotic post-emergence damped-off plants and healthy sunflowers 21 days after inoculation. Height 1: height of sunflowers nine days after inoculation (heights of damped-off plants were taken as zero). Height 2: height of sunflowers 21 days after inoculation (heights of damped-off plants were taken as zero). Treatment: non-treated and treated with mefenoxam (3 mg/kg seed). Isolate: code of *Plasmopara halstedii* isolates used in the experiment (1, 4, 5, 6, 7, 8, 9, and 11) (for more details, see Table 1). Vertical lines represent 95% confidence intervals (95% CI) of the mean values of disease rates and heights.

Plant heights of mefenoxam-treated sunflowers inoculated with *P. halstedii* isolates 1, 4, 5, and 6 were significantly higher than those of the non-treated inoculated plants nine days after inoculation (Figure 1c). On the contrary, there was no significant difference in plant heights between treated and non-treated sunflowers inoculated with isolates 7, 8, 9, and 11. However, by the second recording date, the height of the treated plants was significantly higher than that of the non-treated plants for all isolates except 11 (Figure 1d).

For all parameters tested (Disease 1–2, Height 1–2), the interaction between isolate and treatment was significant (for Disease 1: $F = 12.06$, $p < 0.001$, for Disease 2: $F = 5.36$, $p < 0.001$, for Height 1: $F = 6.61$, $p < 0.001$, for Height 2: $F = 7.37$, $p < 0.001$), i.e., the impact of treatment varied between isolates.

3.2. Microscopic Observations

Sunflower tissue responses to infection by *P. halstedii* in hypocotyl cross-sections are shown in Figure 2. Similar tissue responses were observed in most treated and non-treated plants infected with different isolates, but the intensity of the pathogenic spread and tissue responses were variable (see below). In general, intercellular hyphae and intracellular haustoria were detected in the hypocotyl of non-treated plants in the cortical and the pith parenchyma 21 days after inoculation (Figure 2a). Under UV light, autofluorescence appeared in the intercellular spaces around hyphae, giving the image a dotted appearance (Figure 2b). In contrast, cell browning under normal light (Figure 2c) and an intense autofluorescence of cells showing a hypersensitive-like reaction (Figure 2d) could be detected in cross-sections of several mefenoxam-treated and inoculated sunflowers. Moreover, the development of cellular necrosis by vigorous cell division (Figure 2e) and the strong fluorescent response of surrounding cells (Figure 2f) was also frequently observed in treated and inoculated plants.

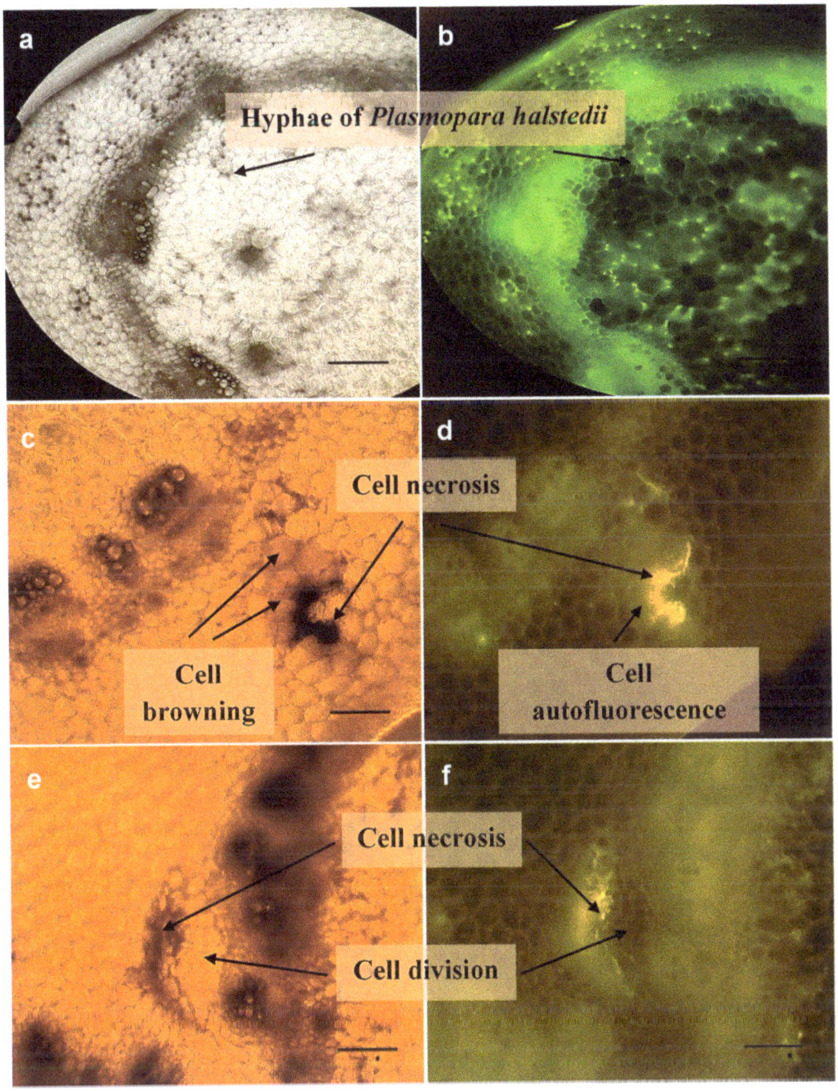

Figure 2. Light micrographs of mefenoxam-activated resistance responses in hypocotyl cross-sections of sunflower. Hyphae of *Plasmopara halstedii* invade cells of non-treated, inoculated susceptible plants (cv. Iregi szürke csíkos) without any host responses in normal (**a**) and in UV light (**b**) (λ = 485 nm) at 21 dpi. Browning (**c**), autofluorescence (hypersensitive reaction) (**d**), and necrosis with intensive cell division (**e**: normal light, **f**: UV light) of cortical parenchyma cells neighboring invaded cells as a host response to the pathogenic attack of mefenoxam-treated inoculated plants at 21 dpi. Scale bar = 100 µm.

The rates of pathogen hyphal spread and tissue responses are shown in Figure 3. Hyphae spread significantly in the cortical and pith parenchyma of non-treated plants inoculated with isolates 1, 4, 5, and 7 compared to mefenoxam-treated plants (Figure 3a,b).

In contrast, more hyphae were found in the cortical and pith parts of mefenoxam-treated sunflowers inoculated with *P. halstedii* isolate 8 than in non-treated ones. The situation was similar to the appearance of hyphae of isolate 11 in the pith. In addition, hyphae were significantly more abundant in the cortical part of non-treated sunflowers inoculated with isolate 9, whereas there was no significant difference in hyphal distribution between treated and non-treated sunflowers for isolate 6 (Figure 3a,b).

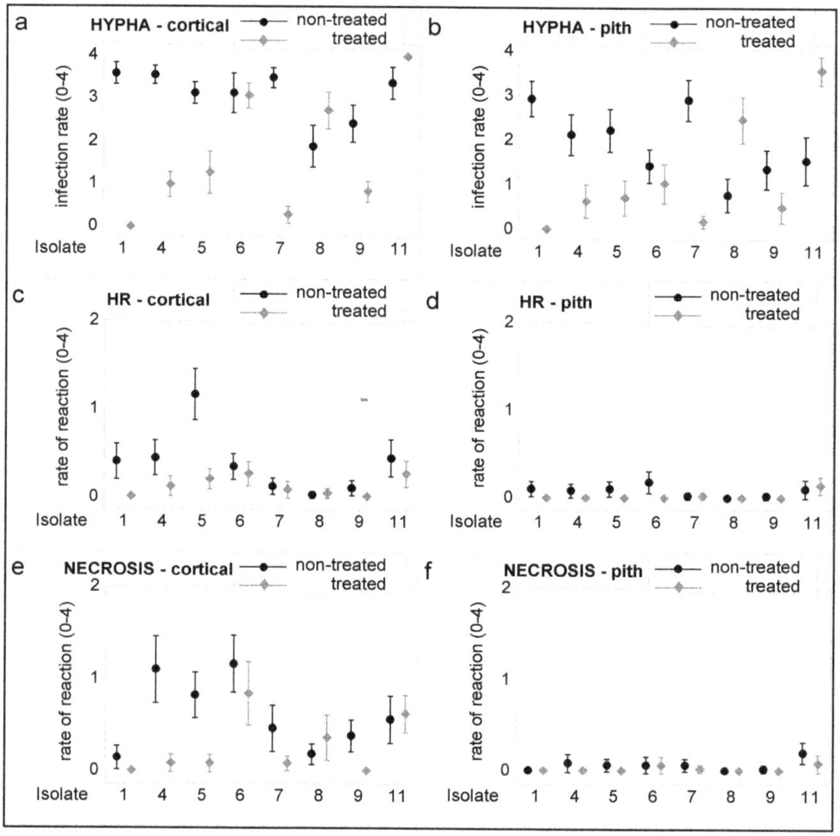

Figure 3. Occurrence of pathogen hyphae (**a**,**b**) and host reactions such as hypersensitive reaction (**c**,**d**) and necrosis (**e**,**f**) in the cortical and pith parenchyma of mefenoxam-treated and non-treated sunflower plants inoculated with *Plasmopara halstedii*. Treatment: non-treated and treated with mefenoxam (3 mg/kg seed). Isolate: code of *Plasmopara halstedii* isolates used in the experiment (1, 4, 5, 6, 7, 8, 9, and 11) (for more details, see Table 1). The infection rate and the rate of the host reaction were measured on a 0–4 scale. Vertical lines represent 95% confidence intervals (95% CI) of the mean values of disease rates and heights.

Generally, fluorescence microscopy of cross-sections of sunflower hypocotyls revealed a relatively higher rate of hypersensitive-like reaction and necrosis (cell death) in the cortical than in the pith parenchyma in this experiment (Figure 3c–f). The hypersensitive reaction was prominent in non-treated plants inoculated with isolate 5 and to a smaller extent in non-treated sunflowers inoculated with isolates 1, 4, 6, and 11 in the cortical parenchyma (Figure 3c,d). However, it was not significant for the latter two compared to mefenoxam-treated plants. The occurrence of cell necrosis in the cortical part was intensive in non-treated plants inoculated with isolates 4, 5, and 6. The latter was not significant compared to mefenoxam-treated sunflowers (Figure 3e). Necrosis in the pith parenchyma cells was minimal in each sample (Figure 3f).

3.3. *Assessing the Sensitivity of Plasmopara halstedii Isolates to Mefenoxam*

Cluster analyses of sunflowers inoculated with different *P. halstedii* isolates based on disease rates and plant heights are shown in Table 2. Four distinct clusters could be identified using macroscopic parameters. Cluster 1 includes non-treated plant samples inoculated with isolates 5, 6, 9, and 11, and mefenoxam-treated plants inoculated with isolate 11, which were found to have high infection levels in both sampling periods. Therefore, the pathogen could penetrate the upper parts of these sunflowers. Plant heights were the lowest in this group. In Cluster 2 are samples of the other parts of non-treated and inoculated plants, where the first infection value (Disease 1) was relatively high, as in Cluster 1. However, unlike the first cluster, the second time point for disease assessment (Disease 2) resulted in much lower infection values and less plant dwarfing in Cluster 2 members (Table 2). In this case, the pathogen could only penetrate to a lesser extent above the hypocotyl.

Table 2. Cluster analyses of sunflowers inoculated with different *P. halstedii* isolates based on disease rates and plant heights.

Variables	Cluster 1	Cluster 2	Cluster 3	Cluster 4
Disease 1 (%)	90.2 ± 6.9 d	72.2 ± 12 c	20.4 ± 12.3 a	38.2 ± 13.3 b
Disease 2 (%)	74.5 ± 10.8 c	29.6 ± 10.4 b	15.9 ± 8.6 a	27.3 ± 10.2 ab
Height 1 (cm)	6.0 ± 0.8 a	7.1 ± 0.4 b	9.7 ± 0.8 c	7.4 ± 0.6 b
Height 2 (cm)	4.0 ± 1.0 a	7.5 ± 0.6 b	11.7 ± 1.1 d	9.3 ± 0.9 c

Data represent the means of variables for each cluster. Values followed by means represent standard deviation. Different letters (a, b, c, d) indicate significant differences based on the Tukey HSD post-hoc test ($p < 0.05$) among clusters, but not comparable between variables. Cluster 1: isolates 1, **5, 6, 9, 11** non-treated, **11** treated. Cluster 2: isolates **1, 4, 7, 8** non-treated. Cluster 3: isolates **1, 4, 5, 6**, 7, 9 treated. Cluster 4: isolates 1, 4, 5, 6, **7, 8, 9** treated, 8, 9 non-treated. Bold isolate numbers indicate dominance of that isolate in that cluster compared to other clusters.

Clusters 3 and 4 mainly include samples of inoculated plants treated with mefenoxam. In contrast to the initial infection rates, there was no significant difference between the two clusters in the second survey. However, the plant height values were significantly higher for Cluster 3 members (Table 2).

Cluster analyses of sunflowers based on the examined microscopic variables inoculated with different *P. halstedii* isolates are presented in Table 3. Three distinct clusters could be identified using microscopic parameters. Samples of non-treated inoculated plants are in the first two clusters, while mefenoxam-treated plants can be found in all three clusters. Moreover, treated plants inoculated with isolates 4 and 5 are equally represented in the first two clusters.

Table 3. Cluster analyses of sunflowers inoculated with different *P. halstedii* isolates based on the examined microscopic variables.

Variables	Cluster 1	Cluster 2	Cluster 3
H_Cort	3.7 ± 0.3 c	3.0 ± 0.5 b	0.2 ± 0.2 a
HR_Cort	0.4 ± 0.4 c	0.2 ± 0.2 b	0 a
NEC_Cort	0.7 ± 0.5 c	0.5 ± 0.4 b	0 a
H_Pith	3.6 ± 0.4 c	0.5 ± 0.3 b	0 a
HR_Pith	0.1 ± 0.2 b	0 a	0 a
NEC_Pith	0.1 ± 0.2 b	0 a	0 a

Data represent the means of variables for each cluster. Values followed by means represent standard deviation. Different letters (e.g., a, b) indicate significant differences based on the Tukey HSD post-hoc test ($p < 0.05$) among clusters, but not comparable between variables. Cluster 1: isolates **1, 4, 5, 7** non-treated, **8, 11** treated. Cluster 2: isolates **4, 5, 6, 8, 9, 11** non-treated, **6** treated. Cluster 3: isolates **1, 4, 5, 7, 9** treated. Bold isolate numbers indicate dominance of that isolate in that cluster compared to other clusters. The underlined isolates were equally represented in the clusters concerned.

For Cluster 1 samples, the pathogen could invade both the cortical and pith parenchyma (Table 3). Not only the spread of hyphae but also the HR and necrosis in different tissue sections were significant in Cluster 1 samples compared to the other two clusters. Treated sunflowers inoculated with *P. halstedii* isolates 8 and 11 are included in the first cluster along with non-treated ones. Unlike the sunflowers in the first cluster, the distribution of hyphae of samples in Cluster 2 (isolates 1, 4, 7, and 8, non-treated) was accompanied by HR and necrosis only in the cortical parenchyma but not in the pith. Most of the treated sunflower samples, except for isolates 6, 8, and 11, are in Cluster 3 (isolates 1, 4, 5, 6, 7, and 9, treated), with few hyphae detected in the cortical tissues. No tissue response was detected in these sunflowers.

3.4. Correlations among Macroscopic Parameters

The results of Pearson correlation based on the examined macroscopic variables (disease rates, plant heights) are shown in Table 4. During the second evaluation, a strong negative correlation was found between the disease rate and plant height values of both non-treated and treated plants. Similarly, the experiment showed a strong negative correlation between the initial disease rates and the final plant height values of treated plants. In contrast, a high positive correlation could be detected between the initial and final plant height data of both treated and non-treated plants. In addition, a strong positive correlation was found between the initial and final disease values of mefenoxam-treated sunflowers.

Table 4. Pearson correlation among the examined variables (disease rates, plant heights).

Variable	Disease 1	Disease 2	Height 1	Height 2
Panel A: Non-treated (n = 80)				
Disease 1	1	0.346 **	−0.465 **	−0.550 **
Disease 2		1	−0.439 **	**−0.713 ****
Height 1			1	**0.737 ****
Height 2				1
Panel B: Treated (n = 80)				
Disease 1	1	**0.701 ****	−0.368 **	**−0.700 ****
Disease 2		1	−0.329 **	**−0.722 ****
Height 1			1	**0.741 ****
Height 2				1

Disease 1: ratio of sporulating damped-off plants and healthy sunflowers nine days after inoculation. Disease 2: ratio of chlorotic damped-off plants and healthy sunflowers 21 days after inoculation. Height 1: height of sunflowers nine days after inoculation (heights of damped-off plants were taken as zero). Height 2: height of sunflowers 21 days after inoculation (heights of damped-off plants were taken as zero). Treatment: non-treated and treated with mefenoxam (3 mg/kg seed). ** Correlation is significant at the 0.01 level (2-tailed). Values in bold indicate a strong correlation between variables.

3.5. Correlations among Microscopic Parameters

The Spearman correlation of the examined microscopic variables is presented in Table 5. There was a strong positive correlation in the occurrence of hyphae in different parenchymatic plant parts (cortical and pith) of both non-treated and treated inoculated sunflowers. Moreover, strong positive correlations were found between the presence of hyphae in the cortical parenchyma tissues and the appearance of hypersensitive reaction and necrosis in treated plants. In addition, a strong positive correlation could be confirmed between the establishment of necrosis in the cortical part and the occurrence of hyphae in the pith of mefenoxam-treated and inoculated sunflowers.

Table 5. Spearman correlation among the examined microscopic variables.

Variable	H_Cort	HR_Cort	NEC_Cort	H_Pith	HR_Pith	NEC_Pith
Panel A: Non-treated (n = 200)						
H_Cort	1	0.211 **	0.291 **	**0.508** **	0.158 **	0.150 **
HR_Cort		1	0.240 **	0.193 **	**0.375** **	0.080
Nec_Cort			1	0.223 **	0.155 **	0.172 **
H_Pith				1	0.156 **	0.240 **
HR_Pith					1	0.106 *
Nec_Pith						1
Panel B: Treated (n = 200)						
H_Cort	1	**0.327** **	**0.488** **	**0.759** **	0.174 **	0.153 **
HR_Cort		1	0.072	0.213 **	0.241 **	0.029
Nec_Cort			1	**0.547** **	0.079	0.180 **
H_Pith				1	0.204 **	0.169 **
HR_Pith					1	0.129 *
Nec_Pith						1

H: hyphae of Plasmopara halstedii, HR: hypersensitive reaction of invaded cells, Nec: necrosis, Cort: cortical parenchyma, Pith: pith parenchyma. ** Correlation is significant at the 0.01 level (2-tailed). * Correlation is significant at the 0.05 level (2-tailed). Values in bold indicate a strong correlation between variables.

4. Discussion

Field isolates of sunflower downy mildew resistant to mefenoxam, a previously effective active ingredient against the pathogen, were found across Europe [15,16,18,20] and in the USA [17] (pp. 79–84). There are no data from Asia and Africa on mefenoxam resistance in the pathogen. Moreover, in a recent study, seven out of ten *P. halstedii* isolates collected in Hungary showed poor to moderate sensitivity to mefenoxam [19]. According to our present study, with more detailed symptom recording and refined statistical analyses than in previous studies, reduced sensitivity could be measured for three out of eight downy mildew isolates.

Cluster analysis based on disease rates and plant heights showed a difference between mefenoxam-treated and non-treated plants in this experiment. The only exception was sunflowers inoculated with *P. halstedii* isolate 11, where the values of mefenoxam-treated and non-treated plants were similar to those of other non-treated plants. In addition, both treated and non-treated plants formed two relatively distinct groups (clusters) based on cluster analysis of disease rates and plant heights. The sunflowers in Cluster 1 (non-treated and inoculated with isolates 5, 6, 9, and 11) had relatively high initial and subsequent infection rates, indicating that the pathogen could penetrate unhindered into the upper parts of the plant. This was associated with significant growth inhibition of these plants. On the other hand, the reaction was similar in mefenoxam-treated plants inoculated with isolate 11; this *P. halstedii* isolate therefore appears to be mefenoxam resistant. Interestingly, in Cluster 2, non-treated plants inoculated with isolates 1, 4, 7, and 8 were characterized by decreased spreading of the pathogen to the aboveground plant parts compared to Cluster 1. The difference in pathogen spread between the two clusters of mainly non-treated plants is likely explained by the different aggressiveness of the *P. halstedii* isolates tested, a common phenomenon indicated by other authors [9].

Nevertheless, the two clusters of mefenoxam-treated and inoculated plants (Clusters 3 and 4) also differed, mainly in the degree of initial disease rate and the development of plant heights. In conclusion, treatment with mefenoxam had different effects on different *P. halstedii* isolates, according to disease rates and plant heights.

Pearson correlation, especially during the second evaluation, showed a strong negative correlation between the disease rate and plant height values of both non-treated and treated plants. This negative correlation is not surprising, as many authors have reported such effects of the pathogen on plant development in susceptible, non-treated sunflowers [8,26]. In the case of treated plants, this negative correlation is presumably related to fungicide resistance since if the pathogen can spread within the plant, the growth-reducing effect is exerted.

The main objective of this study was to investigate the tissue responses of treated plants inoculated with different *P. halstedii* isolates with a fluorescent microscope. Host responses of sunflowers (susceptible, resistant) inoculated with *P. halstedii* have already been examined by several authors [25,27–32]. Mouzeyar et al. [30,31] pointed out that *P. halstedii* could infect susceptible and resistant sunflower lines in a microscopic investigation, although to a lesser extent, even a susceptible plant can react to the pathogen's growth. Our results with fluorescent microscopy of non-treated sunflowers also supported this (Figure 3c–f, Table 3). Mefenoxam treatment alone did not induce autofluorescence in the treated plants in our studies, it was only when the treated plants were inoculated with the pathogen. Autofluorescence, one of the tissue reactions during host–parasite interactions, is mainly associated with the appearance of phenolic compounds (e.g., phytoalexins, lignin), which play an essential role in the plant's defense processes against the pathogen [30].

Moreover, the speed and intensity of host response to *P. halstedii* in a resistant sunflower may vary, and it can appear in the root or different parts of the hypocotyl [30]. Previous authors also described a hypersensitive-like response in the hypocotyl of mefenoxam-treated susceptible sunflowers [23]. They found that all metalaxyl concentrations and application modes provided complete protection against *P. halstedii*. However, only one *P. halstedii* isolate was tested in the latter work that seemed sensitive to the active ingredient.

We first revealed a clear difference in host responses of mefenoxam-treated susceptible sunflowers inoculated with various *P. halstedii* isolates. Treated plants inoculated with some isolates (6, 8, and 11) showed hyphal growth in the cortical and pith parenchyma. A moderate hypersensitive reaction and necrosis could also be detected in the cortical part. This phenomenon was very similar to what usually occurs in non-treated susceptible plants, with the plant response appearing to be a delayed host reaction to a pathogenic attack [8,30]. For other *P. halstedii* isolates, we could detect limited or no mycelial growth in the mefenoxam-treated plants, which was accompanied by a weak or no reaction of the treated sunflowers in their hypocotyls. Because of the lack of massive mycelial growth in the hypocotyl, it is likely that the pathogen was arrested in the root tissues by the chemical.

In our study, cluster analyses of sunflowers based on microscopic variables showed clear differentiation of three groups of mefenoxam-treated sunflowers inoculated with different *P. halstedii* isolates. Those in the first two groups (clusters) showed high (isolates 8 and 11) or moderate resistance (isolate 6) to mefenoxam, while isolates in the third group showed sensitivity. Disease rate and plant height values (macroscopic parameters) of treated and inoculated sunflowers with these resistant isolates also supported this (Table 3). However, only isolate 11 could be defined as having highly decreased sensitivity with the evaluation of visible symptoms (macroscopic parameters) (Figure 3). Hence, examining tissue reactions (e.g., hypersensitive reaction, necrosis) seems more accurate for estimating the sensitivity of *P. halstedii* isolates to mefenoxam than macroscopic symptoms.

In addition to its direct toxic effect on the pathogen, metalaxyl activates the host defense system, which might result in increased sunflower resistance, restricting pathogen development [33,34]. In previous research, histological alterations such as haustoria encapsulation by callose deposits [35] or the development of limited hypersensitive-like lesions were also reported, followed by metalaxyl treatment in some host–parasite interactions where the pathogen was sensitive to the chemical [23,36–38]. However, the question remains whether the direct (fungistatic) or indirect effect (through the host) of metalaxyl is more significant against the sensitive pathogen in different host–parasite relationships.

Examining metalaxyl-sensitive and tolerant *Phytophthora megasperma* isolates in soybean, Cahill and Ward [39] pointed out that metalaxyl enhanced the release of phytoalexin elicitors (glyceollin) in culture fluids of the sensitive isolate but not in those of the tolerant isolate. Releasing elicitors due to metalaxyl treatment could induce host reactions in compatible interactions with the sensitive isolate. In our study, the effective host responses against the sensitive P. halstedii isolates likely occurred at a very early stage of infection in the roots of mefenoxam-treated sunflowers. Despite this, the reaction of mefenoxam-treated plants to resistant isolates could appear later in the hypocotyl, which the delayed stimulation of elicitor activity by the chemical can explain. Our results with the Spearman correlation also demonstrate this. It showed that the spread of the resistant isolates in the cortical parenchyma of treated plants correlated positively with the appearance of HR and necrosis.

Interestingly, more abundant hyphae were found in the pith of treated than non-treated plants inoculated with isolates 8 and 11 (considered resistant). This is in line with the results of Cahill and Ward [39], who reported better growth of metalaxyl-tolerant *Phytophthora megasperma* isolates in the presence of the chemical in vitro and in vivo. Previous authors assumed that metalaxyl could serve as a nutrient and raised the idea of other resistance mechanisms and different interactions with the host (soybean) for those tolerant isolates. In addition, the more significant presence of *P. halstedii* in the pith of sunflowers has been shown to facilitate the spread of the pathogen to the upper parts of the plant (e.g., epicotyl) [40].

Further studies are needed to explore the reasons for the differences in tissue responses to sensitive and resistant isolates of *P. halstedii* in sunflower. In addition, how plant defense mechanisms contribute to the effectiveness of fungicides also has to be elucidated.

Author Contributions: Conceptualization, R.B., N.N. and K.K.; methodology, R.B., N.N., S.V., K.K. and A.B.; software, N.N., R.B. and S.V.; validation, N.N., S.V., K.K., A.B. and R.B.; formal analysis, R.B.; investigation, R.B., N.N., S.V., K.K. and A.B.; resources, N.N. and R.B.; data curation, R.B., S.V. and K.K.; writing—original draft preparation, N.N. and R.B.; writing—review and editing, N.N., S.V., K.K., A.B. and R.B.; visualization, N.N., R.B. and S.V.; supervision, R.B., S.V. and K.K.; project administration, R.B.; funding acquisition, N.N. and R.B. All authors have read and agreed to the published version of the manuscript.

Funding: This research was funded by EU Social Funds (EFOP 3.6.3.-VEKOP-16-2017-00008 project).

Data Availability Statement: All data generated or analyzed during this study are included in this article.

Acknowledgments: Nisha Nisha wishes to thank the Tempus Public Foundation of the Government of Hungary for a doctoral scholarship (Stipendium Hungaricum Scholarship Program Registration Number SHE-15651-001/2017). The authors thank József Kiss for his dedicated help and Rita Baraksó and Andrea Nagy for their technical aid.

Conflicts of Interest: The authors declare no conflict of interest.

References

1. Statista: Production Volume of Sunflower Seed Oil Worldwide from 2012/13 to 2022/22. Available online: https://www.statista.com/statistics/613490/sunflowerseed-oil-production-volume-worldwide/ (accessed on 1 August 2022).
2. Debaeke, P.; Mestries, E.; Desanlis, M.; Seassau, C. Effects of crop management on the incidence and severity of fungal diseases in sunflower. In *Sunflowers: Growth and Development, Environmental Influences and Pests/Diseases*; Arribas, J.I., Ed.; Botanical Research and Practices; Nova Science Publishers: New York, NY, USA, 2014; pp. 201–226. ISBN 978-1-63117-347-9.
3. Bán, R.; Kovács, A.; Nisha, N.; Pálinkás, Z.; Zalai, M.; Yousif, A.I.A.; Körösi, K. New and High Virulent Pathotypes of Sunflower Downy Mildew (*Plasmopara halstedii*) in Seven Countries in Europe. *J. Fungi* **2021**, *7*, 549. [CrossRef] [PubMed]
4. Fawke, S.; Doumane, M.; Schornack, S. Oomycete Interactions with Plants: Infection Strategies and Resistance Principles. *Microbiol. Mol. Biol. Rev.* **2015**, *79*, 263–280. [CrossRef] [PubMed]
5. Viranyi, F.; Spring, O. Advances in sunflower downy mildew research. *Eur. J. Plant Pathol.* **2011**, *129*, 207–220. [CrossRef]
6. Spring, O. Transition of secondary to systemic infection of sunflower with *Plasmopara halstedii*—An underestimated factor in the epidemiology of the pathogen. *Fungal Ecol.* **2009**, *2*, 75–80. [CrossRef]
7. Spring, O. Non-systemic infections of sunflower with *Plasmopara halstedii* and their putative role in the distribution of the pathogen. *J. Plant Dis. Prot.* **2001**, *108*, 329–336.
8. Gascuel, Q.; Martinez, Y.; Boniface, M.-C.; Vear, F.; Pichon, M.; Godiard, L. The sunflower downy mildew pathogen *Plasmopara halstedii*: *Plasmopara halstedii*, sunflower downy mildew. *Mol. Plant Pathol.* **2015**, *16*, 109–122. [CrossRef]
9. Sakr, N. Variation in aggressiveness of *Plasmopara halstedii* (sunflower downy mildew). *J. Plant Dis. Prot.* **2009**, *116*, 247–251. [CrossRef]
10. de Labrouhe, D.T.; Pilorge, E.; Nicolas, P.; Vear, F. Le mildiou du tournesol. In *CETIOM*; INRA: Paris, France, 2000.

11. Gisi, U.; Sierotzki, H. Fungicide modes of action and resistance in downy mildews. *Eur. J. Plant Pathol.* **2008**, *122*, 157–167. [CrossRef]
12. Gisi, U.; Sierotzki, H. Oomycete Fungicides: Phenylamides, Quinone Outside Inhibitors, and Carboxylic Acid Amides. In *Fungicide Resistance in Plant Pathogens*; Ishii, H., Hollomon, D.W., Eds.; Springer: Tokyo, Japan, 2015; pp. 145–174, ISBN 978-4-431-55641-1.
13. Oros, G.; Viranyi, F. Resistance of *Plasmopara halstedii* to metalaxyl in the greenhouse. *Temp. Downy Mildew Newsl.* **1984**, *3*, 22–23.
14. Delen, N.; Onoğur, E.; Yıldız, M. Sensitivity levels to metalaxyl in six *Plasmopara helianthi* Novot. isolates. *J. Turk. Phytopathol.* **1985**, *14*, 31–36.
15. Lafon, S.; Penaud, A.; Walser, P.; De Guenin, M.C.; Molinero, V. Le mildiou du tournesol toujours sous surveillance. *Phytoma Déf. Vég.* **1996**, *484*, 35–36.
16. Albourie, J.-M.; Tourvieille, J.; de Labrouhe, D.T. Resistance to metalaxyl in isolates of the sunflower pathogen *Plasmopara halstedii*. *Eur. J. Plant Pathol.* **1998**, *104*, 235–242. [CrossRef]
17. Gulya, T.J. Metalaxyl resistance in sunflower downy mildew and control through genetics and alternative fungicides. In Proceedings of the 15th International Sunflower Conference, Toulouse, France, 12–16 June 2000; International Sunflower Association: Toulouse, France, 2000; Volume 2, pp. 79–84.
18. Molinero-Ruiz, M.L.; Melero-Vara, J.M.; Gulya, T.J.; Dominguez, J. First Report of Resistance to Metalaxyl in Downy Mildew of Sunflower Caused by *Plasmopara halstedii* in Spain. *Plant Dis.* **2003**, *87*, 749. [CrossRef] [PubMed]
19. Körösi, K.; Kovács, A.; Nisha, N.; Bóta, I.; Perczel, M.; Yousif, A.I.A.; Kiss, J.; Bán, R. New data on pathotype distribution and mefenoxam tolerance of *Plasmopara halstedii* in Hungary. *Plant Prot. Sci.* **2021**, *57*, 31–37. [CrossRef]
20. Iwebor, M.V.; Antonova, T.S.; Araslanova, N.M.; Saukova, S.S.; Pitinova, Y.V.; Eliseeva, K.K. The situation in the population of the sunflower downy mildew pathogen in some regions of the Russian Federation. *Agric. Sci. Euro-North-East* **2022**, *23*, 90–97. [CrossRef]
21. Brent, K.J. Fungicide resistance in crop pathogens: How can it be managed? In *FRAC Monograph*; GIFAP: Brussels, Belgium, 1995; ISBN 978-90-72398-07-9.
22. FRAC: FRAC Code List 2022. Available online: https://www.frac.info/docs/default-source/publications/frac-code-list/frac-code-list-2022--final.pdf?sfvrsn=b6024e9a_2 (accessed on 20 December 2022).
23. Mouzeyar, S.; Vear, F.; de Labrouhe, D.T. Microscopical studies of the effect of metalaxyl on the interaction between sunflower, *Helianthus annuus* L. and downy mildew, *Plasmopara halstedii*. *Eur. J. Plant Pathol.* **1995**, *101*, 399–404. [CrossRef]
24. Trojanová, Z.; Sedlářová, M.; Gulya, T.J.; Lebeda, A. Methodology of virulence screening and race characterization of *Plasmopara halstedii*, and resistance evaluation in sunflower—A review. *Plant Pathol.* **2017**, *66*, 171–185. [CrossRef]
25. Bán, R.; Viranyi, F.; Komjáti, H. Benzothiadiazole-Induced Resistance to *Plasmopara halstedii* (Farl.) Berl. et de Toni in Sunflower. In *Advances in Downy Mildew Research—Volume 2*; Spencer-Phillips, P., Jeger, M., Eds.; Developments in Plant Pathology; Springer: Dordrecht, The Netherlands, 2004; Volume 16, pp. 265–273. ISBN 978-90-481-6697-8.
26. Viranyi, F.; Oros, G. Developmental stage response to fungicides of *Plasmopara halstedii* (sunflower downy mildew). *Mycol. Res.* **1991**, *95*, 199–205. [CrossRef]
27. Allard, C. Invasion et colonisation systomique de la plantule de tournesol (*Helianthus annus* L.) par le *Plasmopara halstedii* (Farl.) Berl. et de Toni. *Ann. Phytopathol.* **1978**, *10*, 197–217.
28. Wehtje, G.; Littelfield, L.; Zimmer, D.E. Ultrastructure of compatible and incompatible reactions of sunflower to *Plasmopara halstedii*. *Can. J. Bot.* **1979**, *57*, 315–323. [CrossRef]
29. Gray, A.B.; Sackston, W.E. Early stages of infection of resistant and susceptible sunflower seedlings by three races of *Plasmopara halstedii*. *Can. J. Bot.* **1985**, *63*, 1725–1729. [CrossRef]
30. Mouzeyar, S.; de Labrouhe, D.T.; Vear, F. Histopathological studies of resistance of sunflower (*Helianthus annuus* L.) to downy mildew (*Plasmopara halstedii*). *J. Phytopathol.* **1993**, *139*, 289–297. [CrossRef]
31. Mouzeyar, S.; de Labrouhe, D.T.; Vear, F. Effect of host-race combination on resistance of sunflower (*Helianthus annuus* L.) to downy mildew (*Plasmopara halstedii*). *J. Phytopathol.* **1994**, *141*, 249–258. [CrossRef]
32. Radwan, O.; Bouzidi, M.F.; Mouzeyar, S. Molecular Characterization of Two Types of Resistance in Sunflower to *Plasmopara halstedii*, the Causal Agent of Downy Mildew. *Phytopathology* **2011**, *101*, 970–979. [CrossRef]
33. Ward, E.W.B. Suppression of metalaxyl activity by glyphosate: Evidence that host defence mechanisms contribute to metalaxyl in hibition of *Phytophthora megasperma* f. sp. *glycinea* in soybeans. *Physiol. Plant Pathol.* **1984**, *25*, 381–386. [CrossRef]
34. Cahill, D.M.; Morris, P.F.; Ward, E.W.B. Influence of metalaxyl on abscisic acid levels in soybean hypocotyls infected with *Phytophthora sojae*. *Physiol. Mol. Plant Pathol.* **1993**, *42*, 109–121. [CrossRef]
35. Hickey, E.L.; Coffey, M.D. The effects of Ridomil on *Peronospora pisi* parasitizing *Pisum sativum*: An ultrastructural investigation. *Physiol. Plant Pathol.* **1980**, *17*, 199–204. [CrossRef]
36. Ward, E.W.B.; Lazarovits, G.; Stössel, P.; Barrie, S.D.; Unwin, C.H. Glyceollin production associated with control of Phythophthora rot of soybeans by the systemic fungicide, metalaxyl. *Phytopathology* **1980**, *70*, 738–740. [CrossRef]
37. Lazarovits, G.; Ward, E.W.B. Relationship between localized glyceollin accumulation and metalaxyl treatment in the control of Phytophthora rot in soybean hypocotyls. *Phytopathology* **1982**, *72*, 1217–1221. [CrossRef]
38. Stössel, P.; Lazarovits, G.; Ward, E.W.B. Light and electron microscopy of Phytophthora rot in soybeans treated with metalaxyl. *Phytopathology* **1982**, *72*, 106–111. [CrossRef]

39. Cahill, D.M.; Ward, E.W.B. Effects of metalaxyl on elicitor activity, stimulation of glyceollin production and growth of sensitive and tolerant isolates of *Phytophthora megasperma* f.sp. *glycinea*. *Physiol. Mol. Plant Pathol.* **1989**, *35*, 97–112. [CrossRef]
40. Heller, A.; Rozynek, B.; Spring, O. Cytological and physiological reasons for the latent type of infection in sunflower caused by *Plasmopara halstedii*. *J. Phytopathol.* **1997**, *145*, 441–445. [CrossRef]

microorganisms

MDPI

Article

Nanofungicides with Selenium and Silicon Can Boost the Growth and Yield of Common Bean (*Phaseolus vulgaris* L.) and Control Alternaria Leaf Spot Disease

Naglaa A. Taha [1], Salem Hamden [2], Yousry A. Bayoumi [3], Tamer Elsakhawy [4], Hassan El-Ramady [5,*] and Svein Ø. Solberg [6,*]

1 Plant Pathology Research Institute, Agriculture Research Center, Giza 12619, Egypt
2 Agricultural Botany Department, Faculty of Agriculture, University of Kafrelsheikh, Kafr El-Sheikh 33516, Egypt
3 Horticulture Department, Faculty of Agriculture, University of Kafrelsheikh, Kafr El-Sheikh 33516, Egypt
4 Agriculture Microbiology Department, Soil, Water and Environment Research Institute (SWERI), Sakha Agricultural Research Station, Agriculture Research Center, Kafr El-Sheikh 33717, Egypt
5 Soil and Water Department, Faculty of Agriculture, University of Kafrelsheikh, Kafr El-Sheikh 33516, Egypt
6 Faculty of Applied Ecology, Agriculture and Biotechnology, Inland Norway University of Applied Sciences, 2401 Elverum, Norway
* Correspondence: hassan.elramady@agr.kfs.edu.eg (H.E.-R.); svein.solberg@inn.no (S.Ø.S.)

Abstract: There is an urgent need to reduce the intensive use of chemical fungicides due to their potential damage to human health and the environment. The current study investigated whether nano-selenium (nano-Se) and nano-silica (nano-SiO_2) could be used against the leaf spot disease caused by *Alternaria alternata* in a common bean (*Phaseolus vulgaris* L.). The engineered Se and SiO_2 nanoparticles were compared to a traditional fungicide and a negative control with no treatment, and experiments were repeated during two successive seasons in fields and in vitro. The in vitro study showed that 100 ppm nano-Se had an efficacy rate of 85.1% on *A. alternata* mycelial growth, followed by the combined applications (Se + SiO_2 at half doses) with an efficacy rate of 77.8%. The field study showed that nano-Se and the combined application of nano-Se and nano-SiO_2 significantly decreased the disease severity of *A. alternata*. There were no significant differences among nano-Se, the combined application, and the fungicide treatment (positive control). As compared to the negative control (no treatment), leaf weight increased by 38.3%, the number of leaves per plant by 25.7%, chlorophyll A by 24%, chlorophyll B by 17.5%, and total dry seed yield by 30%. In addition, nano-Se significantly increased the enzymatic capacity (i.e., CAT, POX, PPO) and antioxidant activity in the leaves. Our current study is the first to report that the selected nano-minerals are real alternatives to chemical fungicides for controlling *A. alternata* in common beans. This work suggests the potential of nanoparticles as alternatives to fungicides. Further studies are needed to better understand the mechanisms and how different nano-materials could be used against phytopathogens.

Keywords: biotic stress; phytopathogen; nanofungicide; antioxidants; electrolyte leakage; selenium; silica

Citation: Taha, N.A.; Hamden, S.; Bayoumi, Y.A.; Elsakhawy, T.; El-Ramady, H.; Solberg, S.Ø. Nanofungicides with Selenium and Silicon Can Boost the Growth and Yield of Common Bean (*Phaseolus vulgaris* L.) and Control Alternaria Leaf Spot Disease. *Microorganisms* **2023**, *11*, 728. https://doi.org/10.3390/microorganisms11030728

Academic Editors: Anuj Ranjan and Abhishek Chauhan

Received: 21 February 2023
Revised: 7 March 2023
Accepted: 9 March 2023
Published: 11 March 2023

1. Introduction

The common bean *Phaseolus vulgaris* L. is one of the superior leguminous vegetables worldwide and is used for green pod and dry seed consumption. Brazil and Mexico are the largest producers with productions of 495,100 and 373,750 MT per year, respectively [1]. The common bean is the third most important food legume worldwide, after soybean and peanut. Total harvested area is 34.8 million ha year^{-1}, which produces about 27.5 million tons year^{-1} [2]. The common bean is rich in protein (22%), dietary fiber, fat, and carbohydrates (62%), and the plant contains valuable phytochemicals and antioxidants, as well as acceptable levels of various vitamins and minerals [3]. Growing beans can improve the soil properties and reduce the nitrogen fertilization due to the plant's characteristic of N-fixation, which increases the soil fertility [4]. However, its production faces many challenges, including pests and diseases, that are related to various bacteria [5,6], fungi [1], nematodes and insects [7], as well as abiotic stresses, e.g., salinity [8], drought [9], and heat [10]. Losses in productivity and quality have commonly reached as high as 60% due to biotic and abiotic stresses [9].

Fungal phytopathogens have caused severe symptoms, including leaf and pod spots, leaf blight, rust, and root-rot [11]. *Alternaria* spp. are destructive to crops such as the apple *Malus domestica* Borkh. [12], the tomato *Solanum lycopersicum* L. [13], the potato *Solanum tuberosum* L. [14], and the common bean [11]. *Alternaria alternata* caused leaf spot and leaf blight diseases [15]. Other small-spored *Alternaria* species include *A. arborescens* E.G. Simmons and *A. tenuissima* (Kunze) Wiltshire, which both caused brown leaf spots, as well as *A. solani* Sorauer, which caused early blight [16]. To control these phytopathogens, chemical fungicides have been applied [17]; however, the overuse of these chemicals may represent a real threat to the environment, as well as a serious risk to human health [18]. Although biological control strategies have been developed, they are expensive and, often, take a long time to be effective, so many farmers regard them as infeasible alternatives [19]. Consequently, research has continued for additional alternatives with low environmental and health risks [20]. One such alternative is nano-minerals [21–23].

Searching for alternatives to traditional pesticides is an emerging research topic. Much research has been focused on developing novel "sustainable pesticides", which include nutrient fungicides such as CuO-NPs [24]. The protective role of many elements in nano-form has been confirmed against different plant pathogens, including bacteria, fungi, actinomycetes, and nematodes [25]. These nano-metals/metalloids have included titanium (TiO_2-NPs) [26,27], silver (Ag-NPs) [28], magnesium (MgO-NPs) [29], silicon (SiO_2-NPs) [30], copper (Cu-NPs) [31], zinc (ZnO-NPs) [32], and other nanomaterials [33].

Certain nanoparticles (NPs) of Se and SiO_2 are considered nanofungicides due to their effect against a number of phytopathogens [22,23,30]. Nanofungicides are interesting to explore due to their low dose requirement; low dose-dependent toxicity; high solubility and permeability; targeted delivery; enhanced bioavailability; and controlled release [23]. Nano-selenium and nano-silica are well-known as anti-stressors for various cultivated plants, as confirmed in many studies, and many studies have been carried out to examine their role in supporting plant cultivation under abiotic stress [34], such as drought on strawberry *Fragaria* spp. [35], the toxicity of heavy metals on rice *Oryza sativa* L. [36], and salinity on rice [37]. However, few studies have reported on the combined application of nano-Se and nano-silica to mitigate biotic stress. In one study, their combined application was used against root-rot disease induced by *Fusarium* spp. on bread wheat *Triticum aestivum* L. [38]. Furthermore, the use of nano-selenium was reported in different studies, e.g., against tomato leaf blight caused by *Alternaria alternata* [26] and against tomato late-blight disease [39]. Although nano-silicon has the potential for mitigating biotic stress in plants [40,41], more investigations are required. This includes nano-silicon's effect against different diseases, such as stem canker; stem and leaf blight; leaf and root wilt; leaf spot; and soft rot [42]. The use of nutrients in nano-form has demonstrated their protective role against phytopathogens. Hence, it is worth exploring nanofungicides and the potential of this kind of fungicide. This could provide new directions for the application of sustainable nanofungicides as a potential replacement for traditional chemical fungicides. Therefore, this work aimed to study the role both of nano-selenium (nano-Se) and nano-silica (nano-SiO_2), both individually and combined, in controlling leaf spot disease caused by *Alternaia alternata* in the common bean. Nano-Se and nano-silica were compared to a commercial fungicide and a negative control with no treatment, during two successive seasons. In addition to yield measurements, the chlorophyll content, the enzymatic activities, and the total antioxidants in plant leaves were studied after the applications.

2. Materials and Methods

2.1. Pathogen Isolation, Purification, and Pathogenicity Tests

Six pathogenic fungi were isolated from the infected leaves of common bean plants, showing typical leaf spot symptoms, that were obtained from commercial fields in El Beheira Governorate, Egypt. Infected leaves were washed, cut into small pieces (5 mm), surface sterilized with a sodium hypochlorite solution (0.5%) for 2–3 min, and then washed 3 times using sterilized water. Samples were dried between 2 layers of sterilized filter papers and moved to a potato dextrose agar (PDA) medium in 9 cm Petri dishes at 28 $\pm$ 2 °C and cultivated for 72 h. Pure cultures were collected for each of the six isolates using the hyphal tip technique. The purified isolates were confirmed for pathogenicity on a susceptible cultivar Giza 12, which was obtained from the Horticulture Research Institute (Agricultural Research Center, Sakha, Egypt).

The plants were grown in pots 30 cm in diameter with 4 plants per pot, and they placed in a greenhouse. The number of replicates for each pathogen isolate was five. All fungal isolates were prepared using conidial suspensions in distilled water at a rate of 5×10^5 spores mL^{-1} taken from 10-day-old PDA cultures. The spore suspensions were sprayed on the whole bean plants at 35 days old. Based on the protocol established by Panwar et al. [43], the symptoms were observed two weeks after inoculation, and the disease severity was recorded. The most virulent pathogenic isolate was selected for further in vitro experiments. The most virulent isolate was identified as *Alternaria alternata* based on their morphological features and microscopic parameters, as described by Ozcelik and Ozcelik [44] in the Mycology and Disease Survey Research Department (Plant Pathology Research Institute, ARC, Giza, Egypt).

2.2. Preparing Nanoparticles

Preparing Se-nanoparticles was achieved using isolated microbes, which were obtained from soil samples collected from the experimental farm at the Sakha Agricultural Research Station (Kafr El-Sheikh, Egypt). These isolates were mainly selected based on their potential to tolerate high concentrations of selenium (as a selenite). These isolates were also screened for selenite tolerance using tryptone soy broth (TSB), which was amended with Na_2SeO_3 (300 mg L^{-1}). Nutrient broth medium was prepared, and a sterilized sodium hydrogen selenite ($NaHSeO_3$) solution was supplemented from a 10,000 mg L^{-1} stock solution to reach 300 mg L^{-1} concentration. The isolated microbes were identified as *Bacillus cereus*, and the TAH strain of this species was used for the biosynthesis of nano-selenium. The biologically synthesized selenium ranged 41–102 nm in size and were produced at the Agricultural Microbiology Research Department (SWERI, ARC, Giza, Egypt), according to Ghazi et al. [45]. The high-resolution transmission electron microscopy was used to measure the size of the nanoparticles (HR-TEM, Tecnai G20, FEI, The Netherlands), and this was conducted by the Nanotechnology and Advanced Material Central Laboratory, ARC. Silicon oxide nanoparticles, NPs-SiO_2, were prepared by fine-grinding and purchased from Agricultural Microbiology Laboratory (SWERI, ARC, Giza, Egypt) and with a diameter of 10 nm, a specific surface area of 260–320 m^2 g^{-1}, and a pH of 4–4.5.

2.3. In Vitro Experiments

Different doses of nano-selenium (nano-Se) and nano-silica (nano-SiO_2), alone and together, were tested against the most virulent isolate (A5) of *A. alternata*. This was conducted in vitro. The applied doses of nano-Se were 25, 50, and 100 ppm, whereas nano-SiO_2 were 100 and 200 ppm. The combined treatment had half of the highest doses of each compound (i.e., 50 ppm nano-Se + 100 ppm nano-SiO_2). All applied doses of the nanoparticles were added to 100 mL of PDA media. Each dose had 5 replicates in different Petri dishes (each 9 cm in diameter). When the fungal cultures were 7 days old, 5 mm agar plugs were obtained from the edges (with vigorously growing fungi) and inoculated at the middle of the plates with PDA media, plus the different rates of nanoparticles; this was performed on all five replicates. The plates were incubated for 10 days at 28 $\pm$ 2 °C. Under full growth, the colony diameters were measured and then compared to the control plates. According to Ferreira et al. [46], the inhibiting growth percents were calculated and compared to the negative control (no treatment) using the following formula:

$$Reduction\ rate\ (\%) = \frac{R - r}{R} \times 100$$

where (R) is the radial growth of fungi as a control and (r) is the radial growth of fungi in the treated plates.

2.4. Field Experiments

The field experiments were carried out in the common bean cv. Giza 12, which had been grown under open field conditions during the summer seasons 2021 and 2022 at a private farm in Aljazeera village, El Rahmaniya city, El Beheira Governorate, Egypt (31.1°06′19.4″ N 30.6°37′47.9″ E). This farm was chosen due to its disease history, as *Alternaria alternata* has regularly been causing problems, and its frequent bean cultivation. A randomized complete block design was arranged with four replicates. Each plot was 12 m^2 (4 m long and with 5 rows each 0.6 m broad). Bean seeds were spaced 10 cm on ridges, and the sowing was carried out on the 20th of February and 5th of March in 2021 and 2022, respectively. An overview of the treatments is provided in Table 1.

Table 1. Details about the treatments. All treatments were conducted 4 times a season starting 30 days after sowing and repeated with 10 days intervals between each spraying (treated 30, 40, 50, and 60 days after sowing).

Code	Description of the Treatment (s)
T1	Untreated plants
T2	Foliar application of 100 ppm biologically synthesized nano-Se from Agricultural Microbiology Research Department (Giza, Egypt)
T3	Foliar application of 200 ppm fine-ground nano-Si from Agricultural Microbiology Research Department (Giza, Egypt)
T4	Foliar application of 50% of T2 and T3 (i.e., 50 ppm of nano-Se + 100 ppm of nano-SiO_2)
T5	Spraying of commercial fungicide Score 250 EC (Difenoconazole, dose 1 mL/2 L from Syngenta (Basel, Switzerland)

The severity degree of the leaf spot disease was rated in the field, according to Panwar et al. [43]: 0 = no symptoms of leaf spot; 1 = up to 1% indication of disease in leaf area; 3 = 1–10% indication of disease in leaf area; 5 = 11–25% indication of disease in leaf area; 7 = 26–50% indication of disease in leaf area; and 9 = >50% indication of disease in leaf area. The scoring was conducted 45, 60, 75, and 90 days, from sowing and the disease severity was calculated using the following formula:

Disease severity (%) = [Sum of all disease rating/(Total number of ratings × Maximum disease grade)] × 100.

Efficacy (E%) of treatments against the pathogen was calculated as: E% = [(A − B)/A] × 100,

where E (efficacy percent), A (disease severity of control), and B (disease severity of treatment).

The mean of the area under the disease progress curve (AUDPC) for each replicate was assessed according to Pandy et al. [47]: AUDPC = D [1/2 (Y1 + Yk) + (Y2 + Y3 + … … + Yk − 1)], where D (time interval), Y1 (first disease severity), Yk (last disease severity), Y2, Y3, … and Yk − 1 (intermediate disease severity).

2.5. Vegetative Growth Traits, Yield and Photosynthetic Attributes

Vegetative growth traits, including the number of leaves per plant, stem length, fresh and dry weight of common bean plants, were measured during both seasons. All vegetative growth traits were measured 70 days from sowing during both seasons. Chlorophyll a and b contents (mg 100 g^{-1} FW) were measured in plant leaves using the method of Nagata and Yamashita [48] by spectrophotometer analysis at the wavelengths 645 and 663 nm and calculated by substituting the readings in the following equations:

$$\text{Chl. a} = 0.999 \times A_{663} - 0.0989 \times A_{645}$$

$$\text{Chl. b} = -0.328 \times A_{663} + 1.77 \times A_{645}$$

Chlorophyll fluorescence was measured as an important parameter for photosynthetic performance and its response to stress. This parameter expressed the maximum efficiency of the photosystem PSII (Fv/Fm) and was measured using a portable Optic-Science OS-30p + Fluorometer (Opti-Sciences, Inc., Hudson, NH, USA), according to Maxwell and Johnson [49].

Plants were harvested at the seed maturity stage. Dry seeds were manually extracted and recorded as the weight of seeds per plant (g) and per plot (g). Thereafter, the value was calculated to provide the seed yield in Mg (tons) ha^{-1}.

2.6. Antioxidant Enzymes, Antioxidative Activity and Electrolyte Leakage

To determine the antioxidant enzymatic activities (catalase, peroxidase, and polyphenol oxidase), 0.5 g of fully expanded young leaves were homogenized in liquid nitrogen with 3 mL of extraction buffer (50 mM TRIS buffer (pH 7.8) containing 1 mM EDTA-Na_2 and 7.5% polyvinylpyrrolidone)) using a pre-chilled mortar and pestle. The homogenate was filtered through 4 layers of cheesecloth and centrifuged at 12,000 rpm for 20 min at 4 °C. The supernatant, which was re-centrifuged at 12,000 rpm for 20 min at 4 °C, was then used for the total soluble enzymatic activity assay. The enzymatic activities were measured colorimetrically using a double-beam UV/visible spectrophotometer Libra S80PC (TechnoScientific Company, Nottingham, UK), 70 days from sowing.

Catalase (CAT; EC 1.11.1.6) activity was measured by following the consumption of H_2O_2 at 240 nm, according to Aebi [50]. A total of 1 mL of the reaction mixture contained 20 mg total protein, 50 mM sodium phosphate buffer (pH 7.0), and 10 mM H_2O_2. The reaction was initiated by adding the protein extract. For each measurement, a blank corresponded to the absorbance of the mixture at time zero, and the actual reading corresponded to the absorbance after 1 min. One unit of CAT activity was defined as a 0.01 decrease in absorbance at 240 nm mg^{-1} of protein min^{-1}.

Peroxidase (POX; EC 1.11.1.7) activity was determined according to the procedure proposed by Rathmell and Sequeira [51]. The reaction mixture consisted of 2.9 mL of a 100-mM sodium phosphate buffer (pH 6.0 containing 0.25% (*v/v*) guaiacol (2-methoxy phenol) and 100 mM H_2O_2). The reaction was started by adding 100 mL of crude enzyme extract. Changes in absorbance at 470 nm were recorded at 30 s intervals for 3 min. Enzymatic activity was expressed as an increase in the absorbance $min^{-1} \cdot g^{-1}$ fresh weight.

Polyphenol oxidase (PPO; EC 1.10.3.1) activity was determined according to the method described by Malik and Singh [52]. The reaction mixture contained 3.0 mL of buffered catechol solution (0.01 M) freshly prepared in 0.1 M phosphate buffer (pH 6.0). The reaction was initiated by adding 100 mL of the crude enzyme extract. Changes in the absorbance at 495 nm were recorded at 30 s for 3 min. Enzymatic activity was expressed as an increase in the absorbance $min^{-1} \cdot g^{-1}$ fresh weight.

Antioxidative activity in the plant tissue was determined by a DPPH (2,2-diphenyl-1-picryl hydrazyl) assay, as described by Binsan et al. [53]. In brief, plant samples were extracted, and 1.5 mL was added with 1.5 mL of 0.15 mM DPPH in 95% ethanol. The mixture was stored in the dark for 30 min at room temperature. Using a double-beam UV/visible spectrophotometer Libra S80PC (TechnoScientific Company, Nottingham, UK), the absorbance of the resulting solution was measured at 517 nm. The calibration curve was prepared using Trolox in the range of 12.5 to 100 µM.

Electrolyte leakage was measured using an electrical conductivity meter, according to the method of Whitlow et al. [54] and later modified by Szalai et al. [55]. Twenty leaf discs (1 cm^2) were placed individually into flasks containing 25 mL of deionized water (Milli-Q 50, Millipore, Bedford, MA, USA). Flasks were shaken for 20 h at an ambient temperature to facilitate electrolyte leakage from injured tissues. Initial EC measurements were recorded for each vial using an EC meter (Fistreem Jenway Bench Top EC meter, Model 3510 Medica Scientific Company, Stockport, UK). Flasks were then immersed in a hot water bath (Fisher Isotemp, Indiana, PA, USA) at 80 °C (176 F) for 1 h to induce cell rupture. The vials were again placed on the Innova 2100 platform shaker for 20 h at 21 °C (70 F). Final conductivity was measured for each flask. The percentage of electrolyte leakage for each sample was calculated as the initial conductivity/final conductivity × 100. Chlorophyll florescence, chlorophyll contents, all vegetative growth parameters, antioxidant enzymes, and electrolyte leakage were assessed 70 days from sowing in both growing seasons. The general layout of the entire work is overviewed in Figure 1.

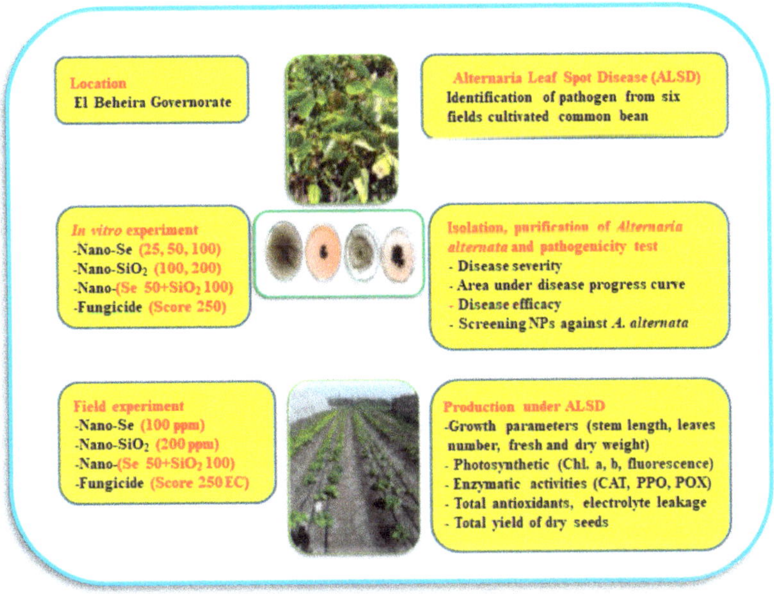

Figure 1. A general overview on the layout of the current study including the location, different experiments, and different measurements. Abbreviations: ALSD: Alternaria leaf spot disease; Chl.: chlorophyll; CAT: catalase; PPO: polyphenol oxidase; POX: peroxidase.

2.7. Microscopic Examination

The fungal culture grown on the PDA media supplemented with selenium nanoparticles was examined by light microscope with a magnification power of 200× by sticky tape touch method after incubation for 7 days. Samples from the control and nano-Se were also examined 5 days from the inoculation on the PDA media with a scanning electron microscope (Model: SEM, JEOL JSM 6510 Iv, Tokyo, Japan) by using the high vacuum mode at the Nanotechnology Institute (Kafrelsheikh University, Kafr El-Sheikh, Egypt). This was conducted to visualize any potential effects of the nano-Se application on *A. alternata*.

2.8. Statistical Analyses

The two field experiments were arranged as randomized complete block designs with four replicates for each treatment in each trial. Statistical analyses were conducted with the CoStat package program (Computer Program Analysis, Version 6.303; CoHort Software, Berkeley, CA, USA) using ANOVA and, thereafter, Duncan's multiple range tests at 5% level of probability to compare the means [56].

3. Results

3.1. Pathogenicity Tests

A greenhouse experiment was conducted to assess the pathogenic ability of the six isolates. The Alternaria isolates were able to infect the susceptible bean cultivar Giza 12 and caused typical leaf spot symptoms, whereas the isolate A5 (*Alternaria alternata*) was the most aggressive one in the experiment (Table 2 and Figure 2). The A5 isolate had the highest disease severity value (78.23%); however, the other isolates varied in their degrees of disease severity. Consequently, the A5 isolate of *A. alternata* was chosen for following studies.

Table 2. Disease severity percentage as the pathogenicity of 6 isolates of *Alternaria alternata* on common bean plants under greenhouse conditions, 21 days after inoculation.

Isolate No.	Disease Severity (%)
A1	41.30 b ± 2.05
A2	23.00 c ± 1.99
A3	27.15 c ± 2.00
A4	14.37 d ± 1.89
A5	78.23 a ± 3.47
A6	30.00 c ± 2.08
F. test	**

Where A1, A2, A3, A4, A5 and A6 are the six isolates of *Alternaria alternata*, which used in the study ** indicates highly significant treatment and values of means in each column followed by the same letter are not significantly at a 95% confidence level (±SD = standard deviation; N = 5).

Figure 2. Images illustrating how the six isolates of *Alternaria alternata* affected the common bean plants. The images were taken 21 days after inoculation under greenhouse conditions.

3.2. In Vitro Antifungal Activity of Nano-Se and Nano-SiO_2

The antifungal effects of nano-Se and nano-SiO_2 alone or in combination were examined in vitro. This was conducted on the most aggressive isolate A5. The highest antagonistic effect was found with the chemical control (95.55% reduction), followed by nano-Se at the highest dose (100 ppm), and the combined application of the nanoparticles (85.1% and 77.78% reduction, respectively). All applied nanoparticles significantly reduced the mycelial growth of the pathogen, as compared to the untreated control (Table 3; Figure 3).

Table 3. Effects of the different treatments on mycelial growth of *Alternaria alternata* and the corresponding reduction in the pathogen under in vitro conditions in the lab.

Treatments	Mycelial Growth (cm)	Reduction (%)
Negative control	9.0 a ± 0.21	0.001 f ± 0.00
Nano-Se (25 ppm)	7.0 b ± 0.17	22.22 e ± 0.95
Nano-Se (50 ppm)	4.5 c ± 0. 08	50.00 cd ± 2.47
Nano-Se (100 ppm)	1.7 d ± 0. 08	85.11 b ± 3. 11
Nano-Si (100 ppm)	6.0 b ± 0. 15	33.33 e ± 2.09
Nano-Si (200 ppm)	3.5 c ± 0. 11	61.11 c ± 3.01
1/2 Nano-(Se + Si) (i.e., 50 + 100 ppm, respectively)	2.0 d ± 0. 05	77.78 bc ± 3.34
Fungicide (Score 250 EC)	0.4 e ± 0. 01	95.55 a ± 4. 37
F. test	**	**

** indicates highly significant effect of treatment. Mean values in each column that are followed by the same letter are not significantly different from each other at a 95% confidence level (±SD = standard deviation; N = 5).

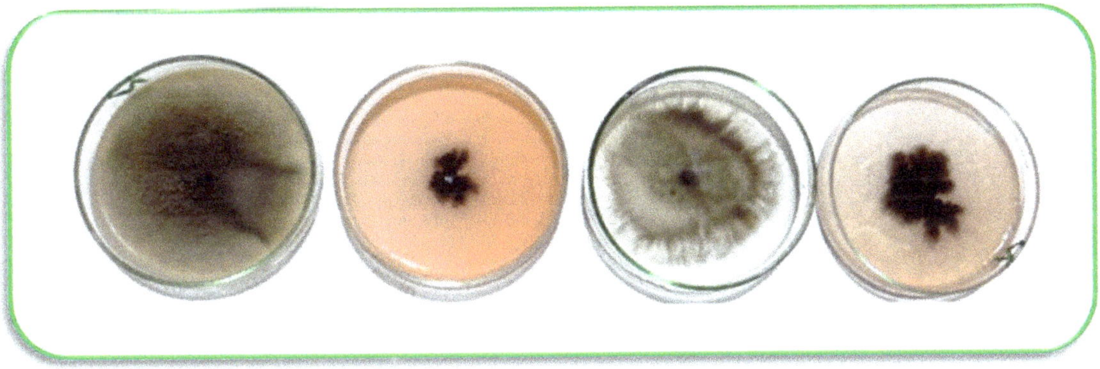

Control Nano-Se (100 ppm) Nano-Si (200 ppm) 1/2 nano-(Se + Si)

Figure 3. Images illustrating of the inhibition of *Alternaria alternata* mycelial growth after some of the treatments in the in vitro experiments. Control = no treatment (negative control).

3.3. Microscopic Investigation

Light microscope images of the Alternaria conidia from the in vitro experiment showed an obvious effect of the nano-Se treatment. A reduction in the number and homogeneity of the conidia, in addition to an occurrence of distortions, and a clear decrease in the conidia size was observed after the nano-Se was compared to the negative controls with no treatment (Figure 4). The change in mycelial growth was examined 5 days after inoculation using a scanning electron microscope apparatus (Figure 5). The nano-Se application (100 ppm) caused morphological alterations in the hyphae and mycelial growth. The mycelium and hyphae of the *A. alternata* fungus was strictly injured in the presence of nano-Se (Figure 5A), as compared to the negative control that was untreated and showed typical mycelial structures for the fungus (Figure 5B).

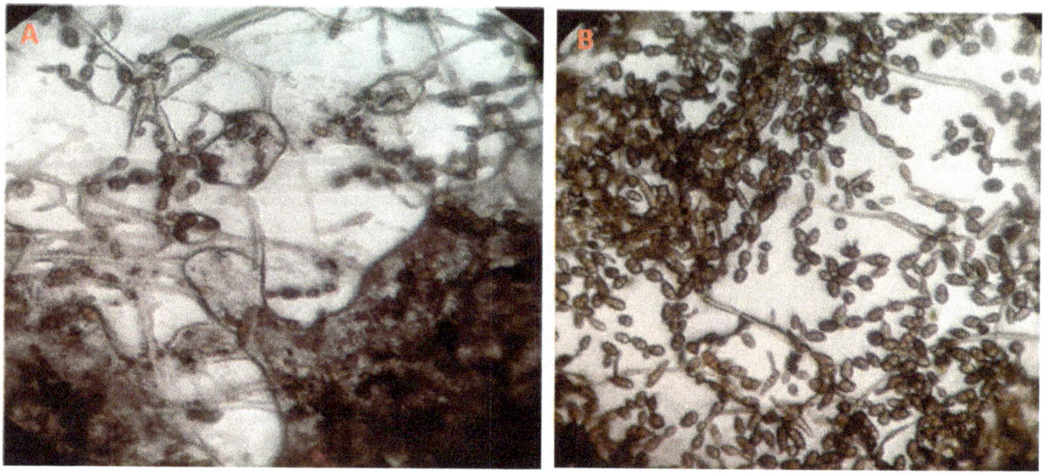

Figure 4. Light microscopic image (200×) of conidia of *Alternaria alternata* grown on PDA for 7 days after incubation, where (**A**) is influenced by Nano-Se (100 ppm) and (**B**) is the negative control with no treatment.

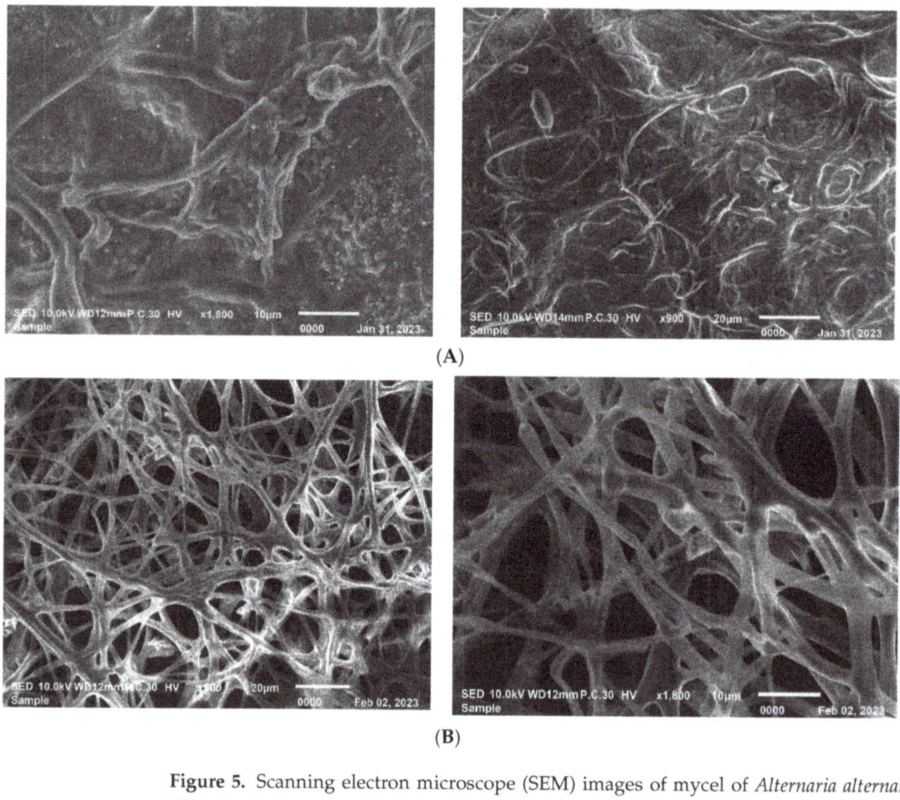

Figure 5. Scanning electron microscope (SEM) images of mycel of *Alternaria alternata* grown on PDA for 7 days after incubation, where (**A**) is influenced by Nano-Se (100 ppm) and (**B**) is the negative control with no treatment. (**A**) Mycelial growth of *Alternaria alternata* influenced by nano-Se application at 100 ppm. (**B**) Normal mycelial growth of *Alternaria alternata* without nano-Se application (Control).

3.4. Disease Severity and Efficacy Percent under Field Conditions

The mean of the area under the disease progress curve (AUDPC) was significantly influenced by the different treatments (Figure 6). When left untreated (negative control), it resulted in the highest AUDPC values, followed by the nano-SiO_2 application, as compared to the other treatments, while the nano-Se application alone or in combination with nano-SiO_2 showed the lowest AUDPC values, and this was shown in both seasons. There were no significant differences between nano Si and the commercial fungicide, and this was found in both seasons (Table 4). The progression of the Alternaria leaf spot disease severity was recorded 4 times: 45, 60, 75, and 90 days after sowing. The disease severity (%) was significantly influenced by the different nano-Se and nano-Si applications at all sampling dates during both seasons. The highest disease severity values were observed with untreated plants (negative control treatment), which were recorded at 29.5%, 42.1%, 61.8%, and 89.6% in the first season and at 31.3%, 49.1%, 66.0% and 94.7% in the second season at the aforementioned days, respectively. Comparatively, the commercial fungicide application (positive control) had a superior effect on the disease severity on all the dates during the growing seasons and showed the lowest disease severity percentages in all cases (Figure 7), as compared to the nanoparticles. Nano-Se alone had a better effect on disease severity (%), as compared to nano-SiO_2 alone or in a combined reduced dose.

The efficacy percentages of all treatments were significantly impacted by the severity percentage of the Alternaria leaf spot disease, as shown in Table 4. The highest efficacy percentage was found from the fungicide treatment followed by nano-Se application, but without significant differences between them. The nano-Si application had lower values than the combined application of nano-Se and Si, which showed intermediate values in both seasons. The AUDPC was significantly influenced by the treatments (Figure 7). The untreated (negative control) group resulted in the highest AUDPC values, followed by the nano-Si application in both seasons, while nano-Se alone or in combination with nano-Si showed the lowest AUDPC values, and the commercial fungicide application showed no significant differences in its results across both seasons.

Table 4. Effects of applied nano-Se and nano-SiO_2 treatments on disease severity and efficacy (%) of Alternaria leaf spot disease under field conditions at 45, 60, 75, and 90 days after the sowing of common beans during two seasons, where T1 = negative control, T2 = 100 ppm nano-Se, T3 = 200 ppm nano-SiO_2, T4 = 1/2 nano-Se + SiO_2, and T5 = fungicide (Score 250 EC at 1 mL/2 L).

Treatments	Disease Severity (%) Days after Sowing Date				Efficacy (%)
	45	60	75	90	
			Season of 2021		
T1	29.5 a ± 1.24	42.1 a ± 2.56	61.8 a ± 3.18	89.6 a ± 3.76	——
T2	1.5 b ± 0.16	5.5 b ± 0.48	14.8 c ± 0.53	20.5 c ± 1.08	77.1 ab ± 2.55
T3	3.3 b ± 0.23	7.5 b ± 0.59	18.1 b ± 0.73	30.3 b ± 1.22	66.1 c ± 2.29
T4	0.0 b ± 0.00	5.8 b ± 0.49	15.1 c ± 0.48	23.5 bc ± 1.18	73.7 b ± 2.51
T5	0.0 b ± 0.00	6.2 b ± 0.61	12.5 d ± 0.67	17.8 c ± 0.79	80.1 a ± 3.05
F. test	**	**	**	**	**
			Season of 2022		
T1	31.3 a ± 3.04	49.1 a ± 4.49	66.0 a ± 4.99	94.6 a ± 4.99	——
T2	2.4 c ± 0.10	10.4 b ± 0.47	17.0 cd ± 1.02	23.0 c ± 1.06	75.7 a ± 4.04
T3	6.0 b ± 0.17	13.0 b ± 0.47	22.1 b ± 1.06	35.0 b ± 2.08	63.0 c ± 3.95
T4	3.5 c ± 0.14	11.7 b ± 0.47	19.1 c ± 1.06	27.8 b ± 2.05	70.5 b ± 4.04
T5	1.9 c ± 0.07	8.3 c ± 0.35	14.7 d ± 1.02	20.9 c ± 1.79	77.9 a ± 4.04
F. test	**	**	**	**	*

Mean values in each column that are followed by the same letter are not significantly different from each other at a 95% confidence level. * and ** indicate significant and highly significant, respectively; (±SD = standard deviation; N = 4).

Figure 6. Effects of the various treatments on the development of leaf spot disease caused by *A. alternata* on common bean plants (Giza 12, cv.) grown in the field and sampled 45, 60, 75, and 90 days after sowing, and where Control = no treatment, Nano-Se = 100 ppm nano-Se, Nano-Si = 200 ppm nano-Si, Nano-(Si + Si) = combined 1/2 dose of nano-Se + nano-Si, and Fungicide = Score 250 EC at 1 mL/2 L.

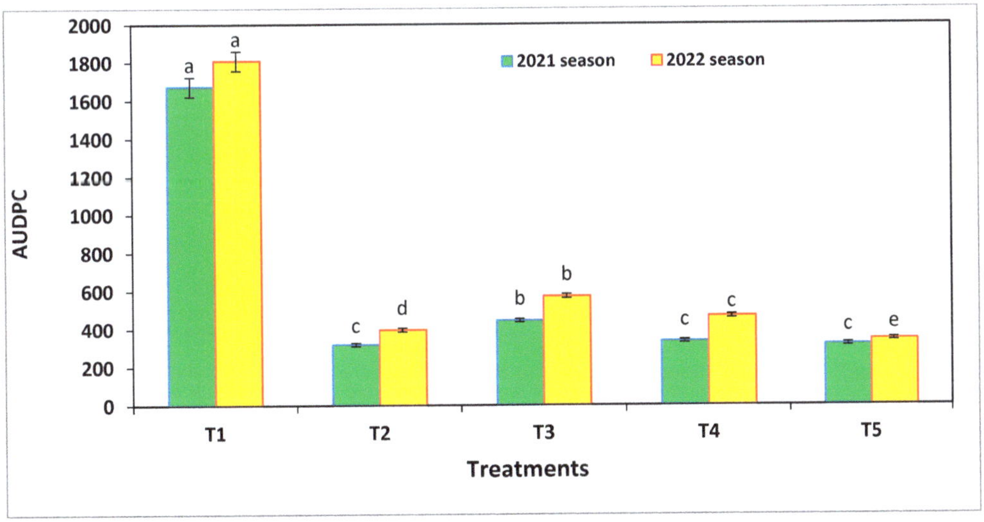

Figure 7. Area under disease progress curve (AUDPC) of the leaf spot disease on common bean in the field experiments, and where T1 = negative control, T2 = 100 ppm nano Se, T3 = 200 ppm nano-Si, T4 = 1/2 nano-Se + Si, and T5 = fungicide (Score 250 EC at 1 mL/2 L). Results from the 2021 and 2022 seasons. Same letter above the error bar indicates no significant differences.

3.5. *Vegetative Growth and Photosynthetic Traits*

Under field conditions, vegetative growth traits, chlorophyll fluorescence and chlorophyll contents were examined 70 days from sowing (Table 5). The stem length, the number of leaves, and the plant fresh and dry masses were significantly enhanced by the different nanoparticle applications, as compared to the fungicide treatment (positive control) and the untreated plants (negative control). The same results were observed in both seasons. Nano-Se produced the tallest plants with the highest number of leaves per plant, which resulted in the highest fresh and dry plant biomass production. The negative controls had the lowest values in both seasons.

In most cases, the combined application of nano-Se and nano-SiO_2 resulted in lower values than nano-Se applied alone. The chlorophyll fluorescence (F_V/F_M) values were significantly increased by all nanoparticles (alone or in combination), as compared to the controls. These were followed by the fungicide treatment, while untreated plants produced the lowest values. The highest values of chlorophyll a and b were obtained by applying nano-Se and fungicide treatments, with no statistical differences in these two, while the negative control plants without any treatment produced the lowest values in both seasons. The combined application of nanoparticles and nano-Si alone had intermediate chlorophyll values.

Table 5. Vegetative growth traits, chlorophyll fluorescence, and chlorophyll content of common bean plants influenced by the treatments in the field experiments and where T1 = negative control, T2 = 100 ppm nano-Se, T3 = 200 ppm nano-SiO_2, T4 = 1/2 nano-Se + SiO_2, and T5 = fungicide (Score 250 EC at 1 mL/2 L).

Treatments	Stem Length (cm)	No. of Leaves/Plant	Plant Fresh Mass (g)	Plant Dry Mass (g)	Chlorophyll Fluorescence (F_V/F_M)	Chl. A	Chl. B
						(mg 100 g^{-1} FW)	
				Season of 2021			
T1	43.7 c	14.2 d	60.5 d	13.66 d	0.724 c	18.22 c	8.14 b
T2	54.0 a	19.1 a	101.5 a	22.14 a	0.798 a	23.99 a	9.23 a

Table 5. *Cont.*

Treatments	Stem Length (cm)	No. of Leaves/Plant	Plant Fresh Mass (g)	Plant Dry Mass (g)	Chlorophyll Fluorescence (F_V/F_M)	Chl. A (mg 100 g^{-1} FW)	Chl. B (mg 100 g^{-1} FW)
T3	51.6 b	17.3 b	93.6 b	19.55 b	0.795 a	22.17 b	8.36 b
T4	53.3 a	17.5 b	95.2 b	20.05 b	0.795 a	23.05 ab	8.55 b
T5	50.6 b	15.7 c	84.6 c	17.99 c	0.771 b	23.99 a	9.18 a
F-test	**	**	**	**	**	**	*
				Season of 2022			
T1	40.4 c	11.8 c	57.1 d	12.45 d	0.733 c	17.09 c	8.02 a
T2	51.8 a	16.7 a	94.7 a	19.77 a	0.778 a	22.55 a	9.02 a
T3	48.9 b	14.7 b	86.3 b	17.56 b	0.776 a	20.08 b	8.34 a
T4	50.1 ab	15.5 ab	85.7 b	17.89 b	0.777 a	20.60 b	8.45 a
T5	48.5 b	14.5 b	79.4 c	16.05 c	0.754 b	21.95 a	8.88 a
F-test	*	*	**	**	*	**	NS

Values of means in each column followed by the same letter are not significant at $p < 0.05$ * or $p < 0.01$ **, respectively.

3.6. *Enzymatic Activities*

Catalase, peroxidase, and polyphenol oxidase were assessed in common bean leaves 70 days after sowing (Figure 8). Overall, the highest values were found after applying nano-Se, followed by fungicide and combined nano-Se and nano-SiO_2. The lowest values were found in the negative control (no treatment).

3.7. *Antioxidant Activity and Electrolyte Leakage*

The application of nano-Se and nano-Si significantly increased the antioxidant activity, as compared to the negative control (no treatment), which resulted the lowest values (27.77 and 31.19 μM 100 g^{-1}) in both seasons, respectively (Figure 9). Nano-Se application alone produced the highest antioxidant values (84.45 and 77.79 μM 100 g^{-1} in the two seasons) followed by the combined application of nano-Se/nano-Si, fungicide treatment, and finally nano-Si spraying alone in both seasons. In contrast, untreated plants had the highest electrolyte leakage percentage (90.66% and 88.48% in the two seasons). All nano- and fungicide treatments resulted in low leakage percentages, as compared to the negative control. Generally, as confirmed by the previous measuring enzymes, the best (lowest) values of electrolyte leakage and the total antioxidants were obtained after applying nano-Se.

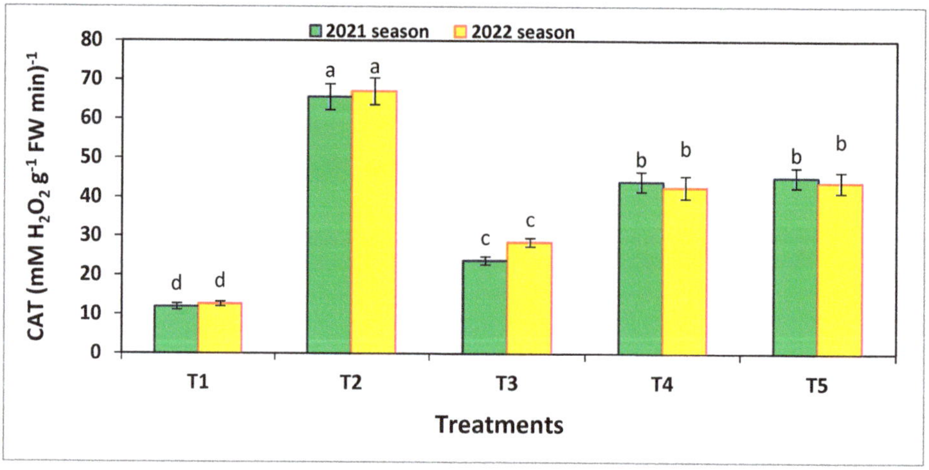

Figure 8. *Cont.*

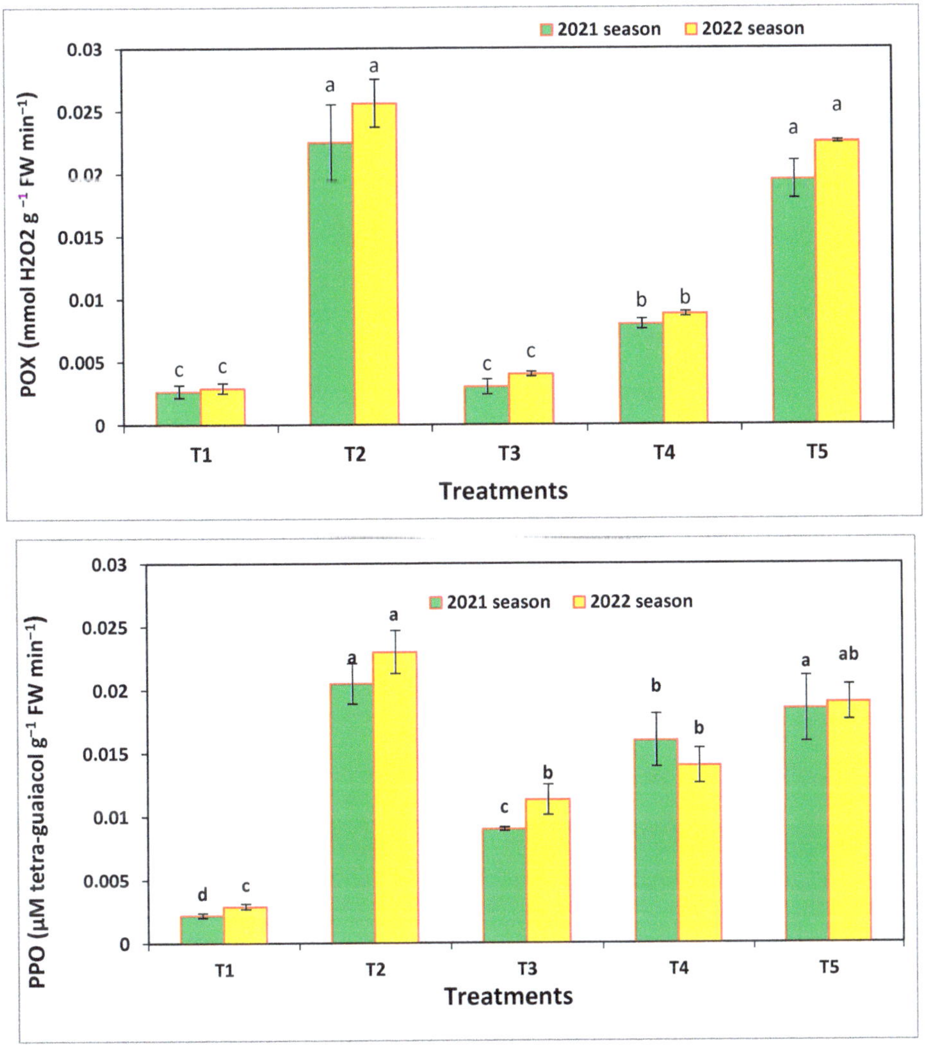

Figure 8. Effects of the different treatments on catalase, peroxidases, and polyphenol oxidase activity in the bean plants measured 70 days after sowing in the field experiments, and where T1 = negative control, T2 = 100 ppm nano-Se, T3 = 200 ppm nano-Si, T4 = 1/2 nano-Se + Si, and T5 = fungicide (Score 250 EC at 1 mL/2 L). Same letter above the error bar indicates no significant differences.

3.8. Response of Total Yield of Dry Seeds to Applied Nanoparticles

The application of nano-Se and nano-SiO_2 alone or in combination significantly increased the dry seed yield for both growing seasons (Table 6). The highest yield was obtained from plants treated with 100 ppm nano-Se (54.22, 45.75 g/plant and 3.07, 2.66 Mg ha^{-1}), followed by the combined application of both nanoparticles at a half dose (49.13, 40.18 g/plant and 2.77, 2.49 Mg ha^{-1}) in both seasons. The untreated plants showed the lowest total seed yield (30.85, 27.19 g/plant and 2.17, 1.77 Mg ha^{-1}) in both seasons. The intermediate values of the total dry seed yield were recorded by the application of fungicide and the 200 ppm nano-SiO_2 treatments in both seasons.

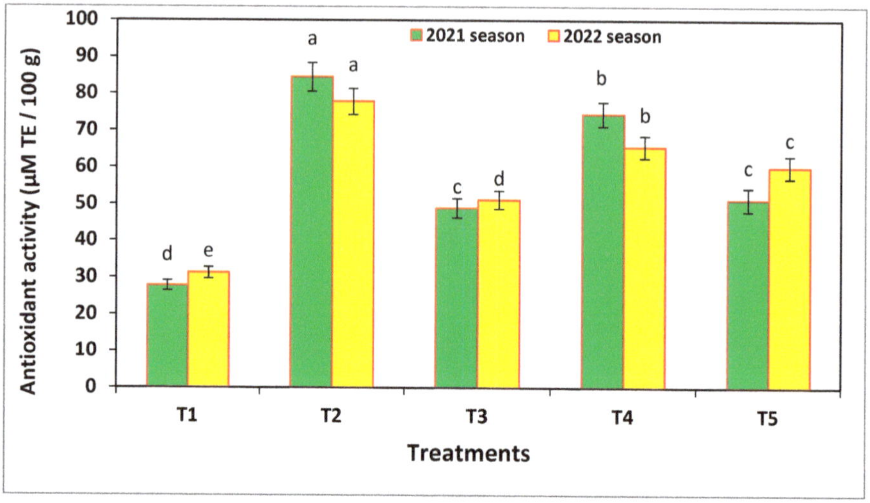

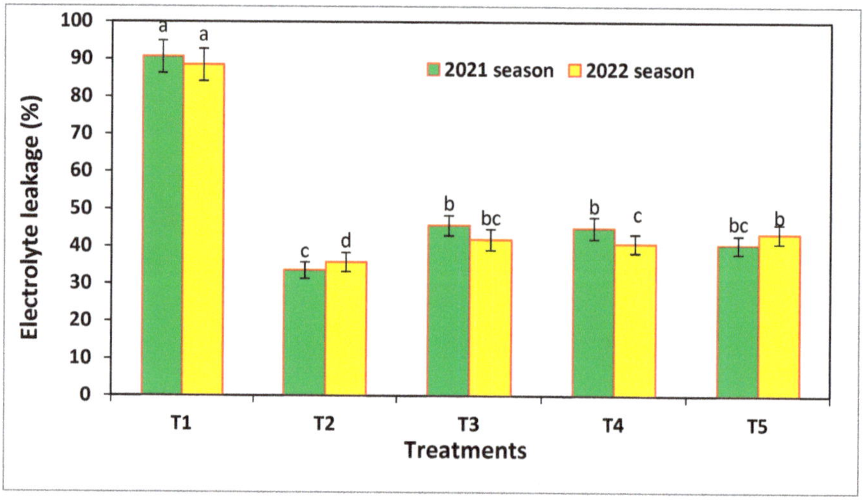

Figure 9. Effects of the different treatments on total antioxidant activity and electrolyte leakage measured 70 days after sowing in the field experiments in 2021 and 2022. Results are given as mean values ± standard error bars and where T1 = negative control, T2 = 100 ppm nano-Se, T3 = 200 ppm nano-Si, T4 = 1/2 nano-Se + Si, and T5 = fungicide (Score 250 EC at 1 mL/2 L). Same letter above the error bar indicates no significant differences.

Table 6. Total dry yield of common bean seeds as influenced by different treatments. Results from the field experiments, and where T1 = negative control, T2 = 100 ppm nano-Se, T3 = 200 ppm nano-SiO_2, T4 = 1/2 nano-Se + SiO_2, and T5 = fungicide (Score 250 EC at 1 mL/2 L).

Treatments	Dry Seed Yield ($g\ plant^{-1}$)	Dry Seed Yield ($Mg\ ha^{-1}$)	Dry Seed Yield ($g\ plant^{-1}$)	Dry Seed Yield ($Mg\ ha^{-1}$)
	Season of 2021		Season of 2022	
T1	30.85 d ± 2.05	2.17 c ± 0.009	27.19 d ± 3.03	1.77 e ± 0.019
T2	54.22 a ± 2.79	3.07 a ± 0.017	45.75 a± 3.55	2.66 a ± 0.025
T3	38.99 c ± 2.48	2.48 b ± 0.025	32.27 c ± 3.35	2.24 d ± 0.033

Table 6. *Cont.*

Treatments	Dry Seed Yield (g plant^{-1})	Dry Seed Yield (Mg ha^{-1})	Dry Seed Yield (g plant^{-1})	Dry Seed Yield (Mg ha^{-1})
	Season of 2021		Season of 2022	
T4	49.13 ab ± 2.75	2.77 a ± 0.032	40.18 b ± 3.44	2.49 b ± 0.029
T5	45.88 b ± 2.57	2.61 b ± 0.044	38.65 b ± 4.01	2.41 c ± 0.018
F-test	**	**	**	**

** indicates highly significant. Mean values in each column followed by the same letter are not significantly different from each other at a 95% confidence level (±SD = standard deviation; N = 4).

4. Discussion

Agriculture faces several challenges that, in the future, may decrease crop production. This includes the pollution from the intensive use of agrochemicals and biotic stresses from pests and diseases. Therefore, the current study was carried out to examine whether nano-sized particles of selenium and silicon could be used to boost plant health and control leaf spot disease caused by the fungus *Alternaria alternata*. Therefore, we sought to answer the following questions: To what extent could nano-Se and nano-Si work against this disease? Could the applied nanoparticles of Se and Si be effective for inhibiting the fugus, as compared to the traditional fungicides? Which nano-treatments would be the most effective? What was the impact of the applied nanofungicides on common bean attributes?

The present study involved two kinds of potential nanofungicides; the first was a biological nano-Se, and the second was nano-SiO_2 that was prepared by a physical method. Our study was conducted to identify the most effective agent against the studied pathogen (*A. alternata*). We selected the phytopathogen of *Alternaria* in our current study. Several species of the *Alternaria* genus are well-known as phytopathogens, which infect many genera of vegetables and fruits. These species have a saprophytic nature, and their pathogenicity actions were due to toxins and cell-wall degrading enzymes [57]. Among those are *Alternaria dauci, A. solani, A. tenuis, A. brassicae, A. brassicicola*, and *A. alternata*. This study focused on *A. alternata* as a common phytopathogen, which could cause severe leaf spot problems in the common bean. *A. alternata* is also considered one of the most important diseases affecting common bean plants under the temperate conditions, as a serious foliar ailment causing massive yield reductions [11,26]. In addition, other crops have been infected, such as faba bean *Vicia faba* L. [15], tomato [57], apple [12], and orange *Citrus* spp. [58].

To what extent could the studied nano-Se and nano-SiO_2 boost the growth and yield of the common bean as well as be used against Alternaria leaf spot disease? From our findings in the current study, it was obvious that the nano-Se and, to some extent, the nano-SiO_2 were effective against this pathogen. The applied in vitro nano-Se alone and/or in combination with nano-SiO_2 recorded an 85.1% and 77.8% reduction in mycelial growth, respectively. This was in line with El-Gazzar and Ismail [26], who reported 81.1% and 89.6% reduction in mycelial growth of *Alternaria* isolated from tomato when using 50 and 100 ppm nano-Se. Therefore, the inhibitory effect could be explained by the congregation and the installation of the particles. The factors of importance included the applied doses, the sizes and shapes of the particles, and the solubility and agglomeration statuses of the particles [59,60]. The results in the current study agreed with the work published by El-Gazzar and Ismail [26], who proved that 100 ppm nano-Se was effective against the leaf blight in tomato plants. Other studies have shown that nano-SiO_2 at 200 ppm was effective against a disease complex in beetroot *Beta vulgaris* L. [32] and on early blight in tomato plants [30]. We included field trials with the examination of various growth parameters, including vegetative traits and yield. All parameters were significantly increased with nano-Se and nano-silica, as compared to the controls. The spraying of nano-Se and/or -SiO_2 alleviated the negative impacts of the fungus stress.

Nano-SiO_2 and/or nano-Se have been reported to increase yields due to their ability to increase the photosynthetic and antioxidative enzymatic activities, as well as their ability to regulate the negative effects of stress, as reported in different species [34]. *A. alternata* could reduce the photosynthesis by inhibiting photosynthetic activity through the necrosis of plant tissues and by damaging structural and stomatal closures of the plants. Our study showed that the photosynthetic attributes, including the chlorophyll content (a and b) and the chlorophyll fluorescence, were significantly influenced by the stress caused by the fungus. This was shown in the positive control, whereas these attributes were less affected by applying nano-Se and/or nano-SiO_2 than by applying a chemical fungicide. Under similar stress conditions, nano-treatments have appeared to promote the absorption and conversion of light, which enabled the treated plants to maintain normal levels of carbon assimilation [61]. The selected nanoparticles showed a significant ability to support photosynthetic attributes under biotic/abiotic stress, as reported in many published studies on other crops (e.g., [32,62–64]). Both nano-Si and nano-Se could increase chlorophyll content under stressful conditions and protect chloroplasts from oxidative damage, potentially because they may act as a cofactor in enzymatic reactions involved in various biosynthetic pathways [35].

Under biotic stresses, plants should be armed with enzymatic and/or non-enzymatic antioxidants. In our study, this included enzymatic activities, electrolyte leakage, and total antioxidants. This defense system, in general, evolves different systemic signaling pathways (i.e., calcium, reactive oxygen species (ROS), and phytohormones). The application of the studied NPs showed that they had the ability to enhance the plant resistance to phytopathogens, probably as a result of producing ROS or phytohormones [64]. We showed that the biological nano-Se recorded the highest values in CAT, PPO, and POX enzymes, as well as the total antioxidant content, followed by the combined application of nano-Se and nano-silica at half doses. This finding confirmed the protective role of both nano-Se and nano-Si against oxidative stress, in this study, as a result of biotic stress. This was shown by the increased activity of the enzymatic antioxidants, which reduced the ROS and decreased the lipid peroxidation [35]. The distinguished role of biological nano-Se was confirmed in previous studies in other crops. The effectiveness of biological nano-Se has been confirmed in different studies and under different stresses, such as cucumber under salinity and heat stress [65], wheat under Fusarium root-rot disease [66], pak choi under heavy metal toxicity [67], and rapeseed under salinity stress [68], whereas the biological silicon-NP improved the common bean yield under heavy metals and salinity stress [38]. In this study, we confirmed its effect against the Alternaria leaf spot disease in the common bean.

Nano-Se and nano-SiO_2 are well-known as anti-stressors under biotic stress (i.e., fungal phytopathogen of *A. alternata*). This antifungal activity was confirmed as inducing plant physiological immunity against phytopathogens such as leaf blight in tomato [26] and early blight disease in eggplant *Solanum melongena* L. [69]. It was reported that spraying 200 ppm SiO_2-NP on tomato plants under bacterial and fungal attacks promoted plant growth parameters, proline, chlorophyll, carotenoid, and enzymatic activities, and the spraying reduced the disease indices [30]. In our study, the biological nano-Se recorded the highest values in most studied parameters, including vegetative attributes, photosynthetic items, enzymatic antioxidants, and dry seed yield, as compared to the controls and nano-SiO_2.

Few recent publications have discussed the combined application of Se and Si (e.g., [70]), whereas the interaction between nano-nutrients such as Se and Si still needs more investigation (e.g., [19,71]). The obtained results indicated that the biological nano-Se had the best results among all treatments, whereas nano-SiO_2 had the lowest results, ranking lower than the combined nano-Se and nano-silica. The biological nano-Se had unique physiochemical characteristics in the plant system [68], which has attracted considerable concern worldwide due to their significant potential for alleviating plant stress [63]. Nano-silica has been effective in reducing serious insect and mechanical damage. Based on the combined application of nano-Se and nano-silica, the effectiveness of nano-Se could have decreased due to a hetero-aggregation: This phenomenon can occur between nanoparticles from different sources, wherein their size increases and then interferes with the needed penetrating power.

5. Conclusions

Through the use of nanotechnology, the production of nano-pesticides with protective effects from elements such as Ag, Cu, Se, Si, TiO_2, and Zn, has the potential to work against several phytopathogens. The current study is the first report on applying bio-nano-Se and nano-silica against the fungus of *A. alternata* in the common bean. The current study showed that biological nano-Se could influence the growth of the common bean positively. This was shown by the increased photosynthetic activity, better antioxidant enzyme efficiencies (CAT, PPO, and POX), and thus, better protection of the plants against this fungus, which also improved the plants' growth and yield. The application of the selected nano-Se and nano-SiO_2 generated beneficial effects by supporting the plant enzymatic systems and decreasing the disease severity of *A. alternata*. Although our findings may contribute to a better understanding of the antifungal effects of the applied nanofungicides, this novel antifungal alternative to chemical control requires additional investigation, particularly related to the interactions between the applied nano-Se and the nano-SiO_2.

Author Contributions: Conceptualization and visualization, Y.A.B. and N.A.T.; methodology, N.A.T. and Y.A.B.; software, T.E.; validation, S.H. and H.E.-R.; formal analysis, S.Ø.S. and Y.A.B.; investigation, N.A.T.; resources, H.E.-R.; data curation, Y.A.B.; writing—original draft preparation, H.E.-R. and S.Ø.S.; writing, review and editing, all authors; visualization, T.E., S.H. and H.E.-R.; supervision, Y.A.B. and S.Ø.S.; project administration, H.E.-R. All authors have read and agreed to the published version of the manuscript.

Funding: This research received no external funding.

Institutional Review Board Statement: Not applicable.

Informed Consent Statement: Not applicable.

Data Availability Statement: Available on request.

Acknowledgments: The authors want to acknowledge the technical and administrative staff at our institutions. We especially acknowledge the support from the staff at the Physiology and Breeding of Horticultural Crops Laboratory at Kafrelsheikh University in conducting the biochemical assays.

Conflicts of Interest: The authors declare no conflict of interest.

References

1. Félix-Pablos, C.M.; Parra-Cota, F.I.; Santoyo, G.; Orozco-Mosqueda, M.C.; Santos-Villalobos, S.D. Draft genome sequence of *Bacillus* sp. strain FSQ1, a biological control agent against white mold in common bean (*Phaseolus vulgaris* L.). *Curr. Res. Microb. Sci.* **2022**, *3*, 100138. [CrossRef] [PubMed]
2. FAOSTAT. *Tomato Production in 2019, Crops/Regions/World List/Production Quantity (Pick Lists)*; Corporate Statistical Database (FAOSTAT); UN Food and Agriculture Organization: Rome, Italy, 2020.
3. Mukankusi, C.; Raatz, B.; Nkalubo, S.; Berhanu, F.; Binagwa, P.; Kilango, M.; Williams, M.; Enid, K.; Chirwa, R.; Beebe, S. Genomics, genetics and breeding of common bean in Africa: A review of tropical legume project. *Plant Breed.* **2019**, *138*, 401–414. [CrossRef] [PubMed]
4. El-Dahshouri, M.F.; Hamouda, H.A.; Anany, T.G. Improving seed production of common bean (*Phaseolus vulgaris* L.) plants as a response for Calcium and Boron. *Agric. Eng. Int. CIGR J.* **2017**, *2017*, 211–219.
5. Girma, F.; Fininsa, C.; Terefe, H.; Amsalu, B. Evaluation of common bean (*Phaseolus vulgaris*) genotypes for resistance to common bacterial blight and angular leaf spot diseases, and agronomic performances. *Heliyon* **2022**, *8*, e10425. [CrossRef] [PubMed]
6. Rahmanzadeh, A.; Khahani, B.S.; Taghavi, M.; Khojasteh, M.; Osdaghi, E. Genome-wide meta-QTL analyses provide novel insight into disease resistance repertoires in common bean. *BMC Genom.* **2022**, *23*, 680. [CrossRef] [PubMed]
7. Adomako, J.S.; Yeboah, J.F.; Asamoah, P.; Amankwaa-Yeboah, E.A.; Adjei, E.A.; Obeng, B.; Sakyiamah, M.; Lamptey, L.; Asibuo, B.J.Y. Survey of plant parasitic nematodes and disease severity of common bean lines evaluated for reaction to root knot nematodes infestation. *Afric. Crop. Sci. J.* **2022**, *30*, 147–154. [CrossRef]
8. Rady, M.M.; Elrys, A.S.; Selem, E.; Mohsen, A.A.A.; Arnaout, S.M.A.I.; El-Sappah, A.H.; El-Tarabily, K.A.; Desoky, E.M. *Spirulina platensis* extract improves the production and defenses of the common bean grown in a heavy metals-contaminated saline soil. *J. Environ. Sci.* **2023**, *129*, 240–257. [CrossRef]

9. Papathanasiou, F.; Ninou, E.; Mylonas, I.; Baxevanos, D.; Papadopoulou, F.; Avdikos, I.; Sistanis, I.; Koskosidis, A.; Vlachostergios, D.N.; Stefanou, S.; et al. The Evaluation of Common Bean (*Phaseolus vulgaris* L.) Genotypes under Water Stress Based on Physiological and Agronomic Parameters. *Plants* **2022**, *11*, 2432. [CrossRef]
10. Zhang, Q.; Geng, J.; Du, Y.; Zhao, Q.; Zhang, W.; Fang, Q.; Yin, Z.; Li, J.; Yuan, X.; Fan, Y.; et al. Heat shock transcription factor (Hsf) gene family in common bean (*Phaseolus vulgaris*): Genome-wide identification, phylogeny, evolutionary expansion and expression analyses at the sprout stage under abiotic stress. *BMC Plant Biol.* **2022**, *22*, 33. [CrossRef]
11. Jin, Y.Z.; Xiong, Y.N.; Xu, C.J.; Ren, J.L.; Guo, Y.X.; Zuo, Y.H.; Zhang, Y.L.; Geng, X.Q. First Report of *Alternaria alternata* Causing Leaf Spot on Kidney Bean in China. *Plant Dis.* **2022**, *106*, 1531. [CrossRef]
12. Armitage, A.D.; Cockerton, H.M.; Sreenivasaprasad, S.; Woodhall, J.; Lane, C.R.; Harrison, R.J.; Clarkson, J.P. Genomics Evolutionary History and Diagnostics of the *Alternaria alternata* Species Group Including Apple and Asian Pear Pathotypes. *Front. Microbiol.* **2020**, *10*, 3124. [CrossRef]
13. Shalaby, T.A.; Taha, N.A.; Taher, D.I.; Metwaly, M.M.; El-Beltagi, H.S.; Rezk, A.A.; El-Ganainy, S.M.; Shehata, W.F.; El-Ramady, H.R.; Bayoumi, Y.A. Paclobutrazol Improves the Quality of Tomato Seedlings to Be Resistant to *Alternaria solani* Blight Disease: Biochemical and Histological Perspectives. *Plants* **2022**, *11*, 425. [CrossRef] [PubMed]
14. Kokaeva, L.Y.; Yarmeeva, M.M.; Kokaeva, Z.G.; Chudinova, E.M.; Balabko, P.N.; Elansky, S.N. Phylogenetic Study of *Alternaria* Potato and Tomato Pathogens in Russia. *Diversity* **2022**, *14*, 685. [CrossRef]
15. Ertoy, N. Morphological and Molecular Characterization of *Alternaria alternata* Causing Leaf Spot in Faba Bean (*Vicia faba* L.) and Determination of The Disease Reactions of Some Faba Bean Varieties Grown in Turkey. *Gesunde Pflanz.* **2022**. [CrossRef]
16. Budde-Rodriguez, S.; Pasche, J.S.; Mallik, I.; Gudmestad, N.C. Sensitivity of *Alternaria* spp. from potato to pyrimethanil, cyprodinil, and fludioxonil. *J. Crop. Prot.* **2022**, *152*, 105855. [CrossRef]
17. Tleuova, A.B.; Wielogorska, E.; Prasad Talluri, V.S.S.L.; Štěpánek, F.; Elliott, C.T.; Grigoriev, D.O. Recent advances and remaining barriers to producing novel formulations of fungicides for safe and sustainable agriculture. *J. Control. Release* **2020**, *326*, 468–481. [CrossRef]
18. Simon-Delso, N.; San Martin, G.; Bruneau, E.; Hautier, L. Time-to-death approach to reveal chronic and cumulative toxicity of a fungicide for honeybees not revealed with the standard ten-day test. *Sci. Rep.* **2018**, *8*, 7241. [CrossRef] [PubMed]
19. Quiterio-Gutiérrez, T.; Ortega-Ortiz, H.; Cadenas-Pliego, G.; Hernández-Fuentes, A.; Sandoval-Rangel, A.; Benavides-Mendoza, A.; la Fuente, M.C.; Juárez-Maldonado, A. The Application of Selenium and Copper Nanoparticles Modifies the Biochemical Responses of Tomato Plants under Stress by *Alternaria solani*. *Int. J. Mol. Sci.* **2019**, *20*, 1950. [CrossRef]
20. Bramhanwade, K.; Shende, S.; Bonde, S.; Gade, A.; Rai, M. Fungicidal activity of Cu nanoparticles against Fusarium causing crop diseases. *Environ. Chem. Lett.* **2016**, *14*, 229–235. [CrossRef]
21. Kutawa, A.B.; Ahmad, K.; Ali, A.; Hussein, M.Z.; Abdul Wahab, M.A.; Adamu, A.; Ismaila, A.A.; Gunasena, M.T.; Rahman, M.Z.; Hossain, M.I. Trends in Nanotechnology and Its Potentialities to Control Plant Pathogenic Fungi: A Review. *Biology* **2021**, *10*, 881. [CrossRef]
22. Sardar, M.; Ahmed, W.; Al Ayoubi, S.; Nisa, S.; Bibi, Y.; Sabir, M.; Khan, M.M.; Ahmed, W.; Qayyum, A. Fungicidal synergistic effect of biogenically synthesized zinc oxide and copper oxide nanoparticles against *Alternaria citri* causing citrus black rot disease. *Saudi J. Biol. Sci.* **2022**, *29*, 88–95. [CrossRef] [PubMed]
23. Periakaruppan, R.; Palanimuthu, V.; Abed, S.A.; Danaraj, J. New perception about the use of nanofungicides in sustainable agriculture practices. *Arch. Microbiol.* **2023**, *205*, 4. [CrossRef] [PubMed]
24. Cordon, G.; Valiño, I.L.; Prieto, A.; Costa, C.; Marchi, M.C.; Diz, V. Effects of the nanoherbicide made up of atrazine-chitosan on the primary events of photosynthesis. *J. Photochem. Photobiol.* **2022**, *12*, 100144. [CrossRef]
25. Shivashakarappa, K.; Reddy, V.; Tupakula, V.K.; Farnian, A.; Vuppula, A.; Gunnaiah, R. Nanotechnology for the detection of plant pathogens. *Plant Nano Biol.* **2022**, *2*, 100018. [CrossRef]
26. El-Gazzar, N.; Ismail, A.M. The potential use of Titanium, Silver and Selenium nanoparticles in controlling leaf blight of tomato caused by *Alternaria alternata*. *Biocatal. Agric. Biotechnol.* **2020**, *27*, 101708. [CrossRef]
27. Perez-Zavala, F.G.; Atriztan-Hernandez, K.; Mart´ınez-Irastorza, P.; Oropeza-Aburto, A.; Lopez-Arredondo, D.; Herrera-Estrella, L. Titanium nanoparticles activate a transcriptional response in *Arabidopsis* that enhances tolerance to low phosphate, osmotic stress and pathogen infection. *Front. Plant Sci.* **2022**, *13*, 994523. [CrossRef]
28. Paul, A.; Roychoudhury, A. Go green to protect plants: Repurposing the antimicrobial activity of biosynthesized silver nanoparticles to combat phytopathogens. *Nanotechnol. Environ. Eng.* **2021**, *6*, 10. [CrossRef]
29. Wang, Z.L.; Zhang, X.; Fan, G.J.; Que, Y.; Xue, F.; Liu, Y.H. Toxicity Effects and Mechanisms of MgO Nanoparticles on the Oomycete Pathogen *Phytophthora infestans* and Its Host Solanum tuberosum. *Toxics* **2022**, *10*, 553. [CrossRef]
30. Parveen, A.; Siddiqui, Z.A. Impact of silicon dioxide nanoparticles on growth, photosynthetic pigments, proline, activities of defense enzymes and some bacterial and fungal pathogens of tomato. *Vegetos* **2022**, *35*, 83–93. [CrossRef]
31. Kamel, S.M.; Elgobashy, S.F.; Omara, R.I.; Derbalah, A.S.; Abdelfatah, M.; El-Shaer, A.; Al-Askar, A.A.; Abdelkhalek, A.; Abd-Elsalam, K.A.; Essa, T.; et al. Antifungal Activity of Copper Oxide Nanoparticles against Root Rot Disease in Cucumber. *J. Fungi* **2022**, *8*, 911. [CrossRef]
32. Khan, M.R.; Siddiqui, Z.A. Use of silicon dioxide nanoparticles for the management of *Meloidogyne incognita*, *Pectobacterium betavasculorum* and *Rhizoctonia solani* disease complex of beetroot (*Beta vulgaris* L.). *Sci. Hortic.* **2020**, *265*, 109211. [CrossRef]

33. Riseh, R.S.; Hassanisaadi, M.; Vatankhah, M.; Soroush, F.; Varma, R.S. Nano/microencapsulation of plant biocontrol agents by chitosan, alginate, and other important biopolymers as a novel strategy for alleviating plant biotic stresses. *Int. J. Biol. Macromol.* **2022**, *222 Pt A*, 1589–1604. [CrossRef]
34. El-Ramady, H.; Alshaal, T.; Elhawat, N.; El-Nahrawy, E.; Omara, A.E.-D.; El-Nahrawy, S.; Elsakhawy, T.; Ghazi, A.; Abdalla, N.; Fári, M. Biological aspects of selenium and silicon nanoparticles in the terrestrial environments. In *Phytoremediation: Management of Environmental Contaminants*; Ansari, A.A., Gill, S.S., Gill, R., Lanza, G., Newman, L., Eds.; Springer International Publishing: Cham, Switzerland, 2018; pp. 235–264.
35. Zahedi, S.M.; Moharrami, F.; Sarikhani, S.; Padervand, M. Selenium and silica nanostructure-based recovery of strawberry plants subjected to drought stress. *Sci. Rep.* **2020**, *10*, 17672. [CrossRef] [PubMed]
36. Hussain, B.; Lin, Q.; Hamid, Y.; Sanaullah, M.; Di, L.; Hashmi, M.L.R.; Khan, M.B.; He, Z.; Yang, X. Foliage application of selenium and silicon nanoparticles alleviates Cd and Pb toxicity in rice (*Oryza sativa* L.). *Sci. Total Environ.* **2020**, *712*, 136497. [CrossRef] [PubMed]
37. Badawy, S.A.; Zayed, B.A.; Bassiouni, S.M.A.; Mahdi, A.H.A.; Majrashi, A.; Ali, E.F.; Seleiman, M.F. Influence of Nano Silicon and Nano Selenium on Root Characters, Growth, Ion Selectivity, Yield, and Yield Components of Rice (*Oryza sativa* L.) under Salinity Conditions. *Plants* **2021**, *10*, 1657. [CrossRef] [PubMed]
38. El-Saadony, M.T.; Desoky, E.M.; Saad, A.M.; Rania, S.M.; Eid, R.S.M.; Selem, S.; Elrys, A.S. Biological silicon nanoparticles improve *Phaseolus vulgaris* L. yield and minimize its contaminant contents on a heavy metals-contaminated saline soil. *J. Environ. Sci.* **2021**, *106*, 1–14. [CrossRef]
39. Joshi, S.M.; De Britto, S.; Jogaiah, S. Myco-engineered selenium nanoparticles elicit resistance against tomato late blight disease by regulating differential expression of cellular, biochemical and defense responsive genes. *J. Biotechnol.* **2021**, *325*, 196–206. [CrossRef] [PubMed]
40. Bhat, K.A.; Batool, A.; Mansoor, M.; Manzoor, M.; Bashir, Z.; Nazir, M.; Zargar, S.M. Silicon- and nanosilicon-mediated disease resistance in crop plants. In *Silicon and Nano-silicon in Environmental Stress Management and Crop Quality Improvement*; Etesami, H., Al Saeedi, A.H., El-Ramady, H., Fujita, M., Pessarakli, M., Hossain, M.A., Eds.; Academic Press: Cambridge, MA, USA, 2022; pp. 193–205.
41. Dhakate, P.; Kandhol, N.; Raturi, G.; Ray, P.; Bhardwaj, A.; Srivastava, A.; Kaushal, L.; Singh, A.; Pandey, S.; Sharma, S.; et al. Silicon nanoforms in crop improvement and stress management. *Chemosphere* **2022**, *305*, 135165. [CrossRef]
42. El-Ramady, H.; Abdalla, N.; El-Mahrouk, M.E.; Bayoumi, Y.; Elsakhawy, T.; Omara, A.E.-D.; Shalaby, T.A. Anatomical Changes of Cultivated Plants under Combined Stress: An Urgent Need for Investigation. *Environ. Biodivers. Soil Secur.* **2021**, *5*, 73–83. [CrossRef]
43. Panwar, V.; Gangwar, R.K.; Javeria, S.; Yadav, R.S. Antifungal efficacy of fungicides and bio-control agents against leaf spot pathogens, *Alternaria alternata*. *Curr. Discov.* **2013**, *2*, 128–133.
44. Ozcelik, N.S.; Ozcelik, S. Investigation on some factors and strains affecting the production of Alternaria-toxins by a thin layer chromatographic method. *Turk. J. Agric. For.* **1997**, *21*, 1–5. [CrossRef]
45. Ghazi, A.A.; El-Nahrawy, S.; El-Ramady, H.; Ling, W. Biosynthesis of Nano-Selenium and Its Impact on Germination of Wheat under Salt Stress for Sustainable Production. *Sustainability* **2022**, *14*, 1784. [CrossRef]
46. Ferreira, J.H.; Mathee, F.N.; Thomas, A.C. Biological control of *Eutypalota* on grapevine by an antagonistic strain of *Bacillus subtilis*. *Phytopathology* **1991**, *81*, 283–287. [CrossRef]
47. Pandy, H.N.; Menon, T.C.M.; Rao, M.V. Simple formula for calculating area under disease progress curve. *Rachis* **1989**, *8*, 38–39.
48. Nagata, M.; Yamashita, I. Simple Method for Simultaneous Determinations of Chlorophyll and Carotenoids in Tomato Fruit. *Nippon. Shokuhin Kogyo Gakkaishi* **1992**, *39*, 925–928. [CrossRef]
49. Maxwell, K.; Johnson, G.N. Chlorophyll Fluorescence—A Practical Guide. *J. Exp. Bot.* **2000**, *51*, 659–668. [CrossRef] [PubMed]
50. Aebi, H. Catalase in Vitro. In *Methods in Enzymology*; Packer, L., Ed.; Academic Press: San Diego, CA, USA, 1984; pp. 121–126.
51. Rathmell, W.G.; Sequeira, L. Soluble Peroxidase in Fluid from the Intercellular Spaces of Tobacco Leaves. *Plant Physiol.* **1974**, *53*, 317–318. [CrossRef]
52. Malik, C.P.; Singh, M.B. *Plant Emynology and Histo-Enzymology*; Kalyani Publishers: New Delhi, India, 1980.
53. Binsan, W.; Benjakul, S.; Visessanguan, W.; Roytrakul, S.; Tanaka, M.; Kishimura, K. Antioxidative activity of Mungoong, an extract paste, from the cephalothorax of white shrimp (*Litopenaeus vannamei*). *Food Chem.* **2008**, *106*, 185–193. [CrossRef]
54. Whitlow, T.H.; Bassuk, N.L.; Ranney, T.G.; Reichert, D.L. An improved method for using electrolyte leakage to assess membrane competence in plant tissues. *Plant Physiol.* **1992**, *98*, 198–205. [CrossRef]
55. Szalai, G.; Janda, T.; Paldi, E.; Szigeti, Z. Role of light in the development of post- chilling symptoms in maize. *J. Plant Physiol.* **1996**, *148*, 378–383. [CrossRef]
56. Snedecor, G.W.; Cochran, W.G. *Statistical Methods*, 8th ed.; Iowa State University Press: Ames, IA, USA, 1989.
57. Yu, S.; Zhen, C.; Zhao, P.; Li, L.; Qiu, Z.; Gao, H. Antifungal mechanisms of γ-aminobutyric acid against the postharvest pathogen *Alternaria alternata*. *LWT* **2023**, *173*, 114314. [CrossRef]
58. Ghooshkhaneh, N.G.; Golzarian, M.R.; Mollazade, K. VIS-NIR spectroscopy for detection of citrus core rot caused by *Alternaria alternata*. *Food Control.* **2023**, *144*, 109320. [CrossRef]
59. Kotzybik, K.; Graf, V.; Kugler, L.; Stoll, D.A.; Greiner, R.; Geisen, R.; Schmidt-Heydt, M. Influence of different nanomaterials on growth and mycotoxin production of *Penicillium verrucosum*. *PLoS ONE* **2016**, *11*, e0150855. [CrossRef] [PubMed]

60. Vera-Reyes, I.; Altamirano-Hernández, J.; Reyes-de la Cruz, H.; Granados-Echegoyen, C.A.; Loera-Alvarado, G.; López-López, A.; Garcia-Cerda, L.A.; Loera-Alvarado, E. Inhibition of phytopathogenic and beneficial fungi applying silver nanoparticles in vitro. *Molecules* **2022**, *27*, 8147. [CrossRef] [PubMed]
61. Yang, Z.X.; Yang, Y.F.; Yu, S.Z.; Wang, R.G.; Wang, Y.; Chen, H.L. Photosynthetic, photochemical and osmotic regulation changes in tobacco resistant and susceptible to *Alternaria alternata*. *Trop. Plant Pathol.* **2018**, *43*, 413–421. [CrossRef]
62. Tripathi, D.; Singh, M.; Pandey-Rai, S. Crosstalk of nanoparticles and phytohormones regulate plant growth and metabolism under abiotic and biotic stress. *Plant Stress* **2022**, *6*, 100107. [CrossRef]
63. El-Badri, A.M.; Ahmed, M.; Hashem, A.M.; Batool, M.; Ahmed Sherif, A.; Nishawy, E.; Ayaad, M.; Hassan, H.M.; Elrewainy, I.M.; Wang, J.; et al. Comparative efficacy of bio-selenium nanoparticles and sodium selenite on morpho-physiochemical attributes under normal and salt stress conditions, besides selenium detoxification pathways in *Brassica napus* L. *J. Nanobiotechnol.* **2022**, *20*, 163. [CrossRef]
64. Wang, L.; Ning, C.; Pan, T.; Cai, K. Role of Silica Nanoparticles in Abiotic and Biotic Stress Tolerance in Plants: A Review. *Int. J. Mol. Sci.* **2022**, *23*, 1947. [CrossRef]
65. Shalaby, T.A.; Abd-Alkarim, E.; El-Aidy, F.; Hamed, E.; Sharaf-Eldin, M.; Taha, N.; El-Ramady, H.; Bayoumi, Y.; dos Reis, A.R. Nano-selenium, silicon and H_2O_2 boost growth and productivity of cucumber under combined salinity and heat stress. *Ecotoxicol. Environ. Saf.* **2021**, *212*, 111962. [CrossRef]
66. El-Saadony, M.T.; Saad, A.M.; Najjar, A.A.; Alzahrani, S.O.; Alkhatib, F.M.; Shafi, M.E.; Selem, E.; Desoky, E.S.M.; Fouda, S.E.; El-Tahan, A.M.; et al. The use of biological selenium nanoparticles to suppress *Triticum aestivum* L. crown and root rot diseases induced by *Fusarium* species and improve yield under drought and heat stress. *Saudi J. Biol. Sci.* **2021**, *28*, 4461–4471. [CrossRef]
67. Zhu, Y.; Dong, Y.; Zhu, N.; Jin, H. Foliar application of biosynthetic nano-selenium alleviates the toxicity of Cd, Pb, and Hg in *Brassica chinensis* by inhibiting heavy metal adsorption and improving antioxidant system in plant. *Ecotoxicol. Environ. Saf.* **2022**, *240*, 113681. [CrossRef] [PubMed]
68. El-Badri, A.M.; Batool, M.; Mohamed, I.A.A.; Wang, Z.; Wang, C.; Tabl, K.M.; Khatab, A.; Kuai, J.; Wang, J.; Wang, B.; et al. Mitigation of the salinity stress in rapeseed (*Brassica napus* L.) productivity by exogenous applications of bio-selenium nanoparticles during the early seedling stage. *Environ. Pollut.* **2022**, *310*, 119815. [CrossRef] [PubMed]
69. Albalawi, M.A.; Abdelaziz, A.M.; Attia, M.S.; Saied, E.; Elganzory, H.H.; Hashem, A.H. Mycosynthesis of Silica Nanoparticles Using *Aspergillus niger*: Control of *Alternaria solani* Causing Early Blight Disease, Induction of Innate Immunity and Reducing of Oxidative Stress in Eggplant. *Antioxidants* **2022**, *11*, 2323. [CrossRef] [PubMed]
70. Kapoor, B.; Kumar, P.; Gill, N.S.; Sharma, R.; Thakur, N.; Irfan, M. Molecular mechanisms underpinning the silicon-selenium (Si-Se) interactome and cross-talk in stress-induced plant responses. *Plant Soil* **2022**. [CrossRef]
71. Saffan, M.M.; Koriem, M.A.; El-Henawy, A.; El-Mahdy, S.; El-Ramady, H.; Elbehiry, F.; Omara, A.E.-D.; Bayoumi, Y.; Badgar, K.; Prokisch, J. Sustainable Production of Tomato Plants (*Solanum lycopersicum* L.) under Low-Quality Irrigation Water as Affected by Bio-Nanofertilizers of Selenium and Copper. *Sustainability* **2022**, *14*, 3236. [CrossRef]

microorganisms

MDPI

Article

Influence of Probiotic Feed Supplement on *Nosema* spp. Infection Level and the Gut Microbiota of Adult Honeybees (*Apis mellifera* L.)

Ivana Tlak Gajger [1,*], Srebrenka Nejedli [2] and Luka Cvetnić [3]

1 Department for Biology and Pathology of Fish and Bees, Faculty of Veterinary Medicine, University of Zagreb, 10 000 Zagreb, Croatia
2 Department for Anatomy, Histology and Embryology, Faculty of Veterinary Medicine, University of Zagreb, 10 000 Zagreb, Croatia
3 Department for Bacteriology and Parasitology, Croatian Veterinary Institute, 10 000 Zagreb, Croatia
* Correspondence: ivana.tlak@vef.hr; Tel.: +385-91-2390-041

Abstract: Honeybees' gut microbiota can provide new valuable access into the pathogenesis-related factors included in infections. Hence, we researched the presence and comparison of gut microbiota groups in control and *Nosema* spp.-infected honeybee colonies through high-throughput sequencing of the 16S rRNA. As the newest approach in apiary management, we hypothesize that the EM® probiotic for bees could have an important role in therapeutic and immunomodulatory effects on honeybee colonies. The aim of this study was to estimate its impact on the gut microbiota composition of adult honeybees. The major genera were detected, where Lactobacillus was the most abundant genus, followed by Gilliamela, Snodgrassella, and Bifidobacterium. Inoculation with *Nosema* spp. spores made the relative proportions of Bifidobacterium lower, which was ameliorated by EM® for bees' application. In addition, EM® for bee applied treatments suppressed the increase in the number of *Nosema* spp. spores. This result points out that continuous EM® for bees treatment shall change bees' gut microbiome composition and mitigate the influence of *Nosema* spp. infection. *Snodgrassella alvi* was a major member of the honeybee gut microbiota and may be significantly increased by long-term treatment with EM® for bees. Toward these results, it is possible that EM® for bees treatment will protect honeybees from herbicide glyphosate negative effects in agricultural fields by improving microbiome and immune functions.

Keywords: honeybee colony; *Apis mellifera* L.; bacteria; probiotics; EM® for bees; gut microbiota; *Nosema* spp.; colony strength

Citation: Tlak Gajger, I.; Nejedli, S.; Cvetnić, L. Influence of Probiotic Feed Supplement on *Nosema* spp. Infection Level and the Gut Microbiota of Adult Honeybees (*Apis mellifera* L.). *Microorganisms* **2023**, *11*, 610. https://doi.org/10.3390/microorganisms11030610

Academic Editors: Vishnu D. Rajput, Anuj Ranjan and Abhishek Chauhan

Received: 30 January 2023
Revised: 15 February 2023
Accepted: 26 February 2023
Published: 28 February 2023

1. Introduction

Honeybee colonies (*Apis mellifera* L.) represent the most important social insects. They have a crucial role in plant pollination [1] and are closely linked to global food production and natural biodiversity balance maintaining. Therefore, their economic, ecological, and social values are enormous [2,3]. Due to their specific nutrition physiology and flight behavior patterns, they show complex interactions with environmental ecosystems and consequently with a diverse range of microorganisms [4]. In recent years, large-scale colony declines have been reported. These losses have been associated with unfavorable pedoclimatic and forage frames, parasites and pathogen infections, pesticide intoxications, and nutritional stress of different origins [5–9]. All these factors are often combined by the beekeeper's management practices [10–12]. Consequently, they appear in a meaningful and synergistic package approach, possibly causing different disturbances of the adult honeybee gut microbiota composition.

The core gut microbiota of adult honeybees consists of complex bacteria communities with high genomic diversity whose functionality is dictated by the host, environment, and social exchanges as well as microbial interactions among themselves [4,13,14]. This microbiota is located in various parts of the gastrointestinal tract, including the crop (just a few microorganisms originated from the environment, e.g., *Apilacobacilus kunkei*), midgut (mostly no stable microbiota), ileum (diverse microbial clusters; 10^8 bacterial cells, e.g., *Snodgrasella alvi*, *Gillimella apicola*, *Lactobacillus* Firm-4 and Firm-5, *Bifidobacterium* spp., *Frischella perrara*, *Bartonella apis*, *Parasaccharibacter apium*, *Bombella apis*, *Apibacter adventoris*, *Apibacter mensalis*), and rectum (10^9 bacterial cells, e.g., *Lactobacillus* Firm-4, *Lactobacillus* Firm-5, *Bifidobacterium* spp.) [14–18]. Bacterial symbionts or "good bacteria" in the honeybee gut have important functions for host nutrition, food digestion, metabolism, development, weight gain, reproduction, behavioral physiology, and immunity through pathogen and insecticide resistance [19–25]. It is known that gut microorganisms significantly contribute to pollen digestion, which is the main source of proteins in honeybee nutrition. Therefore, the microbiome composition affects survival rates [26]. In addition, gut microflora have a crucial role in vitamin, fatty acid, and amino acid synthesis [27], in the development and maturation or renewal of enterocytes [28], in the increasing antimicrobial peptide gene expression, pheromone production [29], and in the formation of the biofilm on gut epithelium as a mechanical barrier against disease-causative agents [15].

Dysbiosis in honeybees is often defined as gut–intestinal microbial imbalance linked to a host deficiency, such as deficient development, lower body mass, earlier worker mortality, and general health, metabolism, and fitness status [30]. In addition, in such a situation, the different environmental influences (e.g., the high infection rate of *Nosema ceranae* spores or immune response suppression caused by oxidative stress) could change the composition of gut bacterial phylotypes and remaining microbiome components (e.g., fungi) leading to the appearance of visible clinical signs of opportunistic diseases and colonies weakening [31,32]. Infections of honeybees with the endoparasite *N. ceranae* and with simultaneous exposure to pesticides can significantly contribute to gut dysbiosis [33,34]. In addition, previously published results showed that *N. ceranae* was included in the removal of *Serratia* spp. from bees' guts, and a consequently significant disturbance in *Snodgrassella* spp. and *Bartonella* spp. was determined [35]. Furthermore, glyphosate combined with *N. ceranae* spore infection change the gut microbiome composition by decreasing the average proportions of *Snodgrassella alvi* and *Lactobacillus apis* [36]. It is interesting that after in vivo treatment of honeybee colonies with oxalic acid, it reduces amplicon sequence variant richness and alters the gut microbiome composition, especially in the genus *Bombella* and bacteria *Lactobacillus kunkeei* [37].

Published studies on the impacts of used commercial probiotics and prebiotics on different aspects of honeybee health are incomplete and contradictory. Some studies show that probiotics increase adult bees' mortality and different pathogen loads, whereas others suggest that the application of probiotics has a positive effect on the protection against diseases, apian product gains, and the strength of colonies or wax gland development [38–45].

N. ceranae is a microsporidium that causes a parasitic disease of honeybee colonies called nosemosis type C and adversely affects adult honeybees' health by parasitizing in the midgut epithelium and impairing digestion and absorption of nutrients [46]. For efficient beekeeping management and sustainable alternative strategies for nosemosis type C, control is needed, especially today, because there is not a single registered authorized veterinary medicine product for nosemosis control. To facilitate full therapeutic success, there is a need for the appliance of bio-inspired honeybee colony protection products in form of feed supplements [47–52] and novel technology designs based on natural ingredients active against microsporidia pathogens. The other possibility is to trigger honeybees' immune defense responses [53] and, at the same time, avoid the possibility of detection of hazards residual in honey and wax originating from treated hives.

One of the natural and bio-inspired ways to protect honeybees' health and to produce safe apian products is the use of a mix of effective microorganism cultures and EM® PROBIOTIC FOR BEES (hereinafter, EM® for bees). This commercially available product is a proprietary probiotic formulation owned and managed by the EM Research Organization in Okinawa, Japan. It contains multiple species of lactic acid bacteria, yeast, and photosynthetic bacteria. After promising results for honeybee nosemosis combating in apiary and laboratory-controlled conditions [53], we hypothesized that EM® for bees could have an important effect on gut microbiome content. The aim of this study is to evaluate the therapeutic effects of EM® for bees on *Nosema* spp. infection levels in relation to honeybee colony strength in apiary conditions. In addition, the gut microbiota composition was obtained using next generation sequencing (NGS) analyses to check the differences between experimental and control honeybee colony groups.

2. Materials and Methods

2.1. Apiary Conditions

2.1.1. Field Test Design, Feed Treatments, and Bee Sampling

The field experiment was conducted for 42 consecutive days (beginning on the 7th of July 2020) at the beeyard situated in the continental part of Croatia (45°56′54.71″ N, 16°37′46.06″ E) after the major honey harvesting season. For performing the test in field conditions, approximately 12 homogeneous honeybee colonies (*A. mellifera carnica*, Pollmann, 1879) were placed in standard Langstroth–Root (LR) hives. Colonies were acquired from the same apiary. Experimental groups were composed of colonies: TH1, colonies were naturally infected with *Nosema* spp. spores; TH2, colonies were treated with EM® for bees; TH3, colonies were additionally infected with *Nosema* spp. spores before the start of treatment with EM® for bees; and TH4, control, noninfected colonies without treatment. Detailed descriptions of the feeding regime and samplings are given in Supplementary Materials.

Experimental honeybee colonies (TH2 and TH3) were additionally fed with a total of 300 mL of sugar syrup prepared in 1:1 proportion of water (L) and sugar (kg) (Virosecer, Croatia) supplemented with a 5% EM® for bees solution consecutively for 10 days from the initial sampling (start of experiment; sampling conducted prior to the first application session). Sugar syrup with probiotic as well as pure sugar syrup (TH1) were applied to honeybee colonies by spraying directly on frames (15 mL of diluted supplemented solution per frame) covered with adult honeybees. The pertaining control honeybee colony (TH4) received only 300 mL of pure sugar syrup prepared and administrated in the described manner. The dose of EM® for bees was adapted according to the manufacturer's instructions. Throughout the honeybee colonies' clinical inspection, approximately 60 forager adult bees per colony were sampled from the entrance of the hive for microscope quality and quantity analyses on *Nosema* spp. spores. Adult honeybee samples were collected in clean plastic receptacles by catching bees directly in front of hive entrances using long tweezers. Each sample was taken on the 7th of July, 17th of July, 27th of July, and 20th of August 2020.

At the initial clinical inspection, none of the honeybee colonies showed clinical signs of diseases. In addition, the previously performed treatment against the mites *Varroa destructor* was carried out on the 25th of June 2020 (Apitraz®, a.m. amitraz) to avoid the deleterious effects of mite parasitation on honeybee colony health. To the best of our knowledge, no pesticides were used in the surrounding agricultural area during the field experiment.

2.1.2. Sampling of Adult Honeybees' Guts

The full length of the intestine of each individual adult honeybee (n = 10 specimens per pooled sample) was extracted after exposure to a low temperature (10 min at 4 °C). For extraction, a larger pair of forceps was used to fix the head and the thorax of each specimen and additionally a smaller pair of forceps to support the top of the last segment of the abdomen where the intestines were carefully pulled out. After the described step, the oesophagus, honey sac, and midgut were removed by cutting them off. For examination of the gut microbiome content, gut samples (ileum and rectum) were fixed in the Eppendorf tubes and were cooled directly in a box full of ice during transportation. In the laboratory, gut samples were stored at −80 °C until molecular analyses were performed.

2.1.3. Clinical Inspections of Honeybee Colonies and Strength Estimation

The presence of the honeybee queen, the mortality of the adult bees, as well as clinical signs of brood diseases were inspected during each opening of the hives at the experimental beeyard. The Liebefeld method for visual estimation of the number of adult bees and brood was performed to determine the strength of honeybee colonies [54]. The assessment of honeybee colony strength was conducted on the 1st (7th of July) and the 42nd (20th of August) day of the experiment. The estimation was conducted during the morning hours (around 10:00 a.m.), before massive forage flights of workers.

2.2. Laboratory Examinations

2.2.1. Presence Determination and Quantification of *Nosema* spp. Spore Levels

Honeybees were counted in each sample. Their abdomens were separated, thoroughly crushed, and homogenized in a plastic container loaded with 1 mL of pure water per bee specimen. *Nosema* spp. spores were counted in each sample using a Malassez hemocytometer. Infection levels were calculated according to the World Organization for Animal Health (WOAH) guidelines [55]. Each counting procedure was repeated three times. The counting equipment was carefully cleaned after each sample counting to avoid contamination with spores from the previous sample.

2.2.2. Gut microbiota Processing and Analysis

Extraction

All honeybee gut samples were processed by homogenization for 1 min in sterile plastic tubes with the addition of 0.5 mL 1 × TE (10 mM Tris-HCl, 1 mM EDTA, pH 8). Total DNA was extracted using the DNeasy PowerSoil Kit (12888-50, Qiagen, Hilden, Germany) according to the manufacturer's instructions. The DNA extractions were monitored with electrophoresis on a 1% agarose gel to check the purity and then were quantified using a NanoDrop 1000 spectrophotometer (ThermoScientificTM, Waltham, MA, USA). The final concentration of the DNA sample was adjusted to 20 ng/μL.

Amplicon Sequencing by NGS

The V3-V4 regions of bacterial and archaeal 16S rRNA were amplified using the Pro341F (CCTACGGGNBGCASCAG)/Pro805R (GACTACNVGGGTATCTAATCC) primers and the dual-index method [56,57]. Amplicons were paired-end sequenced on a 2 × 284 bp cycle using the Illumina MiSeq system with MiSeq Reagent Kit ver 3 (600 Cycle) chemistry and barcoded. Paired-end sequencing readings were connected using a fast-q-join program with default lining [58]. Only linked readings that had a quality value shot of ≥20 for more than 99% of the sequence were extracted using the FASTX-Toolkit [59]. The chimeric sequences were deleted with usearch 6.1 [60–62]. Nonchimeric readings were submitted for 16S rDNA-based taxonomic analysis using the Ribosomal Database Project (RDP) Classifier ver. 2.11 (attributing taxon—Phylum 0.8*1) and the TechnoSuruga Lab Microbial Identification database DB-BA ver 13.0 (TechnoSuruga Laboratory, Ltd., Shizuoka, Japan) with homology for ≥97% [63,64].

Sequence Analyses

Secondary analyses were conducted based on the results of the core set of the previously mentioned database. For comparative analysis between samples, the software Megagenome@KIN ver 5.0 (World Fusion, Tokyo, Japan) was used. With the purpose of analyzing sequence similarities among different operational taxonomic units (OTUs), multiple sequence alignment was performed by using the free available metagenome analysis software (MEGAN) ver. 2.5.0 and R package multi-comp software ver 4.0.5 environment for statistical computing and graphics [65]. For group comparison analyses, linear discriminant analysis effect size (LEfSe) software ver. 1.0 with the pertaining website was used: https://huttenhower.sph.harvard.edu/galaxy/ (accessed on 16 June 2022).

2.3. Statistical Analyses of Colonies' Strength and Nosema spp. Infection Levels

Data were analyzed with the Stata 13.1. computer program (Stata Corp, College Station, TX, USA). The number of *Nosema* spp. spores detected on a certain date were compared between four groups using the Kruskall–Wallis nonparametric test. Paired comparisons between groups on the same date were conducted by the Dunn test, and the results are expressed as z and p values. The number of spores was log normalized and checked for normality of distribution using the Shapiro–Wilk test. Log-transformed data were longitudinally compared between different time points of sampling within the same group using a paired t-test. All results are presented as the main values and their standard deviations. Statistical significance testing was carried out with a significance level of $\alpha = 0.05$ to define statistical differences (0.95 confidence interval).

3. Results

3.1. Apiary Conditions

Estimated Strength of Honeybee Colonies

Differences in the average number of honeybees *per* group during two estimation dates are shown in Figure 1. Statistically significant differences in honeybee colony strength between the control and experimental groups were determined on day 42 ($p < 0.001$; $F = 19.71$). A higher number of honeybees was estimated in groups treated with EM® for bees (TH2, TH3) compared to control groups (TH1, TH4). The colony strength of the TH1 and TH4 groups was similar.

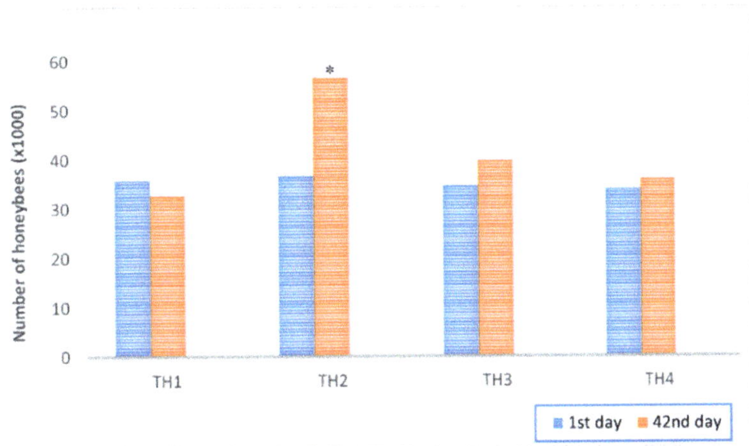

Figure 1. Honeybee colonies' strength variations between control (TH1, TH4) and experimental groups (TH2, TH3) by the estimation days (1st and 42nd day from the initial day of the experiment); statistically significant difference for TH2 on the second day of estimation vs. TH1, TH3, and TH4; * $p < 0.05$.

3.2. Laboratory Examinations

3.2.1. Determination of *Nosema* spp. Infection Levels in Adult Bee Samples

Results of the estimation of *Nosema* spp. infection levels are presented in Table 1. A decline in the number of *Nosema* spp. spores in adult bee samples collected in the colonies fed with EM® for bees (TH2; TH3) in the second ($p < 0.01$; $p < 0.001$) and third ($p < 0.01$; $p < 0.001$) sampling term through a statistically lower number of spores in comparison to the initial sampling before experimental feeding was observed. Only after the last subsequent sampling day (day 44), the increase in the number of spores was estimated for both experimental groups of honeybee colonies. Opposite to the presented results, in the control groups TH1 and TH4, the continuous increase in *Nosema* spp. infection levels was confirmed for each subsequent sampling term, respectively.

Table 1. The number of *Nosema* spp. spores per honeybee during supplemental feeding in apiary conditions; mean values ± SD.

Group		Spores of *Nosema* spp. ($\times 10^6$)			
		7th July	17th July	27th July	20th August
TH1	Mean	2.77	4.26	8.57	15.08
([1]N = 9)	SD	1.23	2.67	6.19	3.65
TH2	Mean	8.02	2.02	0.77	4.66
([1]N = 9)	SD	10.19	2.88	0.42	3.02
TH3	Mean	14.31	8.66	5.84	22.91
([1]N = 9)	SD	8.71	5.68	4.71	11.88
TH4	Mean	15.44	16.42	18.05	59.91
([1]N = 9)	SD	12.38	10.62	10.53	65.53
p (Kruskall–Wallis)		0.001	0.0005	0.0001	0.0001

[1] Sample from each colony was examined in triplicate.

Furthermore, in the experimental group TH2, the decreased number of spores compared to the initial spore count was 25.18% on average on day 10, 96% on average on day 20, and 58.10% on average on day 44. For group TH3, the reduction in spore counts in comparison with the initial sampling term, on average, were as follows: 60.51% on day 10 and 40.81% on day 20. Observed differences between groups (TH1, TH2, TH3, and TH4) are statistically significant in every time point of control (Table 2a–d).

Table 2. Statistics of *Nosema* spp. spore burden differences between the same group for each sampling time point: 7th July (a), 17th July (b), 27th July (c), and 20th August (d).

TH	z	*p*	TH	z	*p*	TH	z	*p*	TH	z	*p*
1:2	−0.581781	0.2804	1:2	1.354283	0.0878	1:2	3.214710	0.0007	1:2	2.350559	0.0094
1:3	−3.020785	0.0013	1:3	−1.544554	0.0612	1:3	0.705668	0.2402	1:3	−1.387949	0.0826
1:4	−3.199795	0.0007	1:4	−2.585449	0.0049	1:4	−1.680162	0.0465	1:4	−2.350559	0.0094
2:3	−2.439005	0.0074	2:3	−2.898836	0.0019	2:3	−2.509042	0.0061	2:3	−3.738509	0.0001
2:4	−2.618014	0.0044	2:4	−3.939731	0.0000	2:4	−4.894872	0.0000	2:4	−4.701119	0.0000
3:4	−0.179010	0.4290	3:4	−1.040895	0.1490	3:4	−2.385830	0.0085	3:4	−0.962610	0.1679
	(a)			(b)			(c)			(d)	

3.2.2. The Estimation of Intestinal Microbiota Composition, Richness, and Diversity

The number of OTUs and readings at the species level for each sampling date and treatment obtained from the amplicon sequence are shown in Figure 2. The number of OTUs and readings on 27th July increased in TH1, TH2, and TH3. However, the number of readings did not increase in TH4. The lifespan of adult summer honeybees is about 42 days, suggesting that most of the July 27th bees were old, and most of the August 20th bees were bees of the next generation.

The bacterial composition ratios (family, genus, and species) determined by the samples of honeybee intestines for each examined sample are shown in Figures 3–5. In all groups of honeybee colonies and additional feeding regimes (TH1, TH2, TH3, TH4), the bacteria with the highest compositional ratios were the most common bacteria in the honeybee gut microbiome.

The diversity index (Shanon index) using family-level data for all 48 samples is shown for each sampling date, which was significantly higher for TH3 compared to TH1 and TH4 for the 17th of July data, indicating that TH3 has a higher diversity of gut microbiota. No significant differences were found between treatments on the other sampling dates (Figure 6).

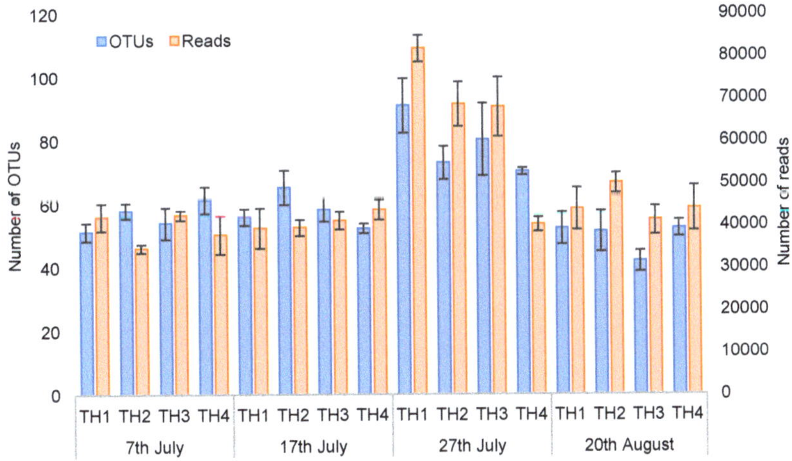

Figure 2. Comparison of the number of OTUs and readings on each treatment and sampling date. Each column in the graph shows the average (n = 3) of the number of OTUs and readings on each treatment and sampling date. Values represent mean + SEM.

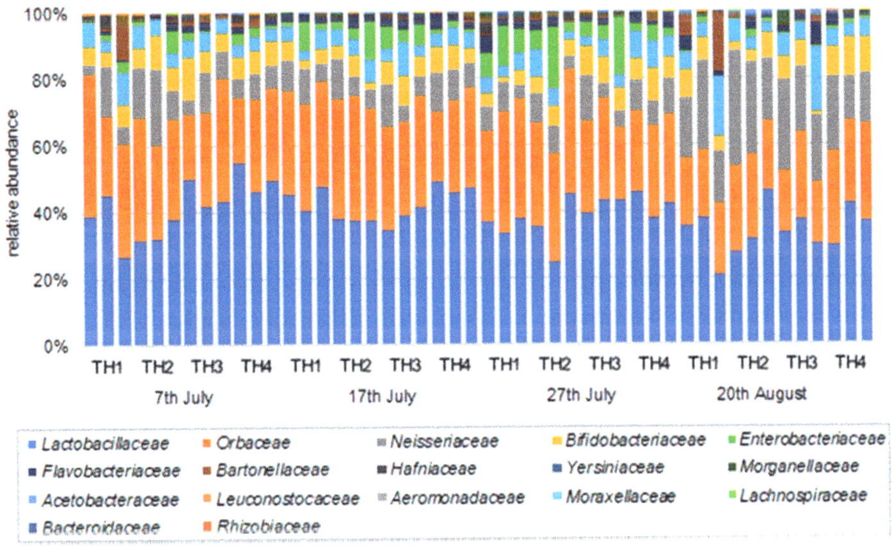

Figure 3. Relative abundance (family) of the gut microbiome in each treatment and on each sampling date.

The results of a principal coordinate analysis (PCoA) visualizing the similarity of complex data (e.g., plotted figures) are shown in Figure 7. The labels of each plot mean the number of *Nosema* spp. spores in the gut of bees. PCoA examined at the species level showed that the plots of the 20th of August samples were away from the other sampling date, further the 27th of July treatments, except for TH4, which was also plotted apart from the other plots. The samples taken on 7th July and 17th July were plotted close together, while TH2 with EM was located slightly further away. In the samples from 27th July, only TH4 was plotted close to the aforementioned samples, while TH2 and TH3 (with EM® for bees treatment) were plotted far from them. This cluster is a relatively low spore count group for *Nosema* spp.

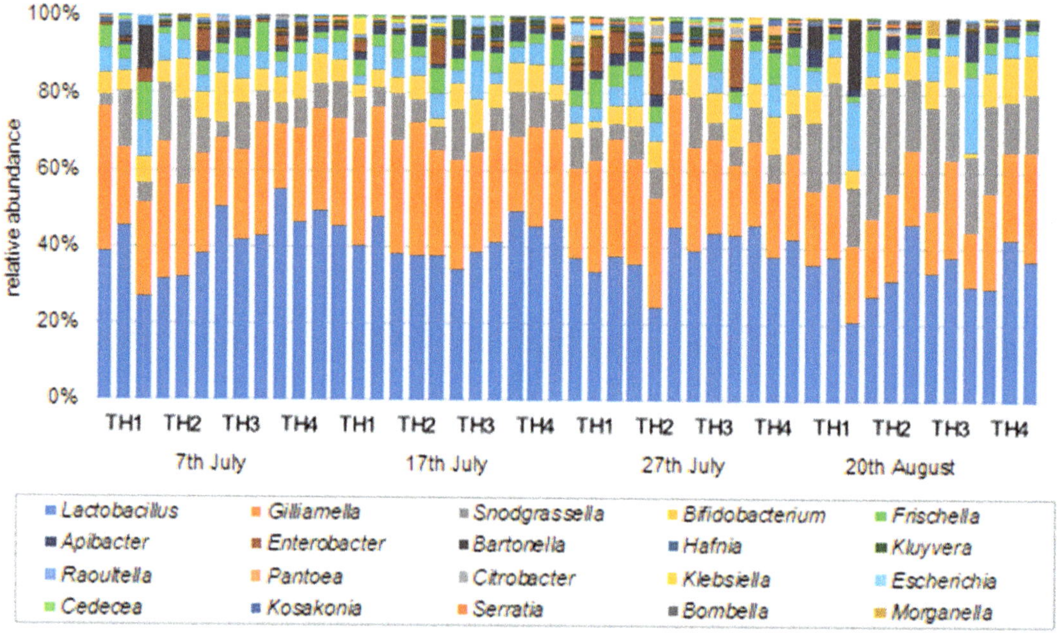

Figure 4. Relative abundance (genus) of the gut microbiome in each treatment on each sampling date.

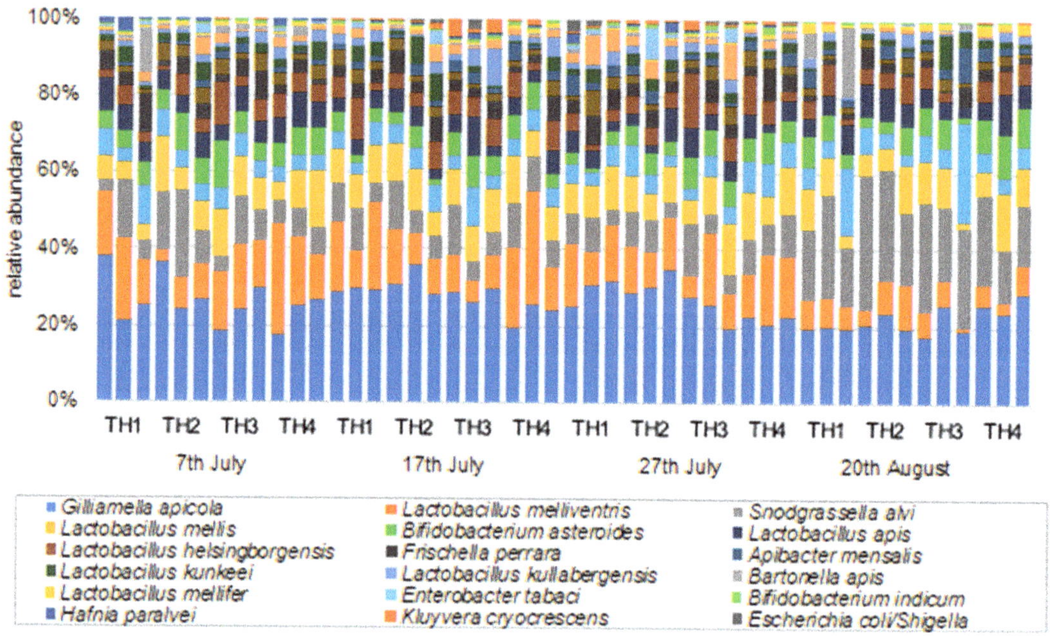

Figure 5. Relative abundance (species) of the gut microbiome in each treatment and on each sampling date.

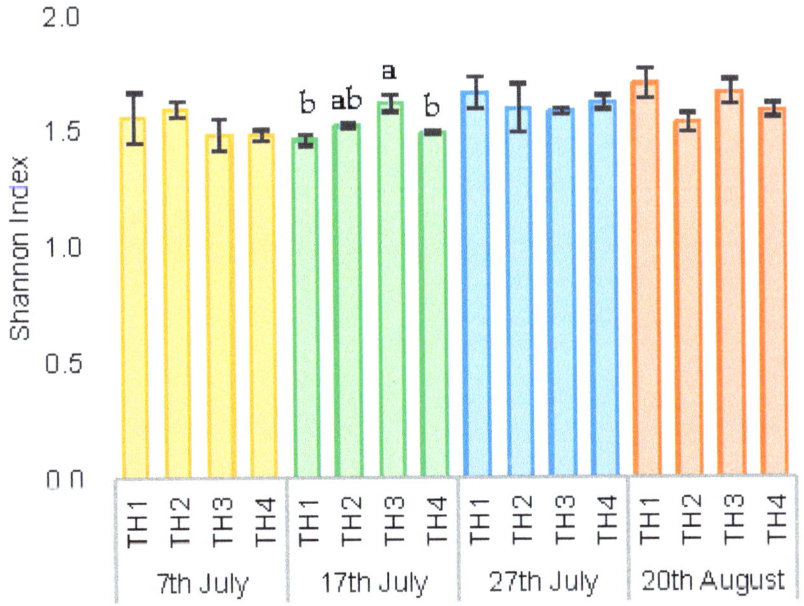

Figure 6. Diversity index (Shannon index) using family data: comparison of treatment intervals and collection dates. There is a significant difference between the days of collection. Values represent mean ± SEM (n = 3). Letters above bars indicate treatments that differ significantly (Tukey HSD test, $\alpha = 0.05$).

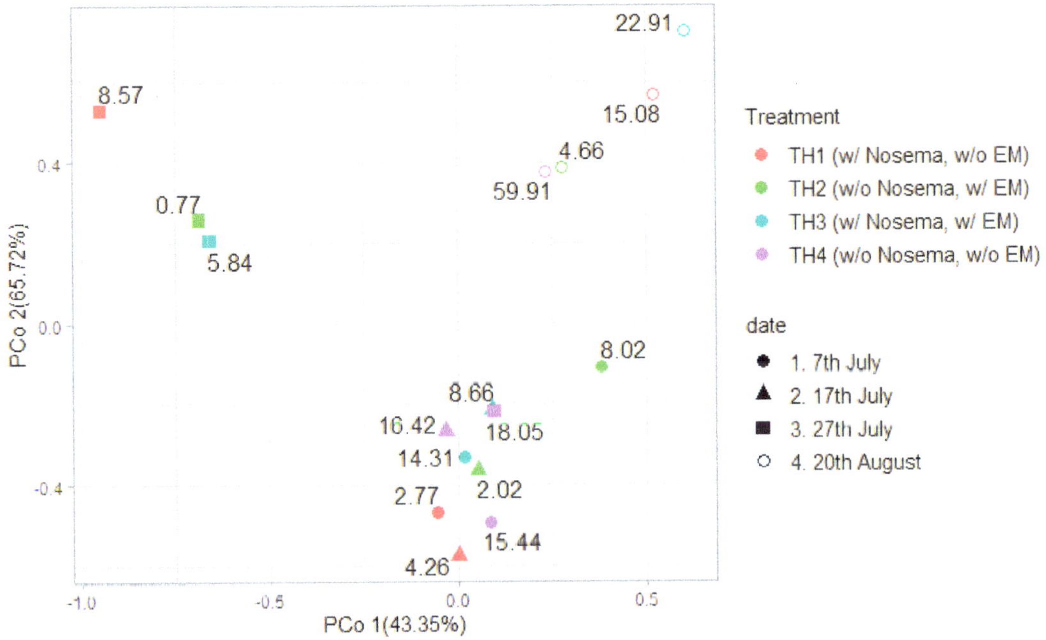

Figure 7. Principal component analysis (PCoA) using the species data comparison of treatment intervals and sample collection dates. Labels in each plot represent the number of *Nosema* spp. spores; w/o—without, w—with.

Table 3 shows the percentage of readings of major species to the total readings for each treatment and day with *Gilliamella apicola*, *Lactobacillus meliventris*, *Snodgrassella alvi*, *Lactobacillus melis*, and *Bifidobacterium asteroides*, accounting for a high percentage. The relative proportions of major species tend to vary between treatments and dates.

Table 3. Relative proportions (%) of the major species in the gut microbiome.

GenBank Identification	7th July				17th July				27th July				20th August			
	TH1	TH2	TH3	TH4	TH1	TH2	TH3	TH4	TH1	TH2	TH3	TH4	TH1	TH2	TH3	TH4
Gilliamella apicola	27.6	28.7	23.6	22.7	28.3	30.7	27.7	22.9	27.8	30.1	23.3	21.2	19.4	20.9	20.2	25.5
Lactobacillus melliventris	16.4	6.7	14.9	19.3	16.6	10.2	8.0	20.4	12.7	11.1	10.8	14.7	7.0	7.9	4.4	5.7
Snodgrassella alvi	7.5	15.6	8.2	6.8	8.5	7.9	7.9	9.4	6.8	7.0	7.7	9.2	20.1	27.0	23.9	17.1
Lactobacillus mellis	5.1	9.1	10.0	9.5	8.7	8.6	9.7	8.9	8.6	8.0	9.4	10.1	7.5	7.7	7.2	10.1
Bifidobacterium asteroides	5.1	7.0	7.3	6.6	3.7	3.5	6.3	5.7	2.4	4.6	6.8	6.3	6.0	4.3	5.2	9.8
Lactobacillus apis	6.7	5.3	5.5	7.4	4.1	4.4	5.9	4.3	4.0	4.4	5.5	4.2	5.7	7.7	4.0	7.2
Lactobacillus helsingborgensis	3.1	3.5	6.4	5.3	5.7	4.7	6.1	5.5	2.7	4.1	7.6	6.3	4.8	4.8	2.6	5.6
Frischella perrara	6.2	3.2	4.9	2.6	3.5	5.0	3.2	3.4	5.7	3.6	3.5	5.1	1.5	3.2	2.6	1.9
Lactobacillus kimbladii	1.4	3.8	3.3	2.5	3.1	2.4	1.7	2.4	3.4	1.9	3.5	2.7	1.7	2.4	1.9	1.9
Apibacter mensalis	0.5	1.2	1.5	1.3	1.7	1.8	2.5	2.7	3.3	2.0	1.4	2.3	2.4	1.6	4.6	2.4
Lactobacillus kunkeei	2.2	2.7	1.7	2.5	2.4	4.9	0.8	1.6	1.8	2.6	1.4	1.5	1.1	1.4	4.4	1.6
Lactobacillus kullabergensis	1.1	1.5	1.9	2.4	2.5	1.5	4.9	2.7	1.9	2.1	2.5	0.8	1.6	1.7	1.0	1.6

The effects of *Nosema* spp. spore inoculation and EM® for bees supplement application on the relative proportions of the major species are represented in Figure 8. *Gilliamella apicola*, which accounted for the highest proportion in all treatments, increased the proportion of the intestinal microbiome by inoculation with *Nosema* spores (TH1). On the other hand, of the treatments that were not inoculated with *Nosema* spores, TH4, where spores increased in the gut after infection with *Nosema*, had the lowest proportion of *G. apicola* until 27th July, while TH2 with EM application had the highest relative proportion of *G. apicola* and moreover kept the lowest number of *Nosema* spores in the gut (a). The relative proportion of *Bifidobacterium asteroides* in TH1 inoculated with *Nosema* spp. spores was lower than in TH4 without *Nosema* inoculation, and the proportion continued to decline until 27th July. Even with inoculation with *Nosema* spores, the relative proportion of *B. asteroides* was not low in TH3 with EM application and remained high until 27th July (b).

The relative proportions of Lactobacillus species in the gut microbiome are shown in Figure 9. *L. melliventris* had a higher relative proportion in group TH4, which had the highest number of *Nosema* spp. spores in the gut in comparison with the other treatments and decreased after 17th July. TH1 decreased similarly, but in group TH2 with only EM® for bees application, it tended to increase, although at a low proportion (a). The relative proportion of *L. mellis* was lower in group TH1 with *Nosema* spp. spore inoculation than in group TH4; however, it was significantly higher in TH3 with in-time EM® for bees application. The proportion changes in *L. mellis* were low (b). The relative proportion of *L. heisingborgensis* was lower in group TH1 than in TH4 but tended to be higher in TH3 fed with EM® for bees. The TH4 proportion increased until 27th July, and TH2 fed with EM® for bees showed the same trend, although at a lower proportion (c). The relative proportion of *L. apis* was high on 7th July and then decreased (TH1 and TH4). In groups TH3 and TH2 treated with EM® for bees, the proportion changes were low until 27th July (d).

The relative proportion of *Frischella perrara* tended to be higher in TH1 inoculated with *Nosema* spp. spores than in TH4. The proportion of *F. perrara* increased until 27th July in treatments not inoculated with *Nosema* spp. spores (TH4 and TH2). In the treatments (TH1 and TH3) inoculated with *Nosema* spp. spores, the proportion tended to be lower in TH3 with EM® for bees (Figure 10a). *Apibacter mensalis* in the treatments with inoculation with *Nosema* spp. spores (TH1 and TH3) increased in proportion until 27th July. In contrast, the treatments without inoculation (TH2 and TH4) tended to be stable (Figure 10b).

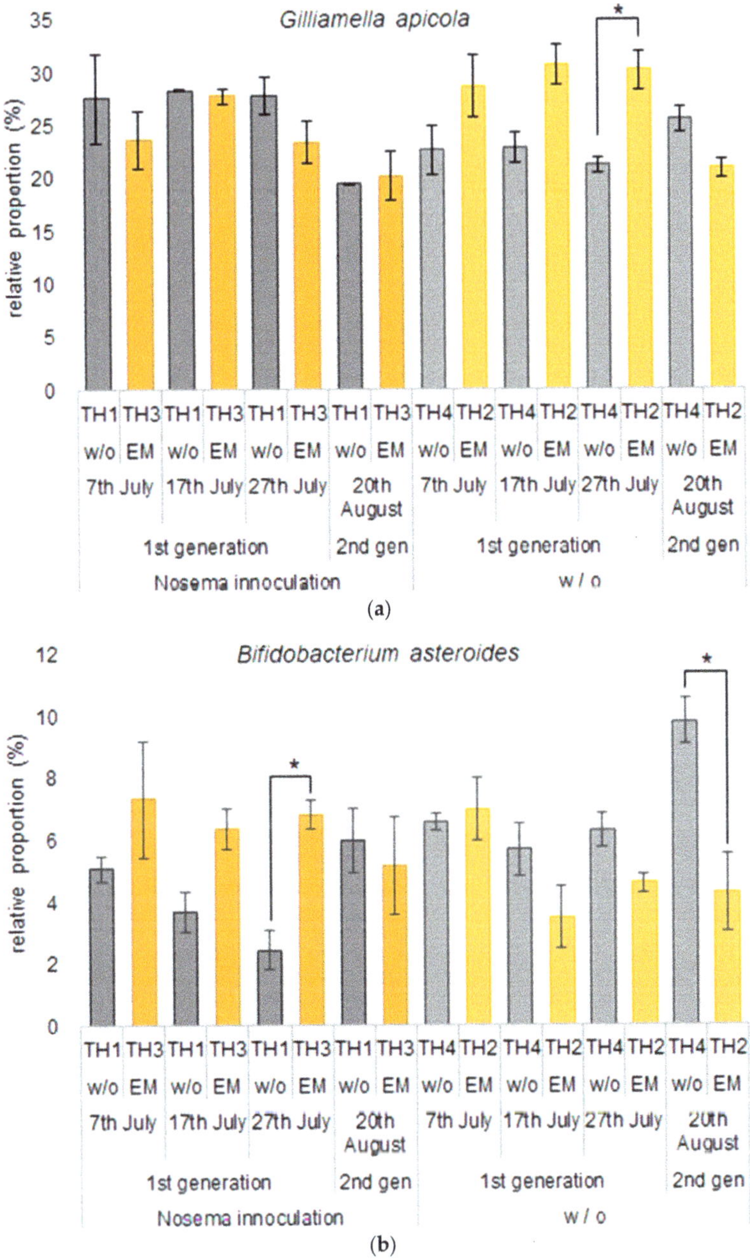

Figure 8. Effect of implemented treatments on the relative proportions of major species: (**a**) *Gilliamella apicola* and (**b**) *Bifidobacterium asteroides*. Values represent mean + SEM (n = 3). Statistical analysis was conducted using Student's *t*-test; * $p < 0.05$; w/o—without, w—with.

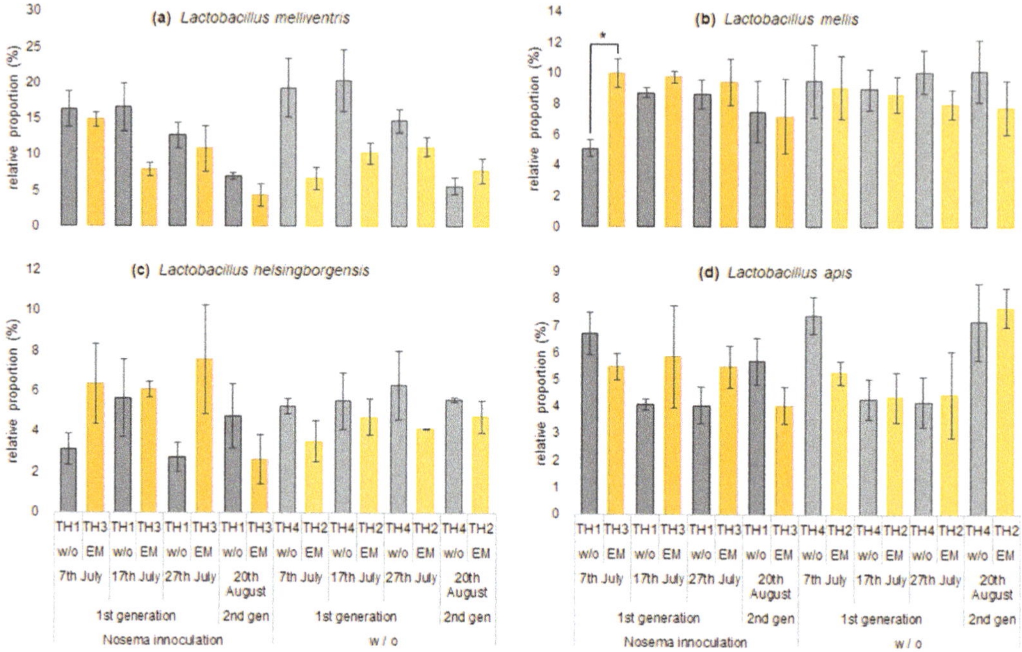

Figure 9. Effect of implemented treatments on the relative proportions of major species: (**a**) *Lactobacillus melliventris*, (**b**) *L. mellis*, (**c**) *L. heisingborgensis*, and (**d**) *L. apis*. Values represent mean ± SEM (n = 3). Statistical analysis was conducted using Student's *t*-test; * $p < 0.05$; w/o—without, w—with.

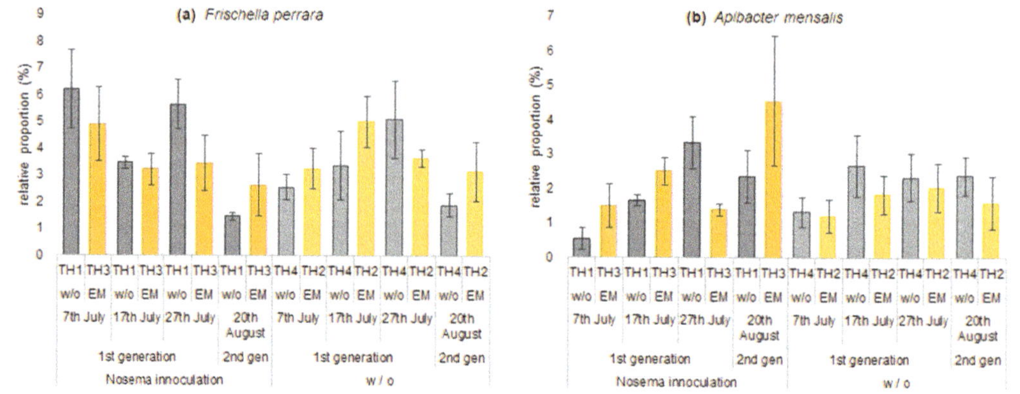

Figure 10. The effect of implemented treatments on the relative proportions of major genera: (**a**) *Frischella perrara* and (**b**) *Apibacter mensalis*. Values represent mean ± SEM (n = 3); w/o—without, w—with.

The results of LEfSe are shown in Figure 11a,b. LEfSe is an analytical method for extracting microbial groups correlated with differences between groups and for searching for microorganisms responsible for certain phenomena, but after analyzing all samples with and without EM® for bees treatment, it did not show any microorganisms characterizing the differences between the experimental and control groups. When LEfSe analysis was performed just on the samples collected on 20th of August, some microbial groups showed significant differences between the groups with and without EM® for bees treatment (TH2 and TH3, TH1 and TH4) (Figure 11).

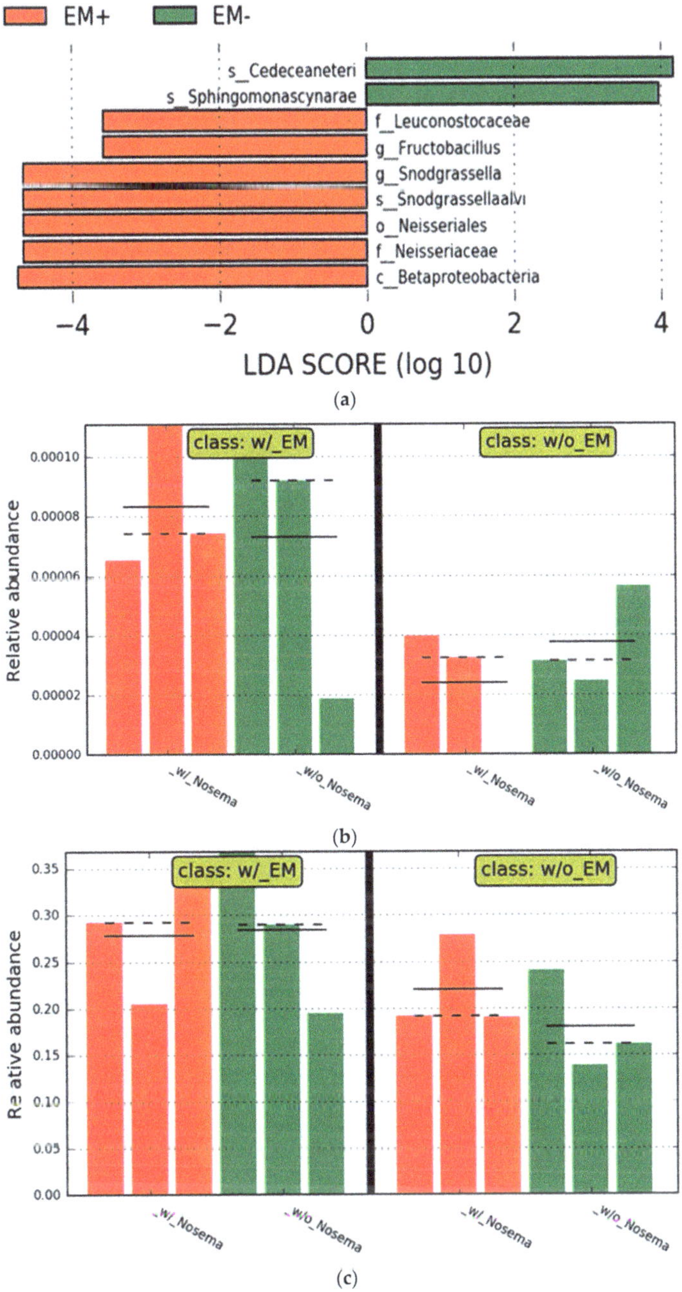

Figure 11. The results of LEfSe (LDA effect size) analysis for samples collected on 20th of August. The histogram of the LDA score shows the biomarkers with statistical differences between the groups (with or without EM® for bees treatment; w/o—without, w—with) (**a**). Histogram of the Fructobacillus (**b**) and *Snodgrassella alvi* (**c**) relative abundance in groups with and without EM® for bee treatments (w/o—without, w—with). Subclasses (with and without *Nosema* spp.) are differentially colored, and the mean and median relative abundance of the biomarkers are indicated with solid and dashed lines, respectively.

The relative abundances of characteristic microorganisms that showed significant differences in the EM® for bees treatment are shown in Figure 11b,c. The relative abundance of Fructobacillus and *Snodgrassella alvi* was significantly increased by supplemental treatment, regardless of the group with and without *Nosema* spp. infection.

4. Discussion

The application of EM® for bees has been confirmed by beekeepers as a good management practice in apiaries. In addition, encouraged by the results of the previous study regarding EM® for bees' impacts on nutritional adult honeybee physiology [53], we considered that it is important to proceed with an investigation on the potential of this complex probiotic to change the gut microbiota composition in adult bees originating from fed colonies.

The strength of colonies was expectedly different between the experimental and control groups of honeybee colonies. In all observation terms, the EM® for bees-supplemented fed colonies were stronger, and the significant difference is most visible on the 42nd day of the experiment. These results are similar to previous scientific records [47,48,66–69]. The field test was chosen since Nosema-diseased adult bees fed with a natural protein food (pollen, especially beebread) show higher microbiota diversity and stability in comparison with those fed only carbohydrates [70]. In this study, it was again observed that EM® for bees applied as a supplement to sugar syrup decreased microsporidium *Nosema* spp. in the honeybee gut. In detail, in the first three sampling terms, lower spore counts in fed colonies compared with controls were determined, which is in accordance with the results of Tlak Gajger et al. (2020) [53]. Only after the last sampling term on day 44, an increase in the number of spores was estimated for both experimental groups of colonies, probably due to an implemented novel feeding regime. In the pertaining control groups (TH1 and TH4), a continuous increase in *Nosema* spp. infection levels was confirmed for each subsequent sampling term, respectively, which was within the expected ranges under the field, environmental, and study conditions.

According to previously published results, the Lactobacillus and Bifidobacterium microorganisms decrease invasion levels of *Nosema* spp. spores in *A. mellifera* [71] and in *A. cerana* [72]. The fed supplementation with *Parasaccharibacter apium* was confirmed to be a good way to improve the resistance of adult bees to Nosema type C invasion [73]. In addition, some other apiculture positive impacts, such as a decrease in varrosis damage, successful honeybee queen production [74], maintaining good colony vitality status [42,75], and increased honey production [74,76], were reported. In this study, *Nosema* spp. spore counts were higher in accidentally naturally infected groups of colonies TH4 than in colonies that were initially naturally infected with spores of *Nosema* spp., TH1 (Table 1), but the relative proportion of Bifidobacteria was lower in TH1, presumably due to longer *Nosema* spp. invasion. It was inferred that when with *Nosema* spp. spores highly infect adult bees that are young enough, the effect on the gut microbiota is greater, and the relative proportion of Bifidobacterium is reduced. EM® for bees treatment was shown to have the potential to mitigate the effects of nosemosis on the gut microbiota content. However, since high levels of *Nosema* spp. spores were detected also in group TH4, where the relative proportion of Bifidobacterium was kept relatively high, it is assumed that Bifidobacterium does not directly inhibit the formation of *Nosema* spp. spores.

When the samples were compared in time series, those taken on the 20th of August, the second generation of adult honeybees, showed a different microbiome content trend from the earlier taken samples (Figure 6). In addition, when the data of each treatment area for the samples taken on the 20th of August were analyzed, there were significant differences in some microbial groups between the samples from the experimental and control groups (Figure 11). For example, *Snodgrassella alvi*, which has been shown to be a characteristic fungus in the honeybee colonies treated with EM® for bees (TH2 and TH3), was higher in the last sampling term (20th of August) in comparison with other treatments (Figure 11). *S. alvi*, which in this study showed significant differences in the experimental and control honeybee colony groups, is a known major member of the honeybee gut microbiota. This bacterium is susceptible to the herbicide glyphosate, and research has shown that glyphosate consistently reduces this bacterium, resulting in lower resistance to pathogens in honeybees [77] and causing metabolic-level perturbation [78]. In addition, the reduction in this bacterium may be a factor that decreases the immunity of honeybees [77].

The number of outs and readings obtained from amplicon sequencing increased on the 27th of July in groups TH1, TH2, and TH3. However, it did not increase in readings for group TH4 (Figure 2). Therefore, the number of intestinal *Nosema* spp. spores suppressed the increase in the number of gut bacteria more than the fresh natural infection with *Nosema* spp. spores. Then, the group TH2 plot on the 7th of July was slightly off, suggesting that the adult honeybee intestinal microbiota was affected by EM® for bees' application. The plots of the sampled bees' gut microbiomes in groups TH1, TH2, TH3, and TH4 on the 20th of August were away from the ones of the other sampling dates. Furthermore, the plots of experimental colonies from groups TH1 and TH3 that were invaded with *Nosema* spp. spores and the ones considered uninvaded (TH2 and TH4) were close together, respectively. Therefore, it is suggested that the gut microbiomes of adult bees sampled on the 20th of August were affected by the invasion with *Nosema* spp. spores more than the level of invasion.

The number of intestinal *Nosema* spp. spores in group TH2 applied only EM® for bees remained lowest, and the relative proportions of Gilliamella were kept higher than in the other groups of colonies. On the other hand, the number of *Nosema* spp. spores remained highest in group TH4, and the relative proportions of Gilliamella were kept lower in the first three sampling terms (Figure 8a). Therefore, EM® for bees' application to honeybees increased the proportion of Gilliamella in the bees' gut microbiomes, indicating that it may inhibit the development of *Nosema* spp. Spores in the midgut.

In groups, TH1 and TH3 invaded with *Nosema* spp. Spores, the level of invasion increased until the 27th of July in TH1 and decreased in TH3 (Table 1). *Bifidobacterium asteroides*, *Lactobacillus mellis*, *L. heisingborgensis*, and *L. apis* were higher in group TH3 than in group TH1 until the 27th of July. We think that probiotics may have suppressed the increase in *Nosema* spp. spore levels in the intestine. Both of these proportions in group TH3 decreased on 20th August, and *Nosema* spores increased, probably as a consequence of the long period passed from initial feeding. On the other hand, group TH2 had the lowest number of *Nosema* spp. spores for all performed sampling terms after the 17th of July and the highest relative proportion of *Gilliamella apicola* until the 27th of July and *Snodgrassella alvi* on the 20th of August.

The relative proportions of *Frischlla perrara* and *Apibacter mensalis* on the 27th of July were higher in group TH1 than in group TH3. Furthermore, the number of *Nosema* spp. spores was also higher. However, on the 20th of August, the proportions of these bacteria were higher in group TH3, and the number of spores was also higher in TH3. Thus, if invaded with *Nosema* spp. spores, these bacteria may be associated with the development of *Nosema* spp. spores in the intestine.

Fructobacillus fructosus, a member of the genus Fructobacillus, has been reported to exhibit antagonistic activity against the pathogen bacterium *Paenibacillus larvae*, the causal agent of American foulbrood [79] However, this species is not known from the results of the LEfSe analysis in this study but represents a good topic for further research. It is known that hive structure, environment, and natural diet influence the assembly and maintenance of honeybee gut microbiota and facilitate future experimental designs [18]. According to our knowledge, this is the first study aimed at investigating the impacts of the probiotic EM® for bees on adult bees' health and immunological conditions by analyzing intestinal microbiome content.

5. Conclusions

There have been many reports from beekeepers that honeybee colonies become more resistant to diseases after using EM® for bees. According to the results of this study, the applied probiotic treatments suppressed the influence of *Nosema* spp. infection and changed bees' gut microbiome composition. In addition, it is probable that the continued use of EM® for bees improves their gut microbiome and, consequently, immunity by maintaining and increasing the level of *S. alvi*. If the higher level of *S. alvi* is supported by feed supplementation with EM® for bees, it may be possible for this beekeeping management practice to reduce the negative effect of glyphosate on honeybees.

Supplementary Materials: The following supporting information can be downloaded at: https://www.mdpi.com/article/10.3390/microorganisms11030610/s1. Table S1 Field test details of feeding regime and sampling.

Author Contributions: Conceptualization, I.T.G.; methodology, I.T.G., S.N.; investigation, I.T.G., S.N. and L.C.; writing—original draft preparation, I.T.G.; writing—review and editing, I.T.G., S.N. and L.C.; visualization, I.T.G.; funding acquisition, I.T.G. All authors have read and agreed to the published version of the manuscript.

Funding: This research received no external funding.

Data Availability Statement: Data are presented in the manuscript.

Acknowledgments: The authors would like to thank B. Kefelja, I. Gajger, G. Husinec, N. Jovanović, S. Nagamoto, H. Tomii, and T. Nishi for their technical assistance.

Conflicts of Interest: The authors declare no conflict of interest. It should be stated that the researched product has been registered and is commercially available in the European market where the field experiment was conducted.

References

1. Klein, A.M.; Vaissiere, B.E.; Cane, J.H.; Steffan-Dewenter, I.; Cunningham, S.A.; Kremen, C.; Tscharntke, T. Importance of pollinators in changing landscapes for world crops. *Proc. R. Soc. B Biol. Sci.* **2007**, *274*, 303–313. [CrossRef] [PubMed]
2. Fontaine, C.; Dajoz, I.; Meriguet, J.; Loreau, M. Functional Diversity of Plant–Pollinator Interaction Webs Enhances the Persistence of Plant Communities. *PLoS Biol.* **2006**, *4*, e1. [CrossRef] [PubMed]
3. Rollin, O.; Benelli, G.; Benvenuti, S.; Decourtye, A.; Wratten, S.D.; Canale, A.; Desneux, N. Weed-insect pollinator networks as bioindicators of ecological sustainability in agriculture. A review. *Agron. Sustain. Dev.* **2016**, *36*, 8. [CrossRef]
4. Ellegaard, K.M.; Engel, P. Genomic diversity landscape of the honey bee gut microbiome. *Nat. Commun.* **2019**, *10*, 446. [CrossRef]
5. Tlak Gajger, I.; Sakač, M.; Gregorc, A. Impact of Thiamethoxam on Honey Bee Queen (*Apis mellifera carnica*) Reproductive Morphology and Physiology. *Bull. Environ. Contam. Toxicol.* **2017**, *99*, 297–302. [CrossRef]
6. Goulson, D.; Nicholls, E.; Botías, C.; Rotheray, E.L. Bee declines driven by combined stress from parasites, pesticides, and lack of flowers. *Science* **2015**, *347*, 6229. [CrossRef] [PubMed]
7. Raymann, K.; Moran, N.A. The role of the gut microbiome in health and disease of adult honey bee workers. *Curr. Opin. Insect Sci.* **2018**, *26*, 97–104. [CrossRef] [PubMed]
8. Alaux, C.; Brunet, J.L.; Dussaubat, C.; Mondet, F.; Tchamitchan, S.; Cousin, M.; Brillard, J.; Baldy, A.; Belzunces, L.P.; Le Conte, Y. Interactions between Nosema microspores and a neonicotinoid weaken honeybees (*Apis mellifera*). *Environ. Microbiol.* **2010**, *12*, 774–782. [CrossRef]
9. Di Pasquale, G.; Salignon, M.; Le Conte, Y.; Belzunces, L.P.; Decourtye, A.; Kretzschmar, A.; Suchail, S.; Brunet, J.-L.; Alaux, C. Influence of Pollen Nutrition on Honey Bee Health: Do Pollen Quality and Diversity Matter? *PLoS ONE* **2013**, *8*, e72016. [CrossRef]
10. Gilioli, G.; Sperandio, G.; Hatjina, F.; Simonetto, A. Towards the development of an index for the holistic assessment of the health status of a honey bee colony. *Ecol. Indic.* **2019**, *101*, 341–347. [CrossRef]
11. Sperandio, G.; Simonetto, A.; Carnesecchi, E.; Costa, C.; Hatjina, F.; Tosi, S.; Gilioli, G. Beekeeping and honey bee colony health: A review and conceptualization of beekeeping management practices implemented in Europe. *Sci. Total Environ.* **2019**, *696*, 133795. [CrossRef]
12. Tlak Gajger, I.; Mañes, A.M.; Formato, G.; Mortarino, M.; Toporcak, J. Veterinarians and beekeeping: What roles, expectations and future perspectives?—A review paper. *Vet. Arhiv* **2021**, *91*, 437–443. [CrossRef]
13. Zheng, H.; Steele, M.I.; Leonard, S.P.; Motta, E.V.S.; Moran, N.A. Honey bees as models for gut microbiome research. *Lab. Anim.* **2018**, *47*, 317–325. [CrossRef] [PubMed]

14. Engel, P.; Martinson, V.G.; Moran, N.A. Functional diversity within the simple gut microbiome of the honey bee. *Proc. Natl. Acad. Sci. USA* **2012**, *109*, 11002–11007. [CrossRef] [PubMed]
15. Vásquez, A.; Forsgren, E.; Fries, I.; Paxton, R.J.; Flaberg, E.; Szekely, L.; Olofsson, T.C. Symbionts as major modulators of insect health: Lactic acid bacteria and honeybees. *PLoS ONE* **2012**, *7*, e33188. [CrossRef]
16. Nowak, A.; Szczuka, D.; Górczyńska, A.; Motyl, I.; Kręgiel, D. Characterization of *Apis mellifera* Gastrointestinal Microbiome and Lactic Acid Bacteria for Honeybee Protection—A Review. *Cells* **2021**, *10*, 701. [CrossRef]
17. Callegari, M.; Crotti, E.; Fusi, M.; Marasco, R.; Gonella, E.; Denoni, I.; Romano, D.; Borin, S.; Tsiamis, G.; Cherif, A.; et al. Compartmentalization of bacterial and fungal microbiomes in the gut of adult honeybees. *npj Biofilms Microbiomes* **2021**, *7*, 42. [CrossRef]
18. Anderson, K.E.; Ricigliano, V.A.; Copeland, D.C.; Mott, B.M.; Maes, P. Social Interaction is Unnecessary for Hindgut Microbiome Transmission in Honey Bees: The Efect of Diet and Social Exposure on Tissue-Specifc Microbiome Assembly. *Microbial Ecol.* **2022**. [CrossRef]
19. Kešnerová, L.; Mars, R.A.T.; Ellegaard, K.M.; Troilo, M.; Sauer, U.; Engel, P. Disentangling metabolic functions of bacteria in the honey bee gut. *PLoS Biol.* **2017**, *15*, e2003467. [CrossRef]
20. Kwong, W.K.; Zheng, H.; Moran, N.A. Convergent evolution of a modified, acetate-driven TCA cycle in bacteria. *Nat. Microbiol.* **2017**, *2*, 17067. [CrossRef]
21. Lee, F.J.; Miller, K.I.; McKinlay, J.B.; Newton, I.L.G. Differential carbohydrate utilization and organic acid production by honey bee symbionts. *FEMS Microbiol. Ecol.* **2018**, *94*, 1–10. [CrossRef] [PubMed]
22. Kwong, W.K.; Moran, N.A. Gut microbial communities of social bees. *Nat. Rev. Microbiol.* **2016**, *14*, 374–384. [CrossRef] [PubMed]
23. Wang, S.; Wang, L.; Fan, X.; Yu, C.; Feng, L.; Yi, L. An insight into diversity and functionalities of gut microbiome in insects. *Curr. Microbiol.* **2020**, *77*, 1976–1986. [CrossRef] [PubMed]
24. Pernice, M.; Simpson, S.J.; Ponton, F. Towards an integrated understanding of gut microbiome using insects as model systems. *J. Insect Physiol.* **2014**, *69*, 12–18. [CrossRef] [PubMed]
25. Jing, T.-Z.; Qi, F.-H.; Wang, Z.-Y. Most dominant roles of insect gut bacteria: Digestion, detoxification or essential nutrient provision? *Microbiome* **2020**, *8*, 38. [CrossRef] [PubMed]
26. Shi, W.; Syrenne, R.; Sun, J.-Z.; Yuan, J.S. Molecular approaches to study the insect gut symbiotic microbiome at the "omics" age. *Insect Sci.* **2010**, *17*, 199–219. [CrossRef]
27. Lee, F.J.; Rusch, D.B.; Stewart, F.J.; Mattila, H.R.; Newton, I.L.G. Saccharide breakdown and fermentation by the honey bee gut microbiome. *Environ. Microbiol.* **2015**, *17*, 796–815. [CrossRef] [PubMed]
28. Parker, A.; Lawson, M.A.E.; Vaux, L.; Pin, C. Host-microbe interaction in the gastrointestinal tract. *Environ. Microbiol.* **2018**, *20*, 2337–2353. [CrossRef]
29. Dillon, R.J.; Vennard, C.T.; Charnley, A.K. A note: Gut bacteria produce components of a locust cohesion pheromone. *J. Appl. Microbiol.* **2002**, *92*, 759–763. [CrossRef]
30. Maes, P.W.; Rodrigues, P.A.P.; Oliver, R.; Mott, B.M.; Anderson, K.E. Diet-related gut bacterial dysbiosis correlates with impaired development, increased mortality and Nosema disease in the honeybee (*Apis mellifera*). *Mol. Ecol.* **2016**, *25*, 5439–5450. [CrossRef]
31. Anderson, K.E.; Ricigliano, V.A. Honey bee gut dysbiosis: A novel context of disease ecology. *Curr. Opin. Insect Sci.* **2017**, *22*, 125–132. [CrossRef] [PubMed]
32. Morimoto, T.; Kojima, Y.; Toki, T.; Komeda, Y.; Yoshiyama, M.; Kimura, K.; Nirasawa, K.; Kadowaki, T. The habitat disruption induces immune-suppression and oxidative stress in honey bees. *Ecol. Evol.* **2011**, *1*, 201–217. [CrossRef] [PubMed]
33. Paris, L.; Peghaire, E.; Moné, A.; Diogon, M.; Debroas, D.; Delbac, F.; El Alaoui, H. Honeybee gut microbiome dysbiosis in pesticide/parasite co-exposures is mainly induced by *Nosema ceranae*. *J. Invertebr. Pathol.* **2020**, *172*, 107348. [CrossRef]
34. Rouzé, R.; Moné, A.; Delbac, F.; Belzunces, L.; Blot, N. The Honeybee Gut Microbiome Is Altered after Chronic Exposure to Different Families of Insecticides and Infection by *Nosema ceranae*. *Microbes Environ.* **2019**, *34*, 226–233. [CrossRef]
35. Panjad, P.; Yongsawas, R.; Sinpoo, C.; Pakwan, C.; Subta, P.; Krongdang, S.; In-on, A.; Chomdej, S.; Chantawannakul, P.; Disayathanoowat, T. Impact of Nosema Disease and American Foulbrood on Gut Bacterial Communities of Honeybees *Apis mellifera*. *Insects* **2021**, *12*, 525. [CrossRef] [PubMed]
36. Castelli, L.; Balbuena, S.; Branchiccela, B.; Zunino, P.; Liberti, J.; Engel, P.; Antúnez, K. Impact of Chronic Exposure to Sublethal Doses of Glyphosate on Honey Bee Immunity, Gut Microbiome and Infection by Pathogens. *Microorganisms* **2021**, *9*, 845. [CrossRef] [PubMed]
37. Cuesta-Mate, A.; Renelies-Hamilton, J.; Kryger, P.; Jansen, A.B.; Sinotte, V.M.; Poulsen, M. Resistance and vulnerability of honeybee (*Apis mellifera*) gut bacteria to commonly used pesticides. *Front. Microbol.* **2021**, *12*, 717990. [CrossRef]
38. Pătruică, S.; Mot, D. The effect of using prebiotic and probiotic products on intestinal micro-flora of the honeybee (*Apis mellifera carpatica*). *Bull. Entomol. Res.* **2012**, *102*, 619–623. [CrossRef]
39. Pătruică, S.; Dumitrescu, G.; Stancu, A.; Bura, M.; Dunea, I.B. The Effect of Prebiotic and Probiotic Feed Supplementation on the Wax Glands of Worker Bees (*Apis mellifera*). *Anim. Sci. Biotechnol.* **2012**, *45*, 268–271.
40. Pătruică, S.; Hutu, I. Economic benefits of using prebiotic and probiotic products as supplements in stimulation feeds administered to bee colonies. *Turk. J. Vet. Anim. Sci.* **2013**, *37*, 259–263.
41. Andrearczyk, S.; Kadhim, M.J.; Knaga, S. Influence of a probiotic on mortality, sugar syrup ingestion and infection of honeybees with *Nosema* spp. under laboratory assessment. *Med. Weter.* **2014**, *70*, 762–765.

42. Arredondo, D.; Castelli, L.; Porrini, M.P.; Garrido, P.M.; Eguaras, M.J.; Zunino, P.; Antunez, K. *Lactobacillus kunkeei* strains decreased the infection by honey bee pathogens *Paenibacillus larvae* and *Nosema ceranae*. *Benef. Microbes* **2018**, *9*, 279–290. [CrossRef] [PubMed]
43. Ptaszyńska, A.A.; Borsuk, G.; Mułenko, W.; Wilk, J. Impact of vertebrate probiotics on honeybee yeast microbiome and on the course of nosemosis. *Med. Weter* **2016**, *72*, 430–434.
44. Daisley, B.A.; Pitek, A.P.; Chmiel, J.A.; Al, K.F.; Chernyshova, A.M.; Faragalla, K.M.; Burton, J.P.; Thompson, G.J.; Reid, J. Novel probiotic approach to counter Paenibacillus larvae infection in honey bees. *ISME J.* **2020**, *14*, 476–491. [CrossRef]
45. Stephan, J.G.; Lamei, S.; Pettis, J.S.; Riesbeck, K.; De Miranda, J.R.; Forsgren, E. Honeybee-specific lactic acid bacterium supplements have no effect on American Foulbrood-infected honeybee colonies. *Appl. Environ. Microbiol.* **2019**, *85*, e00606-19. [CrossRef]
46. Dussaubat, C.; Brunet, J.-L.; Higes, M.; Colbourne, J.K.; López, J.; Choi, J.-H.; Martin-Hernandez, R.; Botías, C.; Cousin, M.; Mcdonnell, C.; et al. Gut pathology and responses to the microsporidium *Nosema ceranae* in the honey bee *Apis mellifera*. *PLoS ONE* **2012**, *7*, e37017. [CrossRef]
47. Tlak Gajger, I.; Vugrek, O.; Pinter, L.; Petrinec, Z. "Nozevit patties" treatment of honeybees (*Apis mellifera*) for the control of *Nosema ceranae* disease. *Am. Bee J.* **2009**, *149*, 1053–1056.
48. Tlak Gajger, I.; Kozaric, Z.; Berta, D.; Nejedli, S.; Petrinec, Z. Effect of the herbal preparation Nozevit on the mid-gut structure of honeybees (*Apis mellifera*) infected with *Nosema* spp. spores. *Vet. Med.* **2011**, *56*, 343–350. [CrossRef]
49. Tlak Gajger, I. Nozevit aerosol application for *Nosema ceranae* disease treatment. *Am. Bee J.* **2011**, *151*, 1087–1090.
50. Tlak Gajger, I.; Smodiš Škerl, M.I.; Šoštarić, P.; Šuran, J.; Sikirić, P.; Vlainić, J. Physiological and Immunological Status of Adult Honeybees (*Apis mellifera*) Fed Sugar Syrup Supplemented with Pentadecapeptide BPC 157. *Biology* **2021**, *10*, 891. [CrossRef]
51. Tlak Gajger, I.; Ribarić, J.; Smodiš Škerl, I.M.; Vlainić, J.; Sikirić, P. Stable gastric pentadecapeptide BPC 157 in honeybee (*Apis mellifera*) therapy, to control *Nosema ceranae* invasions in apiary conditions. *J. Vet. Pharmacol. Therap.* **2018**, *41*, 614–621. [CrossRef] [PubMed]
52. Tlak Gajger, I.; Ribaric, J.; Matak, M.; Svecnjak, L.; Kozaric, Z.; Nejedli, S.; Smodis Skerl, I.M. Zeolite clinoptilolite as a dietary supplement and remedy for honeybee (*Apis mellifera* L.) colonies. *Veternarni Med.* **2015**, *60*, 696–705. [CrossRef]
53. Tlak Gajger, I.; Vlainić, J.; Šoštarić, P.; Prešern, J.; Bubnič, J.; Smodiš Škerl, M.I. Effects on Some Therapeutical, Biochemical, and Immunological Parameters of Honey Bee (*Apis mellifera*) Exposed to Probiotic Treatments, in Field and Laboratory Conditions. *Insects* **2020**, *11*, 638. [CrossRef] [PubMed]
54. Delaplane, K.S.; Van Der Steen, J.; Guzman-Novoa, E. Standard methods for estimating strength parameters of *Apis mellifera* colonies. *J. Apic. Res.* **2013**, *52*, 1–12. [CrossRef]
55. OIE—Office International des Epizooties. Chapter 2.2.4., Nosemosis of Honey Bees. In *Manual of Diagnostic Tests and Vaccines for Terrestrial Animals*; OIE: Pairs, France, 2013. Available online: http://www.oie.int/international-standard-setting/terrestrial-manual/access-online (accessed on 20 May 2022).
56. Takahashi, S.; Tomita, J.; Nishioka, K.; Hisada, T.; Nishijima, M. Development of a Prokaryotic Universal Primer for Simultaneous Analysis of *Bacteria* and *Archaea* Using Next-Generation Sequencing. *PLoS ONE* **2014**, *9*, e105592. [CrossRef]
57. Nguyen, N.P.; Warnow, T.; Pop, M.; White, B. A perspective on 16S rRNA operational taxonomic unit clustering using sequence similarity. *npj Biofilms Microbiomes* **2016**, *2*, 16004. [CrossRef]
58. Hisada, T.; Endoh, K.; Kuriki, K. Inter-and intra-individual variations in seasonal and daily stabilities of the human gut microbiome in Japanese. *Arch. Microbiol.* **2015**, *197*, 919–934. [CrossRef]
59. Aronesty, E. Comparison of sequencing utility programs. *Open Bioinform. J.* **2013**, *7*, 1–8. [CrossRef]
60. Gordon, A.; Hannon, G.J. FASTX-Toolkit FASTQ/A Short-Reads Preprocessing Tools. Available online: http://hannonlab.cshl.edu/fastx_toolkit/index.html (accessed on 30 January 2023).
61. Caporaso, J.G.; Kuczynski, J.; Stombaugh, J.; Bittinger, K.; Bushman, F.D.; Costello, E.K.; Fierer, N.; Peña, A.G.; Goodrich, J.K.; Gordon, J.I.; et al. QIIME allows analysis of high-throughput community sequencing data. *Nat. Methods* **2010**, *7*, 335336. [CrossRef]
62. Edgar, R.C.; Haas, B.J.; Clemente, J.C.; Quince, C.; Knight, R. UCHIME improves sensitivity and speed of chimera detection. *Bioinformatics* **2011**, *27*, 2194–2200. [CrossRef]
63. Wang, Q.; Garrity, G.M.; Tiedje, J.M.; Cole, J.R. Naive bayesian classifier for rapid assignment of rRNA sequences into the new bacterial taxonomy. *Appl. Environ. Microbiol.* **2007**, *73*, 5261–5267. [CrossRef] [PubMed]
64. Kasai, C.; Sugimoto, K.; Moritani, I.; Tanaka, J.; Oya, Y.; Inoue, H.; Tameda, M.; Shiraki, K.; Ito, M.; Takei, Y.; et al. Comparison of the gut microbiome composition between obese and non-obese individuals in a Japanese population, as analyzed by terminal restriction fragment length polymorphism and next-generation sequencing. *BMC Gastroenterol.* **2015**, *15*, 100. [CrossRef]
65. Horthorn, T.; Bretz, F.; Westfall, P. Simultaneous inference in general parametric models. *Biom. J.* **2008**, *50*, 346–363. [CrossRef]
66. Audisio, M.C.; Benítez-Ahrendts, M.R. *Lactobacillus johnsonii* CRL1647, isolated from *Apis mellifera* L. bee-gut, exhibited a beneficial effect on honeybee colonies. *Benef. Microbes* **2011**, *2*, 29–34. [CrossRef] [PubMed]
67. Sabate, D.C.; Cruz, M.S.; Benitez-Ahrendts, M.R.; Audisio, M.C. Beneficial effects of *Bacillus subtilis* subsp. *subtilis* Mori2, a honey-associated strain, on honeybee colony performance. *Probiot. Antimicrob. Proteins* **2012**, *4*, 39–46. [PubMed]
68. Audisio, M.C.; Sabate, D.C.; Benitez-Ahrendts, M.R. Effect of *Lactibacillus johansonii* CRL1647 on different parameters of honeybee colonies and bacterial populations of the bee gut. *Benef. Microbes* **2015**, *25*, 1–10.

69. Alberoni, D.; Baffoni, L.; Gaggìa, F.; Ryan, P.M.; Murphy, K.; Ross, P.R.; Stanton, C.; Di Gioia, D. Impact of beneficial bacteria supplementation on the gut microbiota, colony development and productivity of *Apis mellifera* L. *Benef. Microbes* **2018**, *9*, 9269–9278. [CrossRef] [PubMed]
70. Huang, S.K.; Ye, K.T.; Huang, W.F.; Ying, B.H.; Li, X.S.; Lin, H.; Li, J.H.; Chen, Y.P.; Li, J.L.; Bao, X.L.; et al. Influence of Feeding Type and *Nosema ceranae* Infection on the Gut Microbiota of *Apis cerana* Workers. *mSystems* **2018**, *3*, e00177-18. [CrossRef] [PubMed]
71. Baffoni, L.; Gaggìa, F.; Alberoni, D.; Cabbri, R.; Nanetti, A.; Biavati, B.; Di Gioia, D. Effect of dietary supplementation of Bifidobacterium and Lactobacillus strains in *Apis mellifera* L. against *Nosema ceranae*. *Benef. Microbes* **2016**, *7*, 45–51. [CrossRef]
72. Wu, Y.; Zheng, Y.; Chen, Y.; Chen, G.; Zheng, H.; Hu, F. *Apis Cerana* Gut Microbiota Contribute to Host Health Though Stimulating Host Immune System and Strengthening Host Resistance to *Nosema ceranae*. *R. Soc. Open Sci.* **2020**, *7*, 192100. [CrossRef]
73. Corby-Harris, V.; Snyder, L.; Meador, C.; Naldo, R.; Mott, B.; Andersom, K. *Parasaccharibacter apium*, gen. Nov., sp. Nov., improves honey bee (Hymenoptera: Apidae) resistance to *Nosema*. *J. Econ. Entomol.* **2016**, *109*, 537–543. [CrossRef] [PubMed]
74. Audisio, M.C. Gram-positive bacteria with probiotic potential for the *Apis mellifera* L. honey bee: The experiance in the northwest of Argentina. *Probiot. Antimicrob. Proteins* **2017**, *9*, 22–31. [CrossRef] [PubMed]
75. Yoshiyama, M.; Wu, M.; Sugimura, Y.; Takaya, N.; Kimoto-Nira, H.; Suzuki, C. Inhibition of *Paenibacillus larvae* by lactic acid bacteria isolated from fermented materials. *J. Invertebr. Pathol.* **2013**, *112*, 62–67. [CrossRef] [PubMed]
76. Fanciotti, M.N.; Tejerina, M.; Benítez Ahrendts, M.; Audisio, M.C. Honey yield of different commercial apiaries treated with *Lactobacillus salivarius* A3iob, a new bee-probiotic strain. *Benef. Microbes* **2018**, *9*, 291–298. [CrossRef]
77. Motta, E.V.S.; Raymann, K.; Moran, N.A. Glyphosate perturbs the gut microbiota of honey bees. *App. Biol. Sci.* **2018**, *115*, 10305–10310. [CrossRef]
78. Wang, B.; Habermehl, C.; Jiang, L. Metabolomic analysis of honey bee (*Apis mellifera* L.) response to glyphosate exposure. *Mol. Omics.* **2022**, *18*, 635–642. [CrossRef]
79. Zeid, A.A.A.; Khattaby, A.M.; El-Khair, I.A.A.; Gouda, H.I.A. Detection Bioactive Metabolites of *Fructobacillus fructosus* Strain HI-1 Isolated from Honey Bee's Digestive Tract Against *Paenibacillus larvae*. *Probiot. Antimicrob. Proteins* **2022**, *14*, 476–485. [CrossRef]

microorganisms

Article

Improvement of the Nutraceutical Profile of Brewer's Spent Grain after Treatment with *Trametes versicolor*

Anđela Zeko-Pivač [1], Anja Bošnjaković [1], Mirela Planinić [1], Jelena Parlov Vuković [2], Predrag Novak [3], Tomislav Jednačak [3] and Marina Tišma [1,*]

[1] Faculty of Food Technology Osijek, Josip Juraj Strossmayer University of Osijek, F. Kuhača 18, HR-31 000 Osijek, Croatia
[2] NMR Centre, Ruđer Bošković Institute, Biljenička 54, HR-10000 Zagreb, Croatia
[3] Department of Chemistry, Faculty of Science, University of Zagreb, Horvatovac 102a, HR-10000 Zagreb, Croatia
* Correspondence: mtisma@ptfos.hr

Abstract: Brewer's spent grain (BSG) is an important secondary raw material that provides a readily available natural source of nutraceuticals. It finds its largest application as animal feed and part of the human diet, while the future perspective predicts an application in the production of value-added products. In order to investigate a sustainable BSG treatment method, two BSG samples (BSG1 and BSG2) were evaluated as substrates for the production of hydrolytic (xylanase, β-glucosidase and cellulase) and lignolytic enzymes (laccase, manganese peroxidase and lignin peroxidase) by solid-state fermentation (SSF) with *Trametes versicolor* while improving BSG nutritional value. The biological treatment was successful for the production of all hydrolytic enzymes and laccase and manganese peroxidase, while it was unsuccessful for the production of lignin peroxidase. Because the two BSGs were chemically different, the *Trametes versicolor* enzymes were synthesized at different fermentation times and had different activities. Consequently, the chemical composition of the two BSG samples at the end of fermentation was also different. The biological treatment had a positive effect on the increase in protein content, ash content, polyphenolic compounds, and sugars in BSG1. In BSG2, there was a decrease in the content of reducing sugars. Cellulose, hemicellulose, and lignin were degraded in BSG1, whereas only cellulose was degraded in BSG2, and the content of hemicellulose and lignin increased. The fat content decreased in both samples. The safety-related correctness analysis showed that the biologically treated sample did not contain any harmful components and was therefore safe for use in nutritionally enriched animal feed.

Keywords: brewer's spent grain; solid-state fermentation; *Trametes versicolor*; enzyme cocktail; phenolic compounds

Citation: Zeko-Pivač, A.; Bošnjaković, A.; Planinić, M.; Parlov Vuković, J.; Novak, P.; Jednačak, T.; Tišma, M. Improvement of the Nutraceutical Profile of Brewer's Spent Grain after Treatment with *Trametes versicolor*. *Microorganisms* **2022**, *10*, 2295. https://doi.org/10.3390/microorganisms10112295

Academic Editors: Vishnu D. Rajput, Anuj Ranjan and Abhishek Chauhan

Received: 5 October 2022
Accepted: 15 November 2022
Published: 19 November 2022

1. Introduction

Brewer's spent grain (BSG) is the main solid waste stream of the brewing industry. A total of 20 kg of BSG is produced per 1 hL of brewed beer. It is estimated that about 36.4 million tons of BSG are available in the world per year [1]. Chemically, BSG is a lignocellulosic material, mainly composed of hemicellulose, cellulose, and lignin, but also contains proteins, lipids, vitamins, minerals, and polyphenolic compounds [1–3]. BSG can be, therefore, considered as a potential raw material for various purposes, rather than a waste [4]. BSG is widely used as animal feed, although it has limited digestibility for animals due to the high content of dietary fiber. Improving the nutrient profile of BSG could lead to its larger integration into feed and human nutrition systems and provide several health benefits [5]. Moreover, with the help of appropriate processing methods, BSG can be used not only for the production of higher-quality feed and food, but also for bioenergy and fertilizers, as well as in waste management [1].

Nutraceuticals are compounds derived from natural sources that are becoming a growing trend in health and nutrition, replacing various drugs (non-steroidal anti-inflammatory drugs, analgesic [6], antihypertensive drugs [7], and dietary supplements). Among the wide variety of nutraceuticals such as carbohydrate derivatives, fatty acids, and structural lipids, phenolic compounds attract the most research interest [8]. The importance of this group of compounds is evidenced by the increasing number of reviews [9–11] describing nutraceuticals from structural, medical, and biotechnological aspects. Nutraceuticals also include enzymes, organic acids, vitamins, and minerals. In general, nutraceuticals play an important role as anti-aging agents, antioxidants, anticancer agents, hypoglycemic and hypocholesterolemic agents, antidepressants, etc. The spread of nutraceuticals is increasing worldwide because they are considered safe and free of side effects. The discovery of nutraceuticals is progressing inexorably, but the problem lies in the methods of their production since they are often based on the use of various environmentally harmful chemicals. In addition, there are no regulatory measures for their consumption [12].

The challenge of improving the quality of BSG for use in the food and feed industry is the fact that some components, which are potential nutraceuticals, are bound to lignin, cellulose, and/or hemicellulose or are entrapped inside of those molecules. Different treatments affect the nutritional value of foods and the bioavailability of biologically active substances. Therefore, it is desirable to choose an appropriate technique to increase or maintain the biologically active properties of certain substances [13]. Further to that, biological processing, e.g., solid-state fermentation (SSF), is one of the most important, while being considered a "green chemistry" approach, due to the reduced use of energy and chemicals, and the avoidance of the formation of toxic compounds [14]. The success of SSF is influenced by several factors depending on the nature of the substrate and the microorganism, as well as the scale-up of the process. The most commonly used microorganisms in SSF are fungi, because SSF simulates the natural conditions of the fungal habitat [15,16]. Production of enzymes by SSF at a higher level for commercial interest can be achieved with a selected type of microorganism and substrate for the corresponding enzymes in the fermentation process.

BSG was used as a substrate for the cultivation of various microorganisms for the production of both pure enzymes (laccase, xylanase, cellulase, α-amylase, β-amylase, glucanases) and enzyme cocktails, as recently reviewed [1,17–19]. Because of its lignocellulosic structure and availability of free sugars, it is a source of nutrients and a solid support for the growth of the selected microorganism [20]. During the fungal-based SSF process, which very often includes white-rot fungi, enzymes are synthesized and break down the structure of the substrate and release the trapped nutrients, improving the nutritional value of BSG [21]. White-rot fungi are a physiological group of basidiomycetes with an extracellular lignolytic enzymatic system, which causes wood decay in nature. Fungi and invertebrates are the dominant eukaryote taxa colonizing dead wood, in terms of both abundance and species richness, and they are the key agents of wood decomposition. However, with the exception of termites, the direct effect of invertebrates on wood decay seems to be minor relative to that of fungi [22].

Trametes versicolor is white-rot fungus. Based on the four classic division (Homobasidiomycetes, Heterobasidiomycetes, Urediniomycetes, and Ustilaginomycetes), it belongs to Homobasidiomycetes [23]. According to the Index Fungorum [24], there are 892 species of *Trametes*. *Trametes versicolor* (L.) Lloyd (1921) is one of the most widespread species of *Trametes*.

T. versicolor is heterotrophic organism. It is a saprophyte and uses absorption as the mechanism for feeding, and forms non-motile mycelium of hyphae. When found in nature, *T. versicolor* does not have a stalk, only a cup that attaches directly to the tree or log on which it lives. It has a series of multicolored stripes across the conk. The darker stripes are covered in very small hairs which can help separate *T. versicolor* from similar fungi. Its texture is also very tough and leathery compared with many other fungi which have more delicate skins. It does not have gills, but rather pores [25].

There is a lot of research dealing with the cultivation of *T. versicolor* in different ways (solid-state and submerged fermentation) and with different purposes, as reviewed in our recent review paper [14]. In the last decade, the number of research has increased rapidly, mainly because of ecology and the possibility of reusing widely available lignocellulosic materials as substrates for *T. versicolor* growth for the purpose of producing biofuels and/or value-added products.

In this work, SSF was used for the transformation of two chemically different types of BSG originating from two different breweries. This is the continuation of our work where we have proven that *T. versicolor* produces laccase during cultivation on BSG and influence the liberation of total phenolic compounds [26]. In this work, a complex enzymatic system of *T. versicolor* hydrolytic and lignolytic enzymes was analyzed and a complete chemical analysis of the BSG was done, for each day of fermentation, supported by FTIR and NMR spectra measurements. Analysis and evaluation of the safety-related correctness of treated BSG for a possible use as a feed was done at the end.

2. Materials and Methods

The schematic overview of the experimental set-up is presented in Scheme 1.

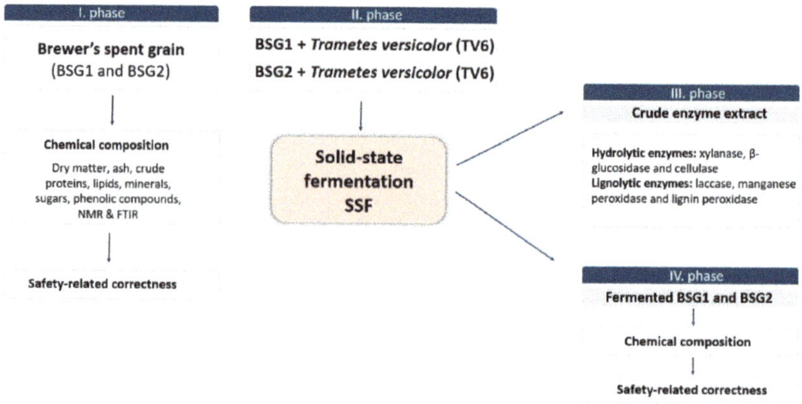

Scheme 1. Schematic representation of the experimental setup.

2.1. Substrates and Microorganisms

Samples of BSG were kindly provided by two local breweries from Croatia, and are referred to in the paper as BSG1 and BSG2. On the day of preparation, a fresh BSG sample with a moisture content of 75% was collected and dried for 8 h at 45 °C in a ventilated oven. Until use, the dry sample was stored at 25 °C. *T. versicolor* TV-6 (MZKI, Ljubljana, Slovenia) was cultivated on potato dextrose agar (PDA) medium at 27 °C for 14 days.

2.2. Biological Treatment of Brewer's Spent Grain by Trametes versicolor

SSF fermentation conditions were performed according to our previous work [26] with applied modifications. In the first step of SSF, 30 g of BSG was added to 720 mL-laboratory jars and mixed with 50 mL of distilled water. It was autoclaved at 121 °C for 15 min and then cooled. The substrate was sterilized (T = 121 °C/15 min), cooled, and inoculated. The inoculum contained 5 mycelial discs (Ø = 1 cm) of *T. versicolor* and 10 mL of sterile water. The height of the inoculated substrate in the laboratory jar was 4.5 cm with a moisture content of 70%. After each day of fermentation, the sample was removed from the laboratory jar and weighed to determine enzyme activity, and the fermentation residue was sterilized. The sterilized residue was dried at room temperature for 48 h and ground on an ultracentrifugal mill (Retsch ZM200, Haan, Germany) to a particle size of ≤1 mm.

2.3. *Enzyme Activities Measurements*

2.3.1. Preparation of Crude Enzyme Extract

For each day of SSF, 2 g of the homogenized sample was weighed from a laboratory jar (in duplicate) and extracted in 10 mL of the appropriate buffer for each enzyme. Extraction was performed on a vortex for 30 min followed by centrifugation at 10,000× *g* (Z 326 K, Hermle Labortechnik GmbH, Wehingen, Germany). Enzyme activities were measured from the supernatant according to the tests described in Section 2.3.2.

2.3.2. Measurements of Enzyme Activities

All enzyme activities were measured in triplicate for each extract using a spectrophotometer (UV-1280, Shimadzu, Kyoto, Japan). Xylanase (endo-1,4-β-xylanase) and cellulase (endoglucanases and exoglucanases) activities were determined by the DNS method [27,28], and β-glucosidase activity was determined according to the study by Karpe et al. (2016) [29]. Manganese peroxidase activity was expressed as the difference between the total and laccase activities, which were also measured under the same conditions [30]. The assay for laccase activity was done according to the study by Tišma et al. (2012) [31]. The assay method for lignin peroxidase was performed following the work by Tien et al. (1988) [32].

2.3.3. pH Measurement

A total of 2 g of fermented BSG was suspended in 10 mL of distilled water. The mixture was vortexed for 30 min and pH was measured with a pH meter (HI 2211 pH/ORP Meter, Hanna instruments).

2.4. *Chemical Composition of BSG*

The chemical composition of untreated samples (BSG1 and BSG2) and treated samples for each day of fermentation was taken according to Section 2.4.1–Section 2.4.7 NMR and FTIR spectra were completed for untreated samples (BSG1 and BSG2) and treated samples BSG1 (day 10) and BSG2 (day 15).

2.4.1. Dry Matter Content

The percentage of dry matter was determined with a rapid moisture analyzer (HR-73, Mettler Toledo, Zürich, Switzerland) according to the thermogravimetric method [33].

2.4.2. Ash Content

The ash content was analyzed by the gravimetric method AACC-08-03 [34].

2.4.3. Crude Proteins

Protein was determined by the Kjeldahl method using the Kjeldahl digestion unit (Behr Labor-Tecnik, Behrotest K12, Düsseldorf, Germany) and water steam distillation system (Gerhard, Vapodest 1, Königswinter, Germany) [35].

2.4.4. Lipids

Lipid content was determined according to the Soxhlet method [35] using an extraction principle (Büchi B-811 LSV, Flawil, Switzerland).

2.4.5. Neutral Detergent Fibers (NDF), Acid Detergent Fibers (ADF), and Acid Detergent Lignin (ADL)

NDF, ADF, and ADL were determined according to the Van Soest method [36] using a fiber analyzer (FIWE 3, VELP Scientifica, Usmate Velate, Italy). The dried and ground sample was weighed (1.0000 ± 0.0001 g) and placed in glass crucible that had previously been dried at 105 °C for 1 h and cooled in a desiccator. For the determination of NDF, the samples were refluxed for 60 min, adding 100 mL of NDF solution, 0.5 g of Na_2SO_3, and a few drops of n-octanol. Samples were then filtered with boiling water and acetone, dried at 105 °C for 8 h, cooled, and weighed. For ADF, 100 mL of ADF solution and a few drops of n-octanol were added and the procedure was repeated as for NDF. To perform ADL analysis, the ADF analysis was performed first and 25 mL of 72% H_2SO_4 was added to the glass wells containing the filtered samples and washed with boiling water and cold acetone. Cold extraction was then performed for 3 h, stirring the samples every hour. At the end of the extraction, the samples were washed with boiling water until the acid was no longer present. The samples were also dried at 105 °C for 8 h, cooled in a desiccator, and weighed. NDF, ADF, and ADL were calculated and corrected for ash by burning glass crucibles with samples in a muffle furnace at 550 °C for 2 h, cooling, and weighing. Hemicellulose content was calculated as the difference between NDF and ADF, cellulose content as the difference between ADF and ADL, and ADL represents lignin content.

2.4.6. Content of Total Reducing Sugars

Total reducing sugars were measured spectrophotometrically (UV-1280, Shimadzu, Japan) at 540 nm according to the DNS method [37].

2.4.7. Content of Total Phenolic Compounds, Total Flavonoids and Total Extractable Proanthocyanidins and Individual Phenolic Compounds

The determination of total phenolic compounds was performed spectrophotometrically using the Folin–Ciocalteu method [38]. Total flavonoids were also determined spectrophotometrically by the aluminum chloride method [39]. Total extractable proanthocyanidins were determined using an acid-butanol solution by a modified spectrophotometric method [40]. All analyses were carried out in duplicate. Analysis of individual phenolic compounds in BSG was determined by ultra-high liquid performance chromatography according to Šelo et al. (2022) [33].

2.4.8. FTIR and NMR Spectra

The FTIR spectra of BSG samples were recorded on a Shimadzu Tracer 100 spectrometer using the ATR (attenuated total reflectance) technique in the single reflection configuration between 4000–400 cm^{-1} with a resolution of 4 cm^{-1} and 128 scans. Solution state 1 H NMR spectra were performed on a Bruker Avance 300 NMR spectrometer using a C/H dual 5 mm probe in D_2O (99.8%, TCI) at room temperature. 1H NMR spectra were recorded with 10 s, 7.6 μs $\pi/2$ pulse length, 16 K time domain, and 100 scans.

Solid-state NMR spectra of BSG samples were measured on a Bruker Avance NEO 400 spectrometer using a broad-band magic angle spinning (MAS) probe. The samples were spun in a 4 mm rotor with a 10 kHz spinning rate. 1H MAS spectra were recorded with 10 scans, 3 s relaxation delay, 0.40 s acquisition time, and 2.5 μs excitation pulse length. 13C cross-polarization (CP) MAS NMR experiments were performed with a standard CP MAS pulse sequence and high-power decoupling during acquisition. The spectra were recorded with 10,000 scans, 3 s relaxation delay, 27 ms acquisition time, and 1 ms contact time. During the contact time, variable amplitude CP ramped from 70% to a maximum of 80 kHz. The protons were decoupled using the SPINAL-64 decoupling. Both 1H MAS and 13C CP MAS spectra were externally referenced to glycine.

2.4.9. Analysis and Evaluation of the Safety-Related Correctness of BSG for Feed

All analyses were performed according to the appropriate accredited method. Detection of nitrates and nitrites was performed by HPLC according to the RU-308-01 method, and gross energy was determined calorimetrically according to HRN EN ISO 9831:2004. Phosphorus content was analyzed spectrophotometrically according to HRN ISO 6491:2001 method, calcium, potassium, iron, lead, cadmium, and arsenic according to the HRN EN ISO/IEC 17025:2017, RU-305-05 on ICP-MS. The analysis for undesirable substances such as mycotoxins was carried out according to the RU-287-04 method (LC-MS/MS), ergot alkaloids RU-149-04 (LC-MS/MS), tropane alkaloids RU-418-01 (LC-MS/MS), polychlorinated biphenyls RU-231-04 (GC-MS/MS), and polycyclic aromatic hydrocarbons RU-230-03; RU-256-02 (GC-MS/MS).

3. Results and Discussion

3.1. Cultivation of Trametes versicolor on Brewer's Spent Grain

In this work, the cultivation of *T. versicolor* on brewer's spent grain in SSF was performed for 15 days in laboratory jars. Moisture content is extremely important in the SSF process because the solid substrate must have adequate available moisture to allow microbial growth and metabolism. Water content that is too high would result in the compaction of the solid, which would introduce the possibility of contamination and prevent oxygen transport. On the other hand, low water content limits nutrient transport and enzyme production [41]. The moisture content of BSG1 during the 15-day fermentation ranged from 66.89% at the beginning to 67.49% at the end of fermentation, and for BSG2 from 69.97% to 67.49%, which is the optimal humidity required for *T. versicolor* growth as was already proved in our previous works. The initial moisture content of BSG was 63% [26] while those of corn silage was 75% [42] and grape pomace was 65–75% [33]. In a study by Iqbal et al. (2011) [43] rice straw used as a substrate for *T. versicolor* had 66.6% of moisture content.

There was no need to add an extra amount of water during SSF while the rate of water production during fungal metabolism was slightly faster than the rate of water evaporation in the case of BSG1. With BSG2, the opposite situation occurred, where water evaporation was slightly faster which lead to a decrease in moisture content, but not below the undesired level.

As a result of organic matter degradation, the loss of the total substrate mass after 15 days of fermentation was 28.74% for BSG1 and 31.01% for BSG2.

The pH analysis showed an increase in the value of fermented BSG1 from an initial 5.44 to 6.11 at the end of fermentation. Fermented BSG2 showed an insignificant decrease from the initial 5.91 to 5.75.

3.2. Trametes versicolor Enzymatic Activities during Cultivation on Brewer's Spent Grain

During SSF cultivation, *T. versicolor* produces a variety of enzymes involved in lignocellulose degradation and modification. In this work, the enzymatic activities of hydrolytic enzymes (xylanase, β-glucosidase and cellulase) and lignolytic enzymes (laccase, manganese peroxidase and lignin peroxidase) were measured during 15 days of SSF.

Aside from being involved in the degradation and modification of lignocellulosic materials, some of those enzymes (e.g., laccase) belong to industrially important enzymes. SSF is gaining more and more attention for the possibility to be used for the production of industrially important enzymes, due to a number of advantages, compared to submerged fermentation. Some of them include the simplicity of implementation, lower production costs, high enzyme productivity, and a positive impact on the environment. However, in addition to the advantages, the SSF also encounters obstacles such as a difficult scale-up, problems in ensuring sterility, heterogeneity of the reaction mixture, oxygen transfer limitation and heat accumulation, variability of conditions during the process, and the effect of inducers [14,44].

The results of enzymatic activities during the cultivation of *T. versicolor* on BSG are shown in Figure 1. Concentrations are expressed as mean value ± standard deviation.

The maximum xylanase productivity of 8430.43 U/g_{DM} was reached on day 15 for BSG1, while for BSG2, the maximum activity of 7608.74 U/g_{DM} was reached on day 14. These results are higher if compared with other microorganisms cultivated on BSG. For example, during cultivation of *Penicillium janczewskii* on BSG, a maximum xylanase activity of 371 U/g_{DM} was achieved on day 7 [45], while cultivation of *Aspergillus niger* strains lead to a maximum xylanase activity of 1400.80 U/g_{DM} [46]. According to the literature, the optimized conditions (pH 6, T = 30 °C, m(BSG) = 15 g) for xylanase production from BSG with *Mucor* species resulted in significantly lower activity values of 67 U/gDM, compared to our results [47]. However, other hydrolytic enzymes (β-glucosidase and cellulase) were less active. The maximum activity of β-glucosidase for BSG1 was 72.55 U/g_{DM} on day 12 of SSF and for BSG2 was 58.56 U/g_{DM} on day 13 of SSF. The maximum cellulase activity of 1.18 U/g_{DM} was reached for BSG1 after 11 days of SSF, and for BSG2 it was 0.92 U/g_{DM}, after 13 days. Similar results were obtained with *Aspergillus brasiliensis* when cultivated on BSG. The obtained β-glucosidase activity was 19.02 U/g_{DM} BSG using 1 × 10^6 spores or 6.64 U/g_{DM} BSG using 25 × 10^6 spores) while activities of cellulases were 7.26 U/g_{DM} BSG using 1 × 10^6 spores or 2.92 U/g_{DM} BSG using 25 × 10^6 spores [48].

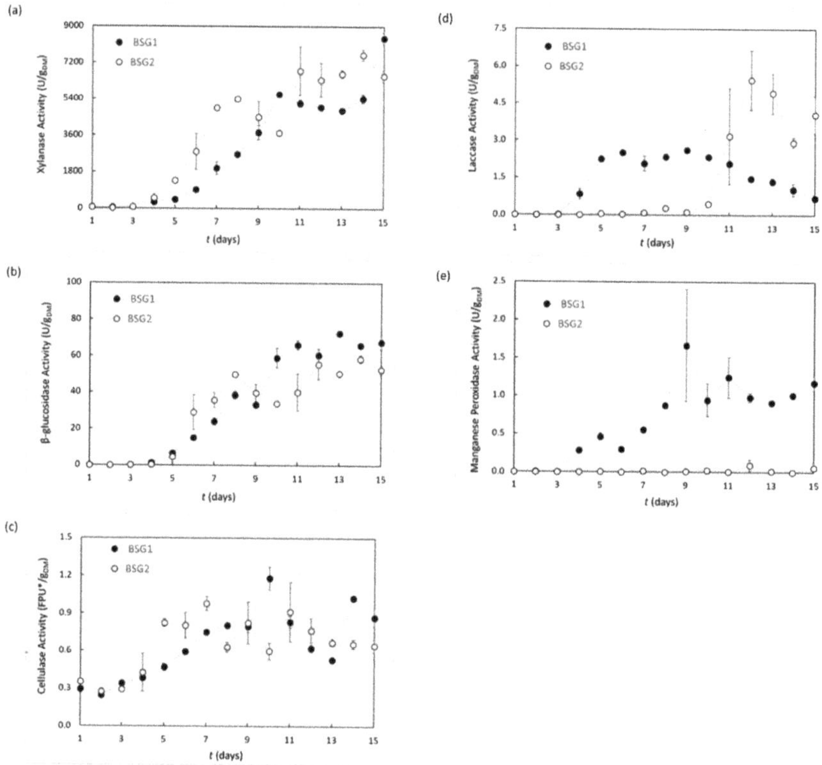

Figure 1. Time course of enzyme activity (**a**) xylanase, (**b**) β-glucosidase, (**c**) cellulase, (**d**) laccase, and (**e**) manganese peroxidase produced by *T. versicolor* on BSG. Activities are expressed as U per gram of dry BSG or * FPU (filter paper unit) per gram of dry BSG.

The combination of *T. versicolor* and BSG under SSF conditions resulted in the successful production of laccase and manganese peroxidase, but proved unsuccessful for lignin peroxidase, while its activity was below the detection limit throughout the whole fermentation process. The maximum activity of laccase (BSG1) was reached after 9 days of fermentation with a value of 2.61 U/g_{DM} and for BSG2 with 5.45 U/g_{DM} after 12 days of fermentation. It was shown that the laccase activities on BSG, produced by *Bjerkandera adusta* and *Schizophyllum commune* reached 1.44 U/g_{DM} and 2.79 U/g_{DM}, respectively [49]. Relatively low activities for manganese peroxidase were obtained in both experiments, with a maximum activity of 1.66 U/g_{DM} after 9 days for BSG1 and for BSG2 of 0.09 U/g_{DM} after 12 days of fermentation.

3.3. Chemical Composition of Brewer's Spent Grain

Since the chemical composition depends on the process of obtaining BSG, i.e., the type of beer and the brewing process, the type of malt used in beer production, the type of barley, and the time of harvest [50], there are variations in the composition of different BSG.According to the literature, BSG is comprised of 15.2–28.7%$_{DM}$ of cellulose, 19–20%$_{DM}$

of hemicellulose, and 3.35–11.41%$_{DM}$ of lignin. It also consists of crude proteins with an average value of 21.25%$_{DM}$, lipids of 8.4%$_{DM}$ and ash of 3.7%$_{DM}$ [1].

3.3.1. Chemical Analysis of the Fibers, Proteins, Lipids and Ash Content

Table 1 shows the chemical composition (cellulose, hemicellulose, lignin, crude proteins, lipids, and ash) of raw samples of BSG1 and BSG2 and fermented BSG1 and BSG2 after 15 days of SSF.

Table 1. Composition of raw and fermented BSG by *T. versicolor* under SSF conditions.

Component [%]	Raw BSG1	Fermented BSG1	Raw BSG2	Fermented BSG2
Cellulose	21.47 ± 1.22	17.82 ± 1.57	28.34 ± 0.46	26.59 ± 0.75
Hemicellulose	30.95 ± 1.55	27.87 ± 0.87	14.89 ± 1.25	20.26 ± 2.39
Lignin	6.94 ± 0.95	5.75 ± 0.34	3.81 ± 0.26	6.90 ± 1.97
Crude Proteins	23.07 ± 0.13	26.42 ± 0.11	18.34 ± 0.01	25.29 ± 0.24
Lipids	8.09 ± 0.17	3.82 ± 0.19	8.85 ± 0.62	3.99 ± 0.55
Ash	3.86 ± 0.03	5.00 ± 0.00	4.24 ± 0.06	5.73 ± 0.11

The analysis of the fiber content results shows that the lignin, cellulose, and hemicellulose content decreased after SSF in BSG1, whereas in BSG2 cellulose content decreased, while the hemicellulose and lignin content increased. The reason for this is that BSG2 contains a higher amount of simple sugars (Section 3.3.2), which *T. versicolor* uses for its growth before other carbon sources. It is assumed that, in this case, a longer incubation period is required for the fungus to begin consuming more complex carbohydrates as a source of growth, with possible degradation. Although xylanase activities were high in both experiments, hemicellulose content decreased only in the BSG1 experiment and increased in the BSG2 experiment. It is likely that other types of enzymes besides those investigated in this study are involved in the modification of BSG. Generally, the use of processed BSG with reduced fiber content in animal feed may improve digestibility in animals that cannot digest fiber [48]. THe protein content increased 1.15-fold for BSG1 and 1.38-fold for BSG2 after 15 days of fermentation, which contributes to the quality of BSG in the form of protein-enriched animal feed. The results of the work by Eliopoulos et al. (2022) [4] demonstrate that SSF has a positive effect on improving protein content of BSG, where the protein content reached 25.01% after 12 days of SSF with *Pleurotus ostreatus*, compared with 16.73% in untreated BSG. SSF with the fungus *Rhizopus* species leads to an increase in protein content from the initial 20.5% to 31.7% after 9 days of SSF [51]. A reduction in lipid content was observed in both BSG1 and BSG2 after 15 days of SSF. Similar results were obtained in the fermentation of BSG with *Rhizopus oligosporus*, where the lipid content decreased from the initial 10.9% to 4.52% after 3 days of SSF [21]. SSF resulted in a 1.30-fold increase in ash content in BSG1 and a 1.35-fold increase in BSG2, with the increase in minerals occurring during organic matter degradation by fermentation. The results of mineral composition are presented in Section 3.3.5.

The results of this study are consistent with the comprehensive literature review [17], which presented that the conversion of BSG involving natural lignocellulosic decomposers by the SSF process significantly improves the nutritional composition of BSG by increasing the content of amino acids, vitamins, and antioxidants while decreasing the content of carbohydrates, fat, and fiber.

3.3.2. Total Reducing Sugars

Various white rot fungi, including *T. versicolor*, are reported in the literature to be good players in removing lignin and hemicellulose with a somewhat lesser effect on cellulose, in lignocellulosic materials. The action of these fungi results in the release of sugars from the mentioned components, e.g., a sugar conversion of 13% was achieved in wheat straw by *P. ostreatus*, *Phanerochaete sordida*, and *Pycnoporus cinnabarin* [52].

The initial concentration of total reducing sugar varies significantly between BSG1 and BSG2 in this work, and the results are shown in Figure 2. Concentrations are expressed as mean value ± standard deviation.

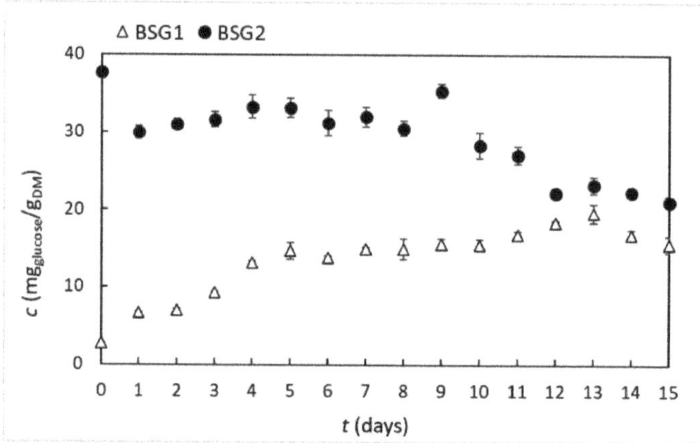

Figure 2. Time course of total reducing sugar concentration during SSF with BSG by *T. versicolor*.

The concentration of total reducing sugar of BSG1 was 2.72 mg/g_{DM}, whereas in BSG2 it was 37.66 mg/g_{DM}. This wide range and difference between BSG1 and BSG2 can possibly be attributed to the type of raw materials used in the process and the applied brewing process. Variations in total reducing sugar content in BSG samples have also been reported in the literature. For example, Llimós et al. (2020) [53] measured 30 mg/g_{DM} in a raw BSG sample, and Fernandes et al. (2021) [20] determined 6.7 mg/g_{DM}.

Regarding the concentration of total reducing sugars, SSF fermentation resulted in an increase in the concentration of reducing sugars in BSG1 after 13 days of fermentation with a maximum concentration of 19.56 mg/g_{DM}, probably due to the action of the enzymes produced and the degradation of other polysaccharide components. In BSG2, the concentration of total reducing sugars decreases to 21.07 mg/g_{DM} at the end of fermentation. In this case, it can be assumed that the fungus prefers to consume fermentable sugars and uses them predominantly as an energy source for its growth [4].

3.3.3. Total and Individual Phenolic Compounds, Total Flavonoids, and Total Extractable Proanthocyanidins

In general, the composition of polyphenolic compounds in raw BSG depends on the production process, especially the brewing temperature and the type of malt used (light and dark) [50]. Due to the large number of enzymes produced during the SSF process, compounds bound to the cell wall are released in the soluble extract, possibly altering the phenolic profile in BSG [54].

The results of the measurement of total polyphenols, flavonoids, and proanthocyanidins are shown in Table 2. for raw and fermented BSG. The individual polyphenolic compounds analyzed by UHPLC are also listed.

The rather low total polyphenol contents in raw BSG are related to the fact that barley does not originally contain a large number of polyphenols and these compounds are associated with lignin and cell wall polysaccharides, so it is necessary to degrade lignocellulosic structures to release polyphenolic compounds [55]. To this purpose, SSF with *T. versicolor* proved successful in the release of total and individual polyphenolic compounds. Total polyphenolic compounds increased 3.75-fold in BSG1 and 1.64-fold in BSG2 by fermentation. Total flavonoids also increased 1.62 fold in BSG1 and 1.14 fold in BSG2. SSF had no positive effect on the concentration of total extractable proanthocyanidins. The increase in total polyphenolic compounds may be attributed to β-glucosidases produced during bioprocessing by the *T. versicolor*. These enzymes hydrolyze β-glucoside bonds to allow free phenolic compounds to react with the Folin–Ciocalteau reagent [15]. In our previous work on BSG with *T. versicolor*, a four-fold increase in total polyphenolic compounds was achieved after 14 days of fermentation, from an initial 2.5 mg/g_{DM} to the final value of 8.7 mg/g_{DM} [26].

Table 2. Concentrations of total phenolic compounds, total flavonoids, total extractable proanthocyanidins, and individual phenolic compounds in raw BSG and fermented BSG with *T. versicolor*.

c_{max} [mg/g_{DM}] *	Raw BSG1	Fermented BSG1	SSF Day	Raw BSG2	Fermented BSG2	SSF Day
Total Polyphenols	4.495 ± 0.270	16.867 ± 0.126	14	9.242 ± 0.068	15.198 ± 0.384	7
Total Flavonoids	1.268 ± 0.025	2.056 ± 0.029	11	1.971 ± 0.044	2.246 ± 0.139	7
Total Proanthocyanidins	0.517 ± 0.431	0.287 ± 0.030	15	0.494 ± 0.009	0.456 ± 0.009	15
Individual Phenols						
p-hydroxybenzoic acid	0.022 ± 0.001	0.105 ± 0.003	3	0.203 ± 0.047	0.324 ± 0.008	6
Catechin	0.148 ± 0.009	1.301 ± 0.025	15	0.188 ± 0.011	1.152 ± 0.047	7
Epicatechin	0.092 ± 0.005	0.137 ± 0.010	11	0.035 ± 0.000	0.149 ± 0.001	3
Vanillic acid	0.002 ± 0.00	0.013 ± 0.000	15	n.d	n.d.	-
Syringic acid	0.031 ± 0.001	0.281 ± 0.000	9	0.015 ± 0.000	0.218 ± 0.000	7
p-coumaric acid	0.008 ± 0.000	0.019 ± 0.000	12	0.004 ± 0.000	0.024 ± 0.001	5
Ferulic acid	n.d.	0.139 ± 0.000	12	0.166 ± 0.027	0.164 ± 0.006	5
Kaempferol	0.023 ± 0.000	0.0230 ± 0.003	10	0.016 ± 0.000	0.040 ± 0.002	7
3,4-dihydroxybenzoic acid	0.462 ± 0.001	6.823 ± 2.159	3	0.025 ± 0.000	0.141 ± 0.008	12
o-coumaric acid	0.012 ± 0.001	0.019 ± 0.001	8	0.012 ± 0.001	0.019 ± 0.001	8
p-hydroxyphenylacetic acid	n.d.	n.d.	-	0.203 ± 0.047	0.324 ± 0.008	6

* concentrations are expressed as mean value ± standard deviation.

The phenolic compounds more commonly found in BSG are hydroxycinnamic acids, especially ferulic acid and *p*-coumaric acid, sinapic acid, caffeic acid, and syringic acid [56]. Our collection includes a total of 27 standards for polyphenolic compounds, 11 of which were detected in this study. For all detected individual polyphenolic compounds, an increase was observed at a particular stage of SSF, with 3,4-dihydroxybenzoic acid in BSG1 standing out the most with a 14.77-fold increase in concentration.

3.3.4. FTIR and NMR

The typical FTIR spectra of raw (black) and fermented (red) are displayed in Figure 3.

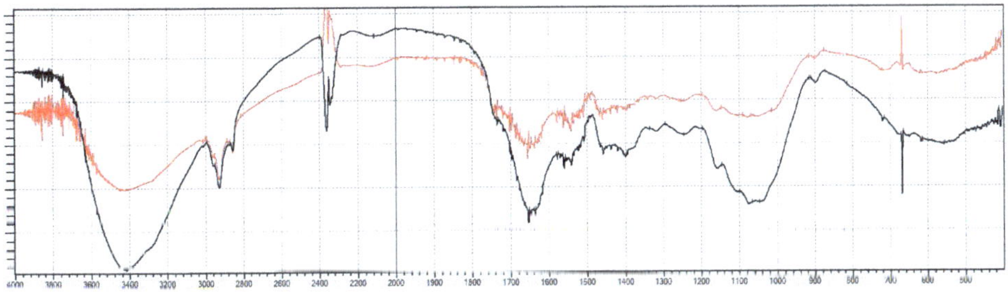

Figure 3. The typical FTIR spectra of raw (red) and fermented (black) BSG samples.

FTIR spectroscopy was used to assess the composition of BSG samples. All samples showed broad adsorption bands in the region between 3000 and 3800 cm^{-1}, belonging to the stretching vibrations of the phenolic and aliphatic –OH groups and between 2800 and 3000 cm^{-1}, attributed to C–H stretching of methyl and methylene groups Furthermore, –C=O (carboxylic acids, and esters) and –C=C stretching bands were observed in the region 1600–1750 cm^{-1}, while C–O–C and C–O vibrations were found in the region between 1000 and 1250 cm^{-1}. C–H deformation, aromatic ring vibration, and –OH bending vibration bands span the region between 1000 and 1500 cm^{-1}. Proton NMR spectra in a solution of raw and fermented BSG samples are presented in Figure 4. The characteristic spectral regions and corresponding functional groups (for all analyzed samples) have been assigned and compared with the data already published in the literature (Table 3) [57–61].

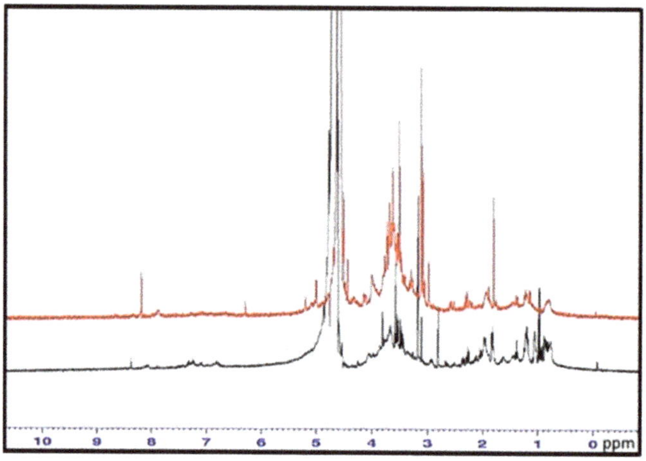

Figure 4. The typical ^{1}H NMR spectra of raw (black) and fermented (red) BSG samples.

Table 3. NMR assignments and percentages of analyzed BSG samples.

δ/ppm	H-Atom	Raw BSG1/% *	Fermented BSG1/% *	Raw BSG2/%	Fermented BSG2/%
0.50–3.00	Aliphatic	29.17	31.33	25.71	30.82
3.00–4.40	Carbohydrate, methoxy	69.45	66.77	71.42	62.89
6.70–7.65	Aromatic	0.70	1.20	2.14	5.66
7.94–8.50	Aromatic, amino acid, formic acid	0.70	0.70	0.73	0.63

* calculated from integrated proton peak intensities.

Despite the fact that BSG components were not separated and the complexity of proton NMR spectra, some significant differences in composition among samples can be noticed. According to the integrated proton peak intensities in NMR spectra, fermented BSG samples seem to show a higher content of aliphatic and aromatic compounds and a lower portion of carbohydrates, amino, and formic acids compared to raw BSG samples.

BSG samples were further analyzed by solid-state NMR using ^{1}H and ^{13}C CP MAS techniques. Carbon and hydrogen atoms were assigned based on already published NMR data and comparison with solution state spectra [58–60].

As can be seen in Figure 5, the most intense signals in the spectra were found in the region between 60 and 110 ppm. This region belongs to the carbohydrates of cellulose and hemicellulose. Aromatic and aliphatic signals are associated with carbon atoms of lignin. The broad signals observed approximately at 55 ppm were attributed to methoxy carbon atoms of lignin. The most deshielded peaks, observed between 169 to 175 ppm, were assigned to carboxylic carbon atoms of lignin and hemicellulose.

The ^{1}H chemical shift assignment in the solid state was made on the basis of the comparison with chemical shifts observed in the solution (Figure 6).

3.3.5. Analysis and Evaluation of the Safety-Related Correctness of BSG for Feed

When BSG is considered a potential addition to animal feed, functional food for humans, or plant food in the form of biofertilizer, it is necessary to examine the parameters that affect the quality of correctness [17]. Research shows that BSG may contain significant amounts of mycotoxins, secondary metabolites originally found in grains that can also be produced by fungi, particularly *Aspergillus*, *Fusarium*, and *Penicillium*. They cause food spoilage and are of significant concern to humans and animals [62,63]. The consumption of feeds containing mycotoxins causes adverse health effects in animals (stunted growth, reduced immunity, chronic and acute diseases, and death) and leads to restrictions in livestock production [64].

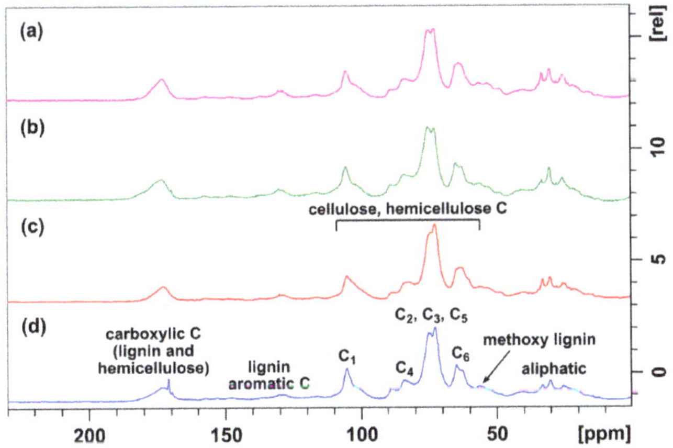

Figure 5. ^{13}C CP MAS spectra of (a) raw BSG1; (b) fermented BSG1; (c) raw BSG2; (d) fermented BSG2.

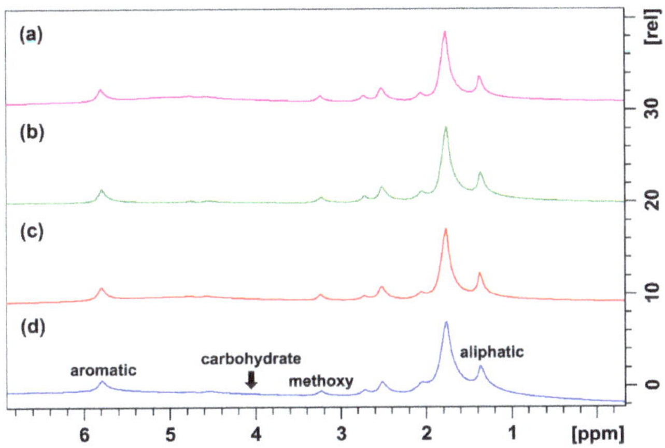

Figure 6. ^{1}H MAS spectra of (a) raw BSG1; (b) fermented BSG1; (c) raw BSG2; (d) fermented BSG2.

The total content of aflatoxins in raw and fermented BSG is <0.003 mg/kg, fumonisins < 0.30 mg/kg, zearalenone < 0.030 mg/kg, T-2 and HT-2 sum < 0.20 µg/kg, and ochratoxin A < 0.001 mg/kg expressed as Limit of Quantification (LOQ).

Other secondary metabolites found in commercially important cereals are ergot alkaloids produced by fungi of the genus *Claviceps* that cause ergot disease. Studies show that ergot alkaloids are often present in animal feed at mean concentrations ranging from 25–96 µg/kg to a maximum of 149–4883 µg/kg [65]. In this work, 12 ergot alkaloids were analyzed with a total concentration of less than 120 µg/kg (LOQ) in all BSG. Cereals may also be contaminated with solanaceous plants (*Solanaceae*), which contain toxic metabolites such as tropane alkaloids. The LOQ for these compounds in feed was 5–25 µg/kg in the study [66], while in this study, atropine and scolopamine were analyzed and their concentration was <1.0 µg/kg LOQ in raw and fermented BSG.

Polychlorinated biphenyls (PCBs) in the feed are a group of complex substances that can affect the hormonal, nervous, and immune systems of animals. Studies show that PCB concentrations in feed were below the LOQ (0.05 mg/kg) in 30 feed samples [67]. The total PCB LOQs in this work were <0.005 mg/kg for raw BSG1 and <0.035 mg/kg for fermented BSG1. For raw and fermented BSG2, the concentration was <0.035 mg/kg. Other hazardous compounds that have carcinogenic and mutagenic properties are polycyclic aromatic hydrocarbons (PAHs), the contaminants from the air, water, and soil. Animals exposed to PAHs experience disorders of the immune system, urinary system, body fluids, and skin and lung damage. In studies of feed mixtures for pigs and cows, PAHs were detected at levels of 0.082 and 0.128 mg/kg, respectively [68], while in this study they were < 0.020 mg/kg LOQ for raw and fermented BSG.

The formulation of animal feed requires a certain amount of minerals necessary to maintain animal health and productivity. Many essential trace elements (Fe, I, Co, Zn, Cu, Mn, Mo, Se) are added to animal feed as dietary supplements, while some (As, Cd, F, Pb, Hg) are considered undesirable substances and have no proven biological functions [69]. The elements analyzed in raw and fermented BSG1 and BSG2, as well as gross energy, are listed in Table 4. For BSG1, the selected analyzes were performed on the sample on the 10th day of fermentation, while for BSG2 the 15th day was selected. The reason for selecting these days was the sum of all the results and enzyme productivity, and the days with the most of the best results were selected.

Table 4. Minerals and heavy metals detected in raw BSG and fermented BSG with *T. versicolor*.

Compound	Raw BSG1	Fermented BSG1	Raw BSG2	Fermented BSG2
Phosphorus [%]	0.48	0.60	0.66	0.80
Calcium [mg/kg]	2073	2456	2763	2994
Potassium [mg/kg]	586	499	722	936
Iron [mg/kg]	173	220	103	113
Lead [mg/kg]	0.049	0.064	0.025	0.033
Cadmium [mg/kg]	0.039	0.048	0.044	0.057
Arsenic [mg/kg]	0.032	0.039	<0.010	0.013

From the results, fermentation releases minerals in the BSG material as the ash content increases, which makes this material attractive for addition to animal feed due to mineral enrichment of the feed composition.

According to the EU DIRECTIVE 2002/32/EC [70], the maximum permissible level of arsenic in feed materials and complementary feed is 2 mg/kg, for lead 10 mg/kg and for nitrites 15 mg/kg. The permissible concentrations of cadmium in feed materials of vegetable origin and complementary feed are 1 and 0.5 mg/kg, respectively. Table 4 shows that the heavy metals in raw as well as in fermented BSG1 and BSG2 are within the permissible concentration limits.

4. Conclusions

The use of *T. versicolor* with BSG under SSF fermentation conditions resulted in the successful production of hydrolytic and lignolytic enzymes, with particular emphasis on xylanase, for which the highest activities were obtained. The production and action of enzymes during the fermentation process resulted in an increase in protein content, ash content, polyphenolic compounds, and sugars in BSG1, while in BSG2 the overall reducing sugar content decreased. The degradation of cellulose, hemicellulose, and lignin as a result of SSF is observed in BSG1, while in BSG2 only cellulose was degraded and the content of hemicellulose and lignin increased. Fat content also decreased in both BSGs during fermentation. These results make SSF a promising alternative for the valorization of BSG as a valuable ingredient for possible use in animal feed. Undesirable substances in animal feed have also been shown to be within acceptable limits for unprocessed and fermented BSG.

Author Contributions: All authors gave their maximum, direct and intellectual contribution during the writing of the paper. Conceptualization, A.Z.-P. and M.T.; Methodology, A.Z.-P., A.B., M.T., M.P., J.P.V., P.N. and T.J.; Validation, A.Z.-P., M.T., and M.P.; Investigation, A.Z.-P., A.B., J.P.V., P.N. and T.J.; Resources, M.T. and M.P.; Writing—Original Draft Preparation, A.Z.-P., M.T. and J.P.V.; Writing—Review and Editing, M.T., M.P. and J.P.V.; Visualization, A.Z.-P., M.T. and M.P.; Supervision, M.T. and M.P.; Project Administration, M.T. and M.P.; Funding Acquisition, M.P. All authors have read and agreed to the published version of the manuscript.

Funding: This research was funded by the European Regional Development Fund (ERDF) (Grant KK.01.1.1.04.0107).

Conflicts of Interest: The authors declare that the research was conducted in the absence of any commercial or financial relationships that could be construed as a potential conflict of interest.

References

1. Zeko-Pivač, A.; Tišma, M.; Žnidaršič-Plazl, P.; Kulisic, B.; Sakellaris, G.; Hao, J.; Planinić, M. The Potential of Brewer's Spent Grain in the Circular Bioeconomy: State of the Art and Future Perspectives. *Front. Bioeng. Biotechnol.* **2022**, *10*, 870744. [CrossRef] [PubMed]
2. Tran, T.T.T.; Ton, N.M.N.; Le, N.D.D.; Le, V.V.M. Addition of brewing spent grains from malt and rice adjunct to the formulation of high fiber biscuit: Effects of particle size of brewing spent grains on the product quality. *Food Res.* **2020**, *4*, 1480–1486. [CrossRef]
3. Yitayew, T.; Moges, D.; Satheesh, N. Effect of Brewery Spent Grain Level and Fermentation Time on the Quality of Bread. *Int. J. Food Sci.* **2022**, *2022*, 8704684. [CrossRef]
4. Eliopoulos, C.; Arapoglou, D.; Chorianopoulos, N.; Markou, G.; Haroutounian, S.A. Conversion of brewers' spent grain into proteinaceous animal feed using solid state fermentation. *Environ. Sci. Pollut. Res.* **2022**, *29*, 29562–29569. [CrossRef] [PubMed]
5. Parchami, M.; Ferreira, J.; Taherzadeh, M. Brewing process development by integration of edible filamentous fungi to upgrade the quality of Brewer's spent grain (BSG). *BioResources* **2021**, *16*, 1686–1701. [CrossRef]
6. Heffernan, S.M.; Conway, G.E. Nutraceutical Alternatives to Pharmaceutical Analgesics in Osteoarthritis. In *Pain Management—Practices, Novel Therapies and Bioactives*; Waisundara, V.Y., Banjari, I., Balkić, J., Eds.; IntechOpen: London, UK, 2021. [CrossRef]
7. Alexander, W. Hypertension: Is it time to replace drugs with nutrition and nutraceuticals? *Pharm. Ther.* **2014**, *39*, 291.
8. Cano y Postigo, L.O.; Jacobo-Velázquez, D.A.; Guajardo-Flores, D.; Garcia Amezquita, L.E.; García-Cayuela, T. Solid-state fermentation for enhancing the nutraceutical content of agrifood by-products: Recent advances and its industrial feasibility. *Food Biosci.* **2021**, *41*, 100926. [CrossRef]
9. Alagawany, M.; Elnesr, S.S.; Farag, M.R.; Abd El-Hack, M.E.; Barkat, R.A.; Gabr, A.A.; Foda, A.M.; Noreldin, A.E.; Khafaga, A.F.; El-Sabrout, K.; et al. Potential role of important nutraceuticals in poultry performance and health—A comprehensive review. *Res. Vet. Sci.* **2021**, *137*, 9–29. [CrossRef]
10. Costagliola, G.; Nuzzi, G.; Spada, E.; Comberiati, P.; Verduci, E.; Peroni, D.G. Nutraceuticals in Viral Infections: An Overview of the Immunomodulating Properties. *Nutrients* **2021**, *13*, 2410. [CrossRef] [PubMed]
11. Taroncher, M.; Vila-Donat, P.; Tolosa, J.; Ruiz, M.J.; Rodríguez-Carrasco, Y. Biological activity and toxicity of plant nutraceuticals: An overview. *Curr. Opin. Food Sci.* **2021**, *42*, 113–118. [CrossRef]
12. Blaze, J. A Comparison of Current Regulatory Frameworks for Nutraceuticals in Australia, Canada, Japan, and the United States. *Innov. Pharm.* **2021**, *12*. [CrossRef] [PubMed]
13. Grgić, J.; Šelo, G.; Planinić, M.; Tišma, M.; Bucić-Kojić, A. Role of the Encapsulation in Bioavailability of Phenolic Compounds. *Antioxidants* **2020**, *9*, 923. [CrossRef] [PubMed]
14. Tišma, M.; Žnidaršič-Plazl, P.; Šelo, G.; Tolj, I.; Šperanda, M.; Bucić-Kojić, A.; Planinić, M. *Trametes versicolor* in lignocellulose-based bioeconomy: State of the art, challenges and opportunities. *Bioresour. Technol.* **2021**, *330*, 124997. [CrossRef] [PubMed]
15. Dulf, F.V.; Vodnar, D.C.; Dulf, E.H.; Pintea, A. Phenolic compounds, flavonoids, lipids and antioxidant potential of apricot (*Prunusarmeniaca* L.) pomace fermented by two filamentous fungal strains in solid state system. *Chem. Cent. J.* **2017**, *11*, 92. [CrossRef] [PubMed]
16. Leite, P.; Silva, C.; Salgado, J.M.; Belo, I. Simultaneous production of lignocellulolytic enzymes and extraction of antioxidant compounds by solid-state fermentation of agro-industrial wastes. *Ind. Crops Prod.* **2019**, *137*, 315–322. [CrossRef]
17. Bianco, A.; Budroni, M.; Zara, S.; Mannazzu, I.; Fancello, F.; Zara, G. The role of microorganisms on biotransformation of brewers' spent grain. *Appl. Microbiol. Biotechnol.* **2020**, *104*, 8661–8678. [CrossRef] [PubMed]
18. Chetrariu, A.; Dabija, A. Brewer's Spent Grains: Possibilities of Valorization, a Review. *Appl. Sci.* **2020**, *10*, 5619. [CrossRef]
19. Mitri, S.; Salameh, S.J.; Khelfa, A.; Leonard, E.; Maroun, R.G.; Louka, N.; Koubaa, M. Valorization of Brewers' Spent Grains: Pretreatments and Fermentation, a Review. *Fermentation* **2022**, *8*, 50. [CrossRef]
20. Fernandes, H.; Moyano, F.; Castro, C.; Salgado, J.; Martinez, F.; Aznar, M.; Fernandes, N.; Ferreira, P.; Gonçalves, M.; Belo, I.; et al. Solid-state fermented brewer's spent grain enzymatic extract increases in vitro and in vivo feed digestibility in European seabass. *Sci. Rep.* **2021**, *11*, 22946. [CrossRef]
21. Cooray, S.T.; Chen, W.N. Valorization of brewer's spent grain using fungi solid-state fermentation to enhance nutritional value. *J. Funct. Foods* **2018**, *42*, 85–94. [CrossRef]

22. Jacobsen, R.M.; Sverdrup-Thygeson, A.; Kauserud, H.; Mundra, S.; Birkemoe, T. Exclusion of invertebrates influences saprotrophic fungal community and wood decay rate in an experimental field study. *Funct. Ecol.* **2018**, *32*, 2571–2582. [CrossRef]
23. Webster, J.; Weber, R. *Introduction to Fungi*, 3rd ed.; Cambridge University Press: Cambridge, UK, 2007; p. 867.
24. Index Fungorum. Available online: http://www.indexfungorum.org/ (accessed on 5 November 2022).
25. Macalester. Turkey Tail Fungus. Available online: https://www.macalester.edu/ordway/biodiversity/inventory/turkeytailfungus/ (accessed on 7 November 2022).
26. Tišma, M.; Jurić, A.; Bucić-Kojić, A.; Panjičko, M.; Planinić, M. Biovalorization of brewers' spent grain for the production of laccase and polyphenols. *J. Inst. Brew.* **2018**, *124*, 182–186. [CrossRef]
27. Bailey, M.J.; Biely, P.; Poutanen, K. Interlaboratory testing of methods for assay of xylanase activity. *J. Biotechnol.* **1992**, *23*, 257–270. [CrossRef]
28. Adney, B.; Baker, J. *Measurement of Cellulase Activities. Laboratory Analytical Procedure*; National Renewable Energy Laboratory (NREL): Golden, CO, USA, 1962.
29. Karpe, A.V.; Dhamale, V.V.; Morrison, P.D.; Beale, D.J.; Harding, I.H.; Palombo, E.A. Winery biomass waste degradation by sequential sonication and mixed fungal enzyme treatments. *Fungal Genet. Biol.* **2016**, *102*, 22–30. [CrossRef] [PubMed]
30. Lueangjaroenkit, P.L.; Kunitake, E.; Sakkam, M.; Kimura, T.; Teerapatsakul, C.; Sakka, K.; Chitradon, L. Light regulation of two new manganese peroxidase-encoding genes in *Trametesolyzona* KU-RNW027. *Microorganisms* **2020**, *8*, 852. [CrossRef] [PubMed]
31. Tišma, M.; Žnidaršič-Plazl, P.; Vasić-Rački, Đ.; Zelić, B. Optimization of laccase production by *Trametes versicolor* cultivated on industrial waste. *Appl. Biochem. Biotechnol.* **2012**, *166*, 36–46. [CrossRef]
32. Tien, M.; Kirk, T.K. Lignin peroxidase of *Phanerochaete chrysosporium*. *Biomass Part B Lignin Pectin Chitin* **1988**, *161*, 238–249. [CrossRef]
33. Šelo, G.; Planinić, M.; Tišma, M.; Grgić, J.; Perković, G.; Koceva Komlenić, D.; Bucić-Kojić, A. A Comparative Study of the Influence of Various Fungal-Based Pretreatments of Grape Pomace on Phenolic Compounds Recovery. *Foods* **2022**, *11*, 1665. [CrossRef] [PubMed]
34. AOAC. AOAC Official Method 2001.11. In *Official Methods of Analysis of AOAC International*, 18th ed.; AOAC International: Gaithersburg, MD, USA, 2005.
35. AOAC. OAC Official Method 945.16. In *Official Methods of Analysis of AOAC International*, 18th ed.; AOAC International: Gaithersburg, MD, USA, 2005.
36. Goering, H.K.; Van Soest, P.J. *Forage Fiber Analyses (Apparatus, Reagents, Procedures, and Some Applications)*; Agricultural Research Service, U.S. Department of Agriculture: Washington, DC, USA, 1970; pp. 1–24.
37. Miller, G.L. Use of Dinitrosalicylic Acid Reagent for Determination of Reducing Sugar. *Anal. Chem.* **1959**, *31*, 426–428. [CrossRef]
38. Marinova, D.; Ribarova, F.; Atanassova, M. Total phenolics and total flavonoids in Bulgarian fruits and vegetables. *J. Chem. Technol. Metall.* **2005**, *40*, 255–260.
39. Planinić, M.; Zelić, B.; Čubel, I.; Bucić-Kojić, A.; Tišma, M. Corn forage biological pretreatment by *Trametes Versicolor* in a tray bioreactor. *Waste Manag. Res.* **2016**, *34*, 802–809. [CrossRef]
40. Škerget, M.; Kotnik, P.; Hadolin, M.; Hraš, A.R.; Simonič, M.; Knez, Ž. Phenols, proanthocyanidins, flavones and flavonols in some plant materials and their antioxidant activities. *Food Chem.* **2005**, *89*, 191–198. [CrossRef]
41. He, Q.; Peng, H.; Sheng, M.; Hu, S.; Qiu, J.; Gu, J. Humidity Control Strategies for Solid-State Fermentation: Capillary Water Supply by Water-Retention Materials and Negative-Pressure Auto-controlled Irrigation. *Front. Bioeng. Biotechnol.* **2019**, *7*, 263. [CrossRef] [PubMed]
42. Bucić-Kojić, A.; Šelo, G.; Zelić, B.; Planinić, M.; Tišma, M. Recovery of Phenolic Acid and Enzyme Production from Corn Silage Biologically Treated by *Trametes versicolor*. *Appl. Biochem. Biotechnol.* **2016**, *181*, 948–960. [CrossRef] [PubMed]
43. Iqbál, H.M.N.; Asgher, M.; Bhatti, H.N. Optimization of physical and nutritional factors for synthesis of lignin degrading enzymes by a novel strain of *Trametes versicolor*. *Bioresources* **2011**, *6*, 1273–1278.
44. Gassara, F.; Brar, S.K.; Tyagi, R.D.; Verma, M.; Surampalli, R.Y. Screening of agro-industrial wastes to produce ligninolytic enzymes by *Phanerochaete chrysosporium*. *Biochem. Eng. J.* **2010**, *49*, 388–394. [CrossRef]
45. Terrasan, C.R.F.; Carmona, E.C. Solid-state fermentation of brewer's spent grain for xylanolytic enzymes production by *Penicillium janczewskii* and analyses of the fermented substrate. *Biosci. J.* **2015**, *31*, 1826–1836. [CrossRef]
46. Moran-Aguilar, M.G.; Costa-Trigo, I.; Calderón-Santoyo, M.; Domínguez, J.M.; Aguilar-Uscanga, M.G. Production of cellulases and xylanases in solid-state fermentation by different strains of *Aspergillus niger* using sugarcane bagasse and brewery spent grain. *Biochem. Eng. J.* **2021**, *172*, 108060. [CrossRef]
47. Hassan, S.S.; Tiwari, B.K.; Williams, G.A.; Jaiswal, A.K. Bioprocessing of brewers' spent grain for production of xylanopectinolytic enzymes by *Mucor* sp. *Bioresour. Technol. Rep.* **2019**, *9*, 100371. [CrossRef]
48. Outeiriño, D.; Costa-Trigo, I.; de Souza Oliveira, R.P.; Pérez Guerra, N.; Domínguez, J.M. A novel approach to the biorefinery of brewery spent grain. *Process Biochem.* **2019**, *85*, 135–142. [CrossRef]
49. Ilić, N.; Davidović, S.; Milić, M.; Rajilić-Stojanović, M.; Pecarski, D.; Ivančić-Šantek, M.; Mihajlovski, K.; Dimitrijević-Branković, S. The agricultural waste utilization for exploration of the extracellular enzyme potential of three novel white-rot Basidiomycetes: *Fomesfomentarius* TMF2, *Schizophyllum commune* TMF3 and *Bjerkandera adusta* TMF1. *Res. Sq.* **2021**. [CrossRef]

50. Liguori, R.; Soccol, R.C.; Vandenberghe, P.L.; Woiciechowski, L.A.; Ionata, E.; Marcolongo, L.; Faraco, V. Selection of the strain *Lactobacillus acidophilus* ATCC 43121 and its application to brewers' spent grain conversion into lactic acid. *BioMed Res. Int.* **2015**, *2015*, 240231. [CrossRef] [PubMed]
51. Ibarruri, J.; Cebrián, M.; Hernández, I. Solid State Fermentation of Brewer's Spent Grain Using *Rhizopus* sp. to Enhance Nutritional Value. *Waste Biomass Valorization* **2019**, *10*, 3687–3700. [CrossRef]
52. Aftab, M.N.; Iqbal, I.; Riaz, F.; Karadag, A.; Tabatabaei, M. Different Pretreatment Methods of Lignocellulosic Biomass for Use in Biofuel Production. In *Biomass for Bioenergy—Recent Trends and Future Challenges*; Abomohra, A.E., Ed.; IntechOpen: London, UK, 2019; pp. 1–24. [CrossRef]
53. Llimós, J.; Martínez-Avila, O.; Marti, E.; Corchado-Lopo, C.; Llenas, L.; Gea, T.; Ponsá, S. Brewer's spent grain biotransformation to produce lignocellulolytic enzymes and polyhydroxyalkanoates in a two-stage valorization scheme. *Biomass Convers. Biorefinery* **2020**, *12*, 3921–3932. [CrossRef]
54. Da Costa Maia, I.; Thomaz dos Santos D'Almeida, C.; Guimarães Freire, D.M.; Cavalcanti, E.; d'Avila, C.; Cameron, L.C.; Dias, J.F.; Larraz Ferreira, M.S. Effect of solid-state fermentation over the release of phenolic compounds from brewer's spent grain revealed by UPLC-MSE. *LWT—Food Sci. Technol.* **2020**, *133*, 110136. [CrossRef]
55. Birsan, R.I.; Wilde, P.; Waldron, K.W.; Rai, D.K. Recovery of polyphenols from brewer's spent grains. *Antioxidants* **2019**, *8*, 380. [CrossRef] [PubMed]
56. Macias-Garbett, R.; Serna-Hernández, S.O.; Sosa-Hernández, J.E.; Parra-Saldívar, R. Phenolic compounds from brewer's spent grains: Toward green recovery methods and applications in the cosmetic industry. *Front. Sustain. Food Syst.* **2021**, *5*, 681684. [CrossRef]
57. Su, Y.; Wenzel, M.; Paasch, S.; Seifert, M.; Böhm, W.; Doert, T.; Weigand, J.J. Recycling of Brewer's Spent Grain as a Biosorbent by Nitro-Oxidation for Uranyl Ion Removal from Wastewater. *ACS Omega* **2021**, *6*, 19364–19377. [CrossRef]
58. El Hariri El Nokab, M.; Habib, M.H.; Alassmy, Y.A.; Abduljawad, M.M.; Alshamrani, K.M.; Sebakhy, K.O. Solid State NMR a Powerful Technique for Investigating Sustainable/Renewable Cellulose-Based Materials. *Polymers* **2022**, *14*, 1049. [CrossRef]
59. Sparrman, T.; Svenningsson, L.; Sahlin-Sjövold, K.; Nordstierna, L.; Westman, G.; Bernin, D. A revised solid-state NMR method to assess the crystallinity of cellulose. *Cellulose* **2019**, *26*, 8993–9003. [CrossRef]
60. Cipriano, D.F.; Chinelatto, L.S.; Nascimento, S.A.; Rezende, C.A.; de Menezes, S.M.C.; Freitas, J.C.C. Potential and limitations of ^{13}C CP/MAS NMR spectroscopy to determine the lignin content of lignocellulosic feedstock. *Biomass Bioenergy* **2020**, *142*, 105792. [CrossRef]
61. Allegretti, C.; Bellinetto, E.; D'Arrigo, P.; Griffini, G.; Marzorati, S.; Rossato, L.A.M.; Ruffini, E.; Schiavi, L.; Serra, S.; Strini, A.; et al. Towards a Complete Exploitation of Brewers' Spent Grain from a Circular Economy Perspective. *Fermentation* **2022**, *8*, 151. [CrossRef]
62. Mastanjević, K.; Šarkanj, B.; Warth, B.; Krska, R.; Sulyok, M.; Mastanjević, K.; Šantek, B.; Krstanović, V. *Fusarium culmorum* multi-toxin screening in malting and brewing by-products. *LWT—Food Sci. Technol.* **2018**, *98*, 642–645. [CrossRef]
63. Lao, E.J. The implications and management strategies of animal feed mycotoxins. *Rev. Environ. Contam. Toxicol.* **2020**, *3*, 53–61. [CrossRef]
64. Santos Pereira, C.C.; Cunha, S.; Fernandes, J.O. Prevalent Mycotoxins in Animal Feed: Occurrence and Analytical Methods. *Toxins* **2019**, *11*, 290. [CrossRef]
65. Babič, J.; Tavčar-Kalcher, G.; Celar, F.A.; Kos, K.; Červek, M.; Jakovac-Strajn, B. Ergot and ergot alkaloids in cereal grains intended for animal feeding collected in Slovenia: Occurrence, pattern and correlations. *Toxins* **2020**, *12*, 730. [CrossRef]
66. Romera-Torres, A.; Romero-González, R.; Martínez Vidal, J.L.; Frenich, A.G. Study of the occurrence of tropane alkaloids in animal feed using LC-HRMS. *Anal. Methods* **2018**, *10*, 3340–3346. [CrossRef]
67. Toman, R.; Pšenková, M.; Tančin, V. Polychlorinated biphenyls in cow's milk, feed and soil in selected areas of Slovakia. *Acta Fytotech. Zootech.* **2020**, *23*, 4. [CrossRef]
68. Şahin, T.; Dalğa, S.; Ölmez, M. Polycyclic Aromatic Hydrocarbons (PAHs) and Their Importance in Animal Nutrition. In *Animal Husbandry*; Kukovics, S., Ed.; IntechOpen: London, UK, 2022. [CrossRef]
69. Hejna, M.; Gottardo, D.; Baldi, A.; Dell'Orto, V.; Cheli, F.; Zaninelli, M.; Rossi, L. Nutritional ecology of heavy metals. *Animal* **2018**, *12*, 2156–2170. [CrossRef]
70. Eur-LEX. Directive 2002/32/EC of the European Parliament and of the Council of 7 May 2002 on Undesirable Substances in Animal Feed. Available online: http://data.europa.eu/eli/dir/2002/32/2019-11-28 (accessed on 20 September 2022).

MDPI AG
Grosspeteranlage 5
4052 Basel
Switzerland
Tel.: +41 61 683 77 34

Microorganisms Editorial Office
E-mail: microorganisms@mdpi.com
www.mdpi.com/journal/microorganisms

www.ingramcontent.com/pod-product-compliance
Lightning Source LLC
Chambersburg PA
CBHW061455210926
51850CB00008B/1724
* 9 7 8 3 7 2 5 8 1 7 0 4 7 *